Charlotte O. Vert

Windows Me
For Me!

Windows ® Me for Me!

Charlotte O. Vert

Windows Millennium Edition Desktop Components

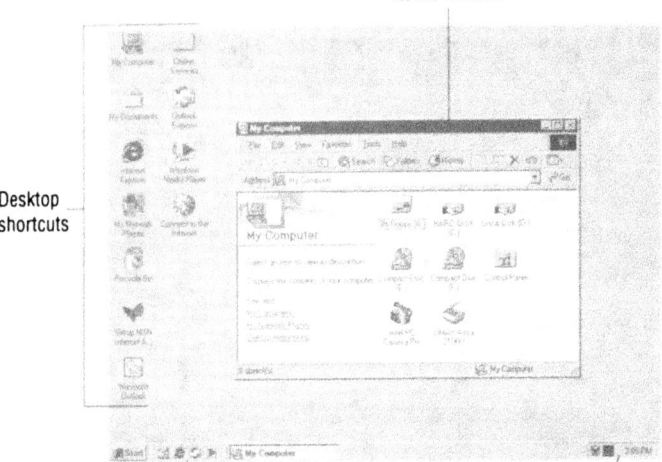

Windows ® Me for Me!

Windows Explorer Components

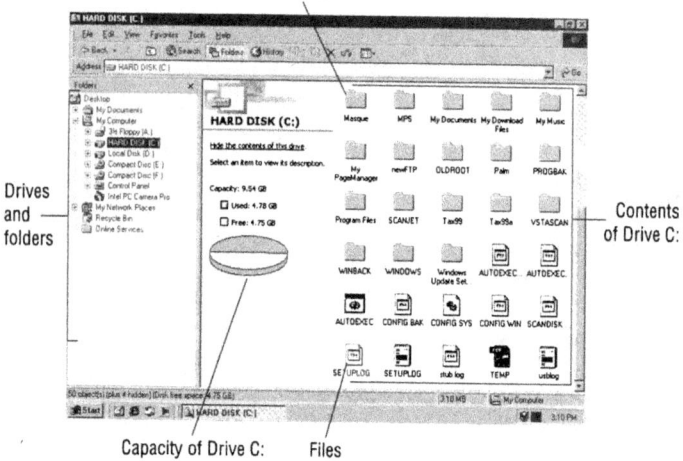

Charlotte O. Vert

Windows® Me For Me!

San Francisco • Paris • Düsseldorf • Soest • London

Windows Me for Me!

Introduction

Windows Millennium Edition (Me) represents the next evolutionary step in the progression of Windows operating systems. It is bigger, faster, slicker, and more capable than anything that has come before. Windows Me supports more different kinds of hardware and includes much more application software than any previous version of Windows.

Windows Me is tailor-made for home and family use. With its easy-to-follow Wizards, you can set up home networking, Internet connection sharing, and more with very little technical expertise needed. It also provides some great multimedia tools for playing, recording, and editing sounds, music, and video.

What's New in Windows Me?

Windows Me is a natural extension of Windows 98, updating the operating system to include support for all the latest hardware standards and improvements that have developed over the last few years. Windows Me fully supports such technologies as IEEE Firewire, USB, and Universal Plug-and-Play.

There are also a host of new or improved features:

Internet Explorer 5.5 Brings the latest in Web browsers to Windows Me. Internet Explorer 5.5 has the latest security and customization features, and the ability to display all the latest Web content.

Home Networking Wizard Walks you through the process of configuring a PC for network use, making a once-complex process very easy.

Windows Media Player A brand-new, greatly enhanced version of this program offers MP3 and CD audio play, as well as Internet radio and a variety of other audio and video formats.

Windows Movie Maker A handy utility for editing digitized video footage and for combining video or still pictures into multimedia shows with musical sound track and narration.

Internet Connection Sharing Allows you to share a modem or other Internet connection (such as cable modem or ISDN connection) with other PCs in your home. (This feature was included in Windows 98 Second Edition too.)

System Restore Provides a way to return to earlier Windows configurations to correct problems. For example, if you installed a program and then Windows started crashing frequently, you could restore your configuration to the way it was before you installed that program.

Games Windows Me comes with some new games for your enjoyment, including a challenging new solitaire variation called Spider and several Internet-based games such as Backgammon and Reversi.

Scanner and Camera Wizard Helps you more easily acquire images from your scanner or digital camera, in most cases allowing you to bypass any proprietary software that came with the device in favor of a standard Windows interface.

How This Book Is Organized

This book presents all the Windows Me features, applications, and configuration options listed in alphabetic order. Most of the entries start with the Windows icon or menu selection you actually use to access the specific feature under discussion. Application toolbars and menu selections are discussed in detail, and configuration options are described in numbered steps for ease of use. A "See also" cross-reference at the end of an entry directs you to other, related entries in the book. If a feature is known by a name other than its official name, you'll find a "See" cross-reference from that name to its official name. You can also access detailed information directly through a complete index.

Conventions Used in This Book

This book uses just a couple of special typographical conventions. Anything that you have to type, such as a command name or a program name, is in **boldface**. Web addresses, also known as URLs, are in a `monospace` font.

Additional information appears throughout the book in the form of Notes, Tips, and Warnings:

NOTES provide additional information about a specific topic.

TIPS give you clues on how to make better use of a Windows feature, or detail shortcut methods you can use to get the same result.

WARNINGS alert you to the potential dangers of using (or abusing) certain features.

Accessibility

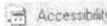 Choose Start ➤ Programs ➤ Accessories ➤ Accessibility to access the Accessibility Wizard, Magnifier, or On-Screen Keyboard.

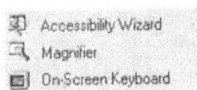

See also Accessibility Options, Accessibility Wizard, Magnifier, On-Screen Keyboard

Accessibility Options

Accessibility Options

Make the computer easier to use for those who are physically challenged—people with reduced vision or hearing impairment, as well as those who have difficulty using the keyboard and the mouse. Choose Start ➤ Settings ➤ Control Panel, and then double-click the Accessibility Options icon to open the Accessibility Properties dialog box.

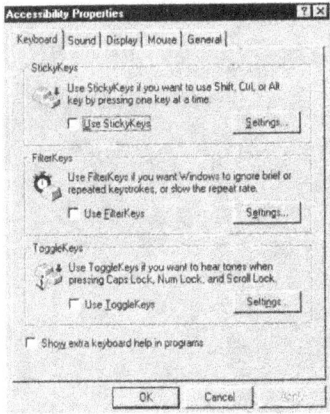

Windows® Me for Me!

ACCESSIBILITY OPTIONS

NOTE Many of the Accessibility Options have an associated shortcut key that allows you to turn them on and off from the keyboard. These are noted in the following sections.

The Accessibility Properties dialog box has five tabs: Keyboard, Sound, Display, Mouse, and General.

Keyboard Tab

Contains the options that facilitate use of the normal keyboard:

StickyKeys Allows you to press one of the modifier keys—Alt, Ctrl, or Shift—and another key one at a time instead of simultaneously. If you cannot press two or three keys at a time, you can still handle these keystrokes. To turn on StickyKeys, check the Use StickyKeys box or press the shortcut key and then click Apply. To set more options for StickyKeys, click the Settings button. To set up the shortcut key, click the Settings button to open the Settings for StickyKeys dialog box. Check the Use Shortcut box, and then click OK.

SHORTCUT KEY Press either Shift key five times.

FilterKeys Desensitizes the keyboard so that it is less likely to give you unwanted keystrokes. It does this by ignoring repeated or short-duration keystrokes or by slowing the rate at which a keystroke is repeated. These features are valuable if you have tremors or tend to "bounce" keys. To turn on FilterKeys, check the Use FilterKeys box or press the shortcut key and then click Apply. To set more options for FilterKeys, click the Settings button. To set up the shortcut key, click the Settings button to open the Settings for FilterKeys dialog box. Check the Use Shortcut box, and then click OK.

SHORTCUT KEY Hold right Shift key for 8 seconds.

ToggleKeys Plays a tone when you turn the Caps Lock, Num Lock, or Scroll Lock keys on or off. You will hear a high-pitched beep when you turn one of them on and a low-pitched beep when you turn one of them off. To turn on ToggleKeys, check the Use ToggleKeys box or press the shortcut key and then click Apply. Click the Settings button to enable or disable the shortcut for ToggleKeys. To set up the shortcut key, click the Settings button to open the Settings for ToggleKeys dialog box. Check the Use Shortcut box and then click OK.

Accessibility Options

SHORTCUT KEY Hold Num Lock key for 5 seconds.

Sound Tab

Contains the options that provide visual aids when Windows generates sounds. These are valuable primarily to people with hearing impairment. The Sound tab has the following options:

SoundSentry Allows you to see when Windows is beeping at you. In place of the audible cue, Windows flashes part of the screen. Click the Settings button to specify which part of the screen you want to flash.

ShowSounds Tells your applications to display a text caption in addition to an audible cue. To select this option, check the Use ShowSounds box.

Display Tab

Allows you to display alternative colors and fonts to make the screens more readable. Check the Use High Contrast box and then click the Settings button to choose the appropriate display options.

SHORTCUT KEY Press Left Alt+Left Shift+Print Screen.

Mouse Tab

Allows you to turn on MouseKeys. When MouseKeys is active, you can use the numeric keypad, which is on the right side of most keyboards, to perform all the functions that can be performed with a mouse. MouseKeys redefines the numeric keypad so that it functions as follows:

- Pressing 5 is the same as clicking the left mouse button once. Pressing 5 twice is the same as right-clicking the mouse.

- Pressing 2, 4, 6, or 8 moves the mouse pointer in the directions indicated by the arrows on the keys.

- Pressing Home, End, Page Up, or Page Down moves the pointer diagonally, according to where each key is positioned on the keypad.

ACCESSIBILITY OPTIONS

- Pressing the minus key (–) and then the 5 key is the same as right-clicking. Pressing the minus key (–) and then the plus key (+) is the same as double-clicking the right mouse button.

- Pressing the plus key (+) is the same as double-clicking the left mouse button.

- Pressing the Ins key is the same as holding down the left mouse button. Press Ins and then press an arrow key to drag a file or a folder.

- Pressing the Del key is the same as releasing the left mouse button after dragging.

- Pressing Ctrl and any number key except 5 jumps the pointer in large increments.

- Pressing Shift and any number key except 5 moves the pointer one pixel at a time.

- Pressing Num Lock switches between MouseKeys and the standard numeric keypad in whatever state (numeric entry or cursor movement) it was in before you started MouseKeys.

SHORTCUT KEY Press Left Alt+Left Shift+Num Lock

General Tab

Provides additional configuration settings that define how and when the Accessibility Options are available:

Automatic Reset Sets the length of time that your computer can be idle before the Accessibility Options turn off. This allows two people with different access needs to use the same computer.

Notification Determines whether the system gives a visual warning, an audio warning, or both when a feature is turned on or off.

SerialKey Devices Connects an alternative input device to your computer using a serial port. This device sends information to the computer, and that information is then treated as keystrokes and mouse events. Check this box to specify that you want to use an alternative input device and then click the Settings button to select the serial port and baud rate.

See also Accessibility Wizard, Control Panel, Display, Internet, Internet Explorer, Magnifier, On-Screen Keyboard

Accessibility Wizard

Gives easy access to several important accessibility features and lets you configure Windows for your level of vision, hearing, and mobility.

Choose Start ≻ Programs ≻ Accessories ≻ Accessibility ≻ Accessibility Wizard to start the Accessibility Wizard. Click Next to begin. Then in the Wizard's first dialog box, press the arrow keys or use the mouse to select the smallest size type that you can read easily and comfortably, and click Next. In the second dialog box, specify the font size you want to use for Windows title bars and menus, and click Next. If you check Use Microsoft Magnifier, a floating window opens at the top of your screen displaying a much enlarged view of the current screen, and you can select the options you want to use with the Magnifier.

In the next dialog box, check those boxes that apply to you, or click the Restore Default Settings button to go back to the original Windows settings. Depending on the choices you make in this dialog box, the Wizard asks you additional questions to help configure your system. Follow the instructions on the screen, and when you complete all the dialog boxes, you will see a final summary dialog box listing your Accessibility selections. Click Finish to put these settings into use, or click Back to refine them.

See also Accessibility Options, Control Panel, Display, Internet Explorer, Magnifier, On-Screen Keyboard

Active Desktop

In Windows, you can use a conventional Windows interface similar to that in earlier versions of Windows, or you can use the Active Desktop. The Active Desktop brings the world of the Web right to the Windows Desktop, allowing you to replace the static Windows wallpaper with a fully configurable, full-screen Web page. The Active Desktop can contain other Web pages, dynamic HTML, and even Java components such as stock tickers and ActiveX controls, and you can add these elements to the Taskbar or to a folder.

To set up your Active Desktop, right-click the Desktop and choose Active Desktop. You will see three options: View As Web Page, Customize My Desktop, and New Desktop Item.

ACTIVE DESKTOP

View As Web Page

Turns on the Active Desktop interface. Selecting this option a second time removes the check mark and turns the Active Desktop off again.

Customize My Desktop

Opens the Display Properties dialog box. You can also right-click the Desktop and select Properties, or if you prefer, choose Start ≻ Settings ≻ Control Panel and select the Display icon. The Display Properties dialog box contains six tabs, but we are concerned with only the two that relate to the active desktop:

Background Lets you choose an HTML document or a picture to use as your Desktop background. In the Wallpaper box, select the background you want to use, or click Pattern to choose or modify the background pattern. You can also click the Browse button to locate a file or to go directly to a Web site to find the HTML document you are interested in using as a background. To cover your entire Desktop with a small wallpaper image, select Tile from the Display box, or choose Center if you prefer to see the image centered. Click the Apply button to see the effect of your changes before you exit the Display Properties dialog box, or click OK to accept the changes and close the dialog box.

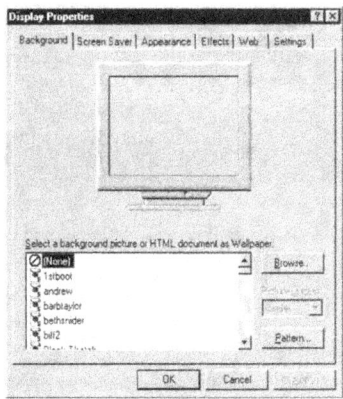

CTIVE DESKTOP

> **TIP** You can also right-click any Web page graphic that takes your fancy and then click Set As Wallpaper

Web Lets you select and organize Active Desktop elements. At the top of the tab, you will see a representation of your Desktop, indicating the location of any Active Desktop elements. These same elements are listed in the box below. Make sure that the Show Web Content on My Active Desktop box is selected, so the active elements will appear.

To add a new element, click the New button to run the new Active Desktop Item Wizard, described in the following section.

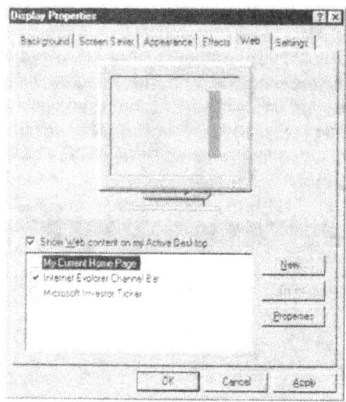

> **NOTE** You can also right-click any link on a Web page, drag it to your Desktop, and then click Create Active Desktop Item Here.

New Active Desktop Item

Runs the New Active Desktop Item Wizard. From here you can enter the URL of a page you want to make into a Desktop item, or you can click the Visit Gallery button to visit the Microsoft Desktop Gallery on the Web, where you can select

ADD NEW HARDWARE

and download active controls such as a stock ticker, a jukebox, or a clock. On that page, select the active item you want and then click Add to Active Desktop.

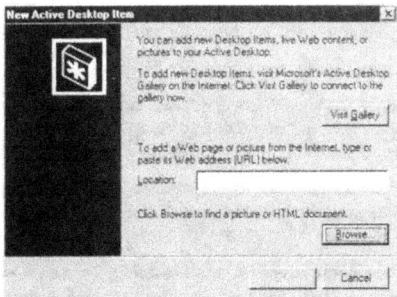

See also Control Panel, Desktop, Display, Folder Options, Internet Explorer, Start, Taskbar

Add New Hardware

Add New Hardware

Guides you through the process of adding new hardware to your system using the Add New Hardware Wizard. This Wizard automatically makes the appropriate changes to the Registry and to the configuration files so that Windows can recognize and support your new hardware. Be sure you have installed or connected your new hardware before you go any further.

To install a Plug-and-Play device on your system, follow these steps:

1. Turn off your computer.
2. Connect or install the new device according to the manufacturer's instructions.
3. Turn your computer back on to restart Windows. Windows will locate your new hardware automatically and install the appropriate software for you.

If Windows does not find your new Plug-and-Play device, check that it is installed properly, and if you can, confirm that the device actually works and is not defective in some way.

ADD NEW HARDWARE

If the device you want to install is not a Plug-and-Play device, follow these steps instead:

1. Turn off your computer.

2. Connect or install the new device according to the manufacturer's instructions.

3. Turn your computer back on to restart Windows.

4. Choose Start ≻ Settings ≻ Control Panel, and then double-click the Add New Hardware icon to open the Add New Hardware Wizard. Click the Next button. Then click Next again to search for Plug-and-Play devices.

5. You must now decide whether you want Windows to attempt to detect your non–Plug-and-Play hardware or if you want to identify the hardware yourself.

    a. Choose Yes if you want Windows to search for your new hardware. You will be warned that Windows may spend several minutes searching, and that your machine could quit functioning during the search. Click Next. A status monitor indicates the progress of the search. Depending on the amount and type of hardware, the detection process could take several minutes.

    b. If you don't want Windows to try to detect the device, click No and then click Next. A dialog box will prompt you to select the new device from a list. Click the hardware type you are installing, and then click Next.

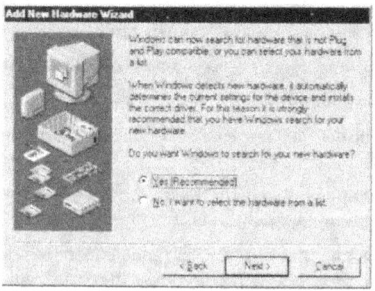

6. From this point on, the dialog boxes depend on the type of hardware you are installing. Simply follow the instructions on the screen to complete the installation.

See also Device Manager, Scanners and Cameras, System

Add/Remove Programs

Add/Remove Programs

Installs or uninstalls individual elements of the Windows operating system itself or certain application programs. Installing or removing application or system software components in this way enables Windows to modify all of the appropriate system and configuration files automatically so that the information in them stays current and correct.

To start Add/Remove Programs, choose Start ➢ Settings ➢ Control Panel, and then double-click the Add/Remove Programs icon to open the Add/Remove Programs Properties dialog box.

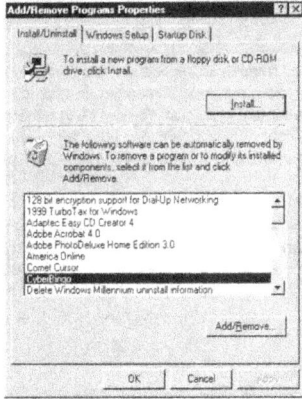

ADD/REMOVE PROGRAMS

Install/Uninstall Tab

To install a new program using the Add/Remove Programs applet, follow these steps:

1. Select the Install/Uninstall tab if it isn't already selected, and then click the Install button.

2. Put the application program CD or floppy disk in the appropriate drive, and click the Next button to display a setup or install message describing the program to be installed.

3. To continue with the installation process, click the Finish button. To make any changes, click Back and repeat the procedure.

To uninstall a program previously installed under Windows, you must follow a different process. The list of programs that have uninstall capability (not all of them do) will appear in the display box of the Install/Uninstall tab. Click the program you want to uninstall, and then click the Add/Remove button. You may see a warning message about removing the application. You will be told when the uninstall is finished. Uninstall may not remove all the files associated with the program. It may leave the program's folder and some data files behind, which you can then delete manually.

NOTE Once you remove an application using Add/Remove Programs, you will have to reinstall it from the original program disks or CD if you change your mind and decide you want to use it again. You can't just restore the files from the Recycle Bin back into their original folders because settings from the Start button and perhaps from the Windows Registry may have been deleted.

WARNING During the Uninstall, you may be asked whether you want to keep or delete a shared file. If in doubt, choose to keep it; it won't hurt anything, and it will ensure that any other programs using that file will still function.

Windows Setup Tab

Some components of the Windows operating system are optional, and you can install or uninstall them as you wish. The Windows Clipboard Viewer is an example. Select the Windows Setup tab to display a list of such components with checkboxes on the left. If the box has a check mark in it, the component is currently installed. If the checkbox is gray, only some elements of that component

ADD/REMOVE PROGRAMS

are installed. To see what is included in a component, click the Details button. Follow these steps to add a Windows component.

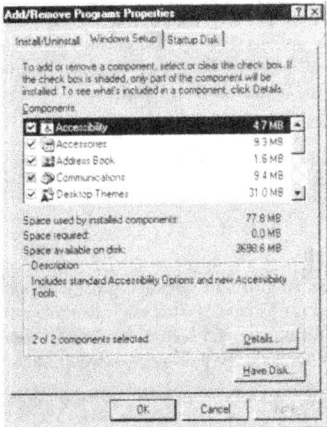

1. Click the feature's checkbox to select it on the Windows Setup tab.

    Or, if the component consists of several elements, click the Details button to display a list of them, and check the boxes you want to install. Then click OK to return to the main listing.

2. Click OK.

3. If prompted, insert the Windows CD and click OK to continue.

To remove a Windows component from your system, follow these steps:

1. Select the component on the Windows Setup tab.

2. Click to remove the checkmark next to the feature.

    Or, if the feature consists of several components, click the Details button, and then remove the checkmark next to the individual pieces of the feature. Click OK to return to the Windows Setup tab.

3. Click OK.

ADDRESS BOOK

Startup Disk Tab

A startup disk is a floppy disk with which you can start (or *boot*) your computer if something happens to your hard drive. When you originally installed Windows, you were asked if you wanted to create a startup disk. If you didn't do it at that time or if the disk you created then is not usable, you can create one now. Simply insert a disk with at least 1.2MB capacity in the appropriate drive, click Create Disk, and follow the instructions on the screen.

> **NOTE** Almost all floppies these days are double-sided, high-density 3.5" disks with 1.44MB capacity.

Network Install Tab

In some cases, you can also install a program directly from a network using the Network Install tab. If the Network Install tab is not present in the Add/Remove Programs Properties dialog box, this feature may not have been enabled on your computer or on your network. See your system administrator for more details.

If the Install/Uninstall tab is present, your system is currently connected to the network, and you can click Install followed by Next to find the setup program for your network. Follow the instructions on-screen.

Address Book

 Address Book Manages your e-mail addresses, as well as your voice, fax, modem, and cellular phone numbers. Once you enter an e-mail address in your Address Book, you can select it from a list rather than type it in every time. To open Address Book, choose Start ➢ Programs ➢ Accessories ➢ Address Book, or click the Address Book icon on the Outlook or Outlook Express toolbar.

Importing an Existing Address Book

Address Book can import information from an existing address book in any of the following formats:

- Windows Address Book
- Microsoft Exchange Personal Address Book
- Microsoft Internet Mail for Windows 3.1 Address Book

ADDRESS BOOK

- Netscape Address Book
- Netscape Communicator Address Book
- Eudora Pro or Lite Address Book
- Lightweight Directory Access Protocol (LDAP)
- Comma-separated text file

To import information from one of these address books, follow these steps:

1. Choose Start ➢ Programs ➢ Accessories ➢ Address Book, or click the Address Book icon on the Outlook Express toolbar.
2. Choose File ➢ Import ➢ Address Book to open the Select Address Book File to Import From dialog box.
3. Select the file you want to import, and click OK.

WARNING If you have Outlook installed, and have the address book set up to automatically share contacts with Outlook, you will not be able to import addresses; the Import command will be unavailable. To stop sharing with Outlook, choose Tools ➢ Options and choose Do Not Share Information Among Microsoft Outlook and Other Applications. Then close the Address Book and reopen it, and you will be able to import.

Address Book Window

The main Address Book window lists the names and e-mail addresses of individuals and groups of individuals, along with business and home phone information if it is available.

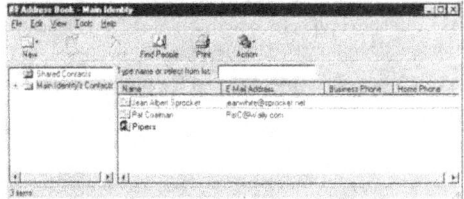

ADDRESS BOOK

The Address Book toolbar contains the following buttons:

New   Opens a menu containing three options: New Contact, New Group, and New Folder.

Properties   Displays the Properties dialog box for the selected entry.

Delete   Removes the selected entry.

Find People   Opens the Find People dialog box so that you can look up people and businesses using the Internet.

Print   Prints the address list.

Action   Opens a menu containing the following options: Send Mail, Send Mail To, Dial, and Internet Call.

Some of these functions are repeated on the Address Book menus, particularly the File, Edit, and Tools menus. The View menu contains options that you can use to configure the toolbar, icons, and entry sort order.

Creating a New Address Book Entry

To add a new entry to your Address Book, click the New button on the Address Book toolbar and choose New Contact, or choose File ➢ New Contact to open the Properties dialog box. This dialog box has the following tabs:

Name   Lets you enter personal information, including the person's first, middle, and last names; a nickname; and an e-mail address.

Home   Allows you to enter additional information about this contact. Enter as much or as little information as makes sense here.

Business   Allows you to enter business-related information. Again, enter as much or as little information as makes sense.

Personal   Allows you to enter personal information, including birthday and anniversary dates.

Other   Offers a chance to store additional information about this contact as a set of text notes.

ADDRESS BOOK

NetMeeting Lets you enter NetMeeting information such as a person's conferencing e-mail address and server name. If NetMeeting is not installed on your system, this tab will be called Conferencing.

Digital IDs Allows you to specify a digital certificate for use with an e-mail address.

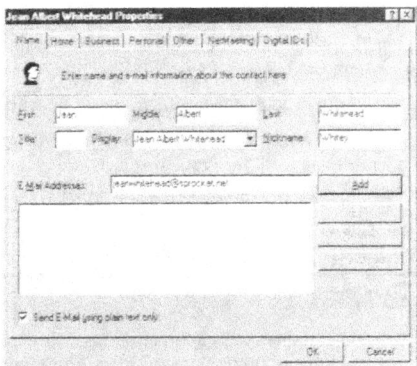

Once you have created a contact entry, when you review an existing contact, you will see essential information presented in a Summary tab in the Properties dialog box.

Setting Up a New Group

You can create groups of e-mail addresses to make it easy to send a message to all the members of the group. You can group people any way you like—by job title, musical taste, or sports team allegiance. When you want to send e-mail to everyone in the group, simply use the group name instead of selecting each e-mail address individually. To create a new group, follow these steps:

1. Click the New button on the Address Book toolbar and select New Group, or choose File ≻ New Group to open the Properties dialog box.

2. In the Group tab, enter the name you want to use for this group into the Group Name field.

ADDRESS TOOLBAR

3. If you want to add a person to this new group who is not yet in your Address Book, click New Contact to open the Properties dialog box. Enter the information, and click OK.

4. Click Select Members to open the Select Group Members dialog box. Add the names of those people who already have entries in your Address Book into this new group. Click OK when you are done to return to the Properties dialog box.

5. In the Group Details tab, use the Notes field for comments about the group. You might note that they meet at the local bookstore on Thursday evenings at 7:30, for example.

6. When you have finished adding members to the group, click OK.

You will now see the name of this new group displayed in the Address Book main window. To send e-mail to all the group members, select the group name, click the Action button on the toolbar, and select Send Mail.

See also Internet Explorer, Outlook Express, Search

Address Toolbar

Shows the location of the page currently being displayed in the main window. This may be a URL on the Internet or on an intranet, or a file or folder stored on your hard disk.

To go to another page, click the arrow at the right end of the Address toolbar to select the appropriate entry, or simply type a new location. When you start to type an address that you have previously entered, the AutoComplete feature recognizes the address and completes the entry for you.

The Address toolbar is available in most Windows file management windows, including Explorer, Internet Explorer, My Computer, Control Panel, and others.

See also Explorer, Internet Explorer

Automatic Updates

Automatic Updates

Keeps your copy of Windows Millennium updated with the most recent files from Microsoft. After initial configuration, the feature works behind the scenes, notifying you when an update is ready and installing it, with your approval.

AUTOMATIC UPDATES

After you install Windows, an Update icon appears in your system tray. Click it to view a license agreement for the feature, and accept it.

Then, whenever you are connected to the Internet, the Automatic Update feature periodically checks for updates and downloads them to your PC. When an update is available for installation, the Automatic Update icon reappears in the system tray. Click on it and then follow the prompts to install the update.

You can configure Automatic Update through its icon in the Control Panel. Double-click its icon to open an Automatic Updates box in which you can choose whether to be notified before the download, or whether you wish to turn the feature off completely.

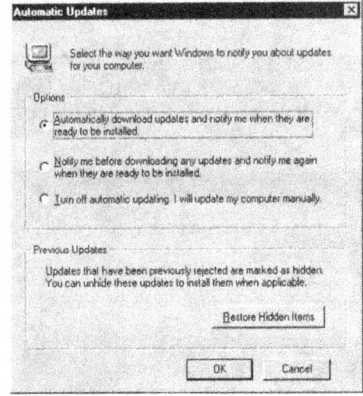

BACKUP

Backup

Previous versions of Windows came with a Backup program, but this version does not. You can, however, buy a third-party backup program, such as BackupExec, at any computer or office supply store.

Briefcase

See My Briefcase

Browse

The Browse button is available in many common dialog boxes when you have to choose or enter a filename, find a folder, or specify a Web address, or URL. Clicking the Browse button or the Find File button opens the Browse dialog box.

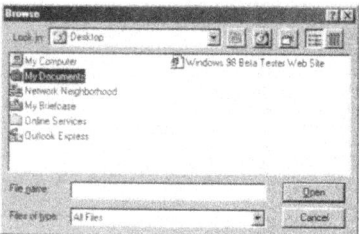

You can look through folders on any disk on any shared computer on the network to find the file you want. When you find the file, folder, computer, or Web site, double-click it to open, import, or enter it in a text box.

CALCULATOR

Calculator

Performs standard arithmetic and scientific or statistical operations. The calculation procedures and the keys available depend on whether you are doing standard or scientific math. Let's take a look.

Calculator Window

Choose Start ➤ Programs ➤ Accessories ➤ Calculator to open the Calculator window. It opens in the view selected when the window was last closed.

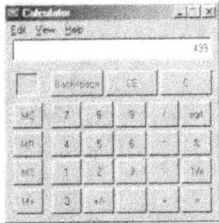

In both Standard and Scientific views, the calculator contains these keys:

Backspace Deletes a single digit.

CE Erases the last entry. You can also press the Delete key on the keyboard.

C Clears a calculation altogether. You can also press the Esc key on the keyboard.

MC Clears a number from the calculator's memory.

MR Recalls the number from the calculator's memory.

MS Stores a number in the calculator's memory, erasing whatever was already in memory. An M appears in the box on the left below the display area.

M+ Adds a number to the number in the calculator's memory.

Windows ® Me for Me!

xxx

FOREWORD

As has occurred in the past, there may come a time in the future where your natural right to resist unjust rule is curtailed by totalitarian governments or foreign invaders to such a degree that simply owning a book dealing with the types of subject matter contained herein is a criminal offense. Indeed, if you're living in a place which is suffering under martial law, such an offense may be punishable by summary execution without trial.

There are two courses of action to ensuring such an issue doesn't become a problem for you:

1. Cooperate completely with whichever ruling junta has its boot on your neck and refrain from owning, reading and diseminating such material;
or

2. APPEAR to cooperate fully with whichever ruling junta you are covertly plotting and/or fighting against, and keep cleverly camouflaged copies of useful books such as this one to hand, hidden in plain sight.

Sandwiched between the dull, but delightfully 90s-retro covers of Charlotte O. Vert's 20 years-out-of-date computer book *Windows Me for Me!* are four classic works by some of the greatest guerrilla warfare theorists and practioners of the 20th Century.

Each of these works, if presented as a traditional book, are so well known that they would be spotted a mile off by even the lowliest jack-booted thug conducting a warrantless search of your premises. That same jack-booted thug, if not completely bored to tears by the title and covers, would, upon flicking through the first and last pages of this publication, disregard it completely, then toss it over his or her shoulder before continuing to rifle through your belongings looking for incriminating evidence against you.

The works reproduced in *Windows Me for Me!* are a quartet of publications written by guerrilla warfare theorists at the height of their fame between the 1930s and 1960s. The authors who wrote the works selected for this particular volume are all communist in orientation, but don't let this deter you if you are opposed to that ideology. These works have inspired and educated freedom fighters of all political stripes around the world since publication. Each of the works reproduced here appears on the "required-reading" lists for Free World special operations personnel whose role is to organise secret guerrilla armies in foreign countries right under the noses of despotic regimes.

These works were included in this volume specially as a hand-picked selection of high quality manuals on the theories and practice of unconventional warfare operations against an oppressive regime. If they are good enough as light reading for a US Special Forces team they are certainly good enough for the rest of us.

The first item, *People's War People's Army* by North Vietnamese Army General Vo Nguyen Giap, was published in 1961, just seven years after the defeat of the French colonial power in Indochina at Dien Bien Phu, and just as the Vietnam War as we know it was beginning to hot up. Ultimately, Giap's strategies as outlined in People's War People's Army, would see the defeat of not only the French, but also the South Vietnamese and the United States. The Vietnamese conflicts, as broken down in Giap's book, remain a textbook example of where a group with minimal equipment, but with an eye on the "long game" can triumph over a technologically-and-numerically superior opponent. It's little wonder that modern enemies of the Western Way of Life such as Al Qaeda and Daesh have extensively studied Giap's teachings and integrated them into their own "long game" strategies.

The second item, Mao Tse Tung's *On Guerrilla Warfare*, was written in 1937 at a time when Mao's Chinese communists had fought Nationalist Generalissimo Chiang

Kai Shek's forces to a standstill, but had joined with their foe to battle the Imperial Japanese who had invaded parts of China from 1931 and who had amped up their operations in China by the time of the publication of mao's book, capturing and sacking the Chinese capital city Nanking in a barbarous orgy of rape, kiling and arson. After the defeat of Japan in 1945, the cmmunists and nationalists again took up arms against each other until the communist victory in 1949. The nationalists fled to Taiwan and the communist's resulting People's Republic of China eventually became an economic and political superpower.

The book is not only a theoretical exploration of the strategy of guerrilla warfare, it's also a "how-to" manual. One of the most influential documents of the 20th Century, Mao Tse Tung's *On Guerrilla Warfare* has become THE basic textbook for waging revolution and for fighting foreign invaders in underdeveloped and emergent areas throughout the world and will continue to do so well into the 21st Century and probably beyond.

The third item is the cult classic by Ernesto "Che" Guevara, *Guerrilla Warfare*. Guevara was a commander in Fidel Castro's anti-Batista guerrilla army in Cuba in the late 1950s. Most guerrilla campaigns against a despotic regime at home begin with a protracted period of discontent among the population and civil unrest which may span decades before breaking out into a guerrilla war. The social and economic conditions in Cuba at the time hadn't even come close to reaching those levels of public discontent, and at that rate the spontaneous creation of a resistance army to fight the Batista regime may not have naturally occurred until the 1970s or later.

Castro and Guevara devised a means to speed up the process, which would not only serve to get the war over and done with quicker, but would also catch the enemy off guard. After all, no one was starving in Cuba before Castro's guerrillas came so as far as the regime was concerned, there was no reason for the Cuban people to take up arms against them. Castro's and Che Guevara's strategy was what they called "foco" (Spanish for "focus"). This meant it wasn't necessary to wait for the conditions to be ripe for guerrilla warfare. Instead they would begin operations and make the guerrilla army itself the focal point of the anti-regime activities, in similar fashion to a snowball rolling down a hill. The act of armed resistance against the oppressor would itself lead to an uprising by the people and this is exactly what happened in Cuba, leading to one-time political dissident and revolutionary Fidel Castro gaining power as Prime Minister of Cuba in 1959.

Che Guevara detailed the salient points of the Cuban Revolution in his book, which was also written to function as a "how-to" manual. Regardless of the applicability and usefulness of the "foco" theory of guerrilla warfare he espouses, Guevara's book remains a valuable resource for the study and conduct of guerrilla warfare.

The final item in this volume is *One Hundred Fifty Questions to a Guerrilla* by Alberto Bayo y Giroud. Born in Cuba, Bayo had been a loyalist Republican guerrilla commander fighting against the Nationalists in the Spanish Civil War (1936-1939). With the victory of the pro-fascist, Nazi-German-supported Nationalists under General Franco, Bayo fled to Mexico, where in 1955 he met Cuban dissident Fidel Castro and an Argentinian doctor-turned wannabe-Marxist-revolutionary called Che Guevara. Bayo was reinfected with revolutionary zeal and set about teaching the pair everything he knew about fighting a guerrilla war, training Castro's cadre of guerrillas in secret training sessions in the Chapultepec park in Mexico City. *One Hundred Fifty Questions to a Guerrilla* was written around the same time to act as an aid de memoir for the Cuban guerrillas and along with Mao Tse Tung's *On Guerrilla Warfare* it became a standard textbook at guerrilla training camps all over Latin America from the 1960s onwards.

Presented in a "question and answer" format, Bayo addresses how guerrilla units should be organized, how to defend a conquered city, the skillset of the "perfect guerrilla" and how to attack a town.

Other volumes in this series cover security, organisation, intelligence, ordnance, supply and guerrilla warfare theory. All are supplied under boring, unrelated covers, but all are "authored" by Charlotte O. Vert.

                C.O. Vert Press,
                          2017

CONTENTS

Introduction (for camouflage)		vii
Foreword		Master Page 1
Contents		Master Page 5
Part 1 -	*People's War People's Army* - Vo Nguyen Giap	Master Page 7
Part 2 -	*On Guerrilla Warfare* - Mao tse Tung	Master Page 249
Part 3 -	*Guerrilla Warfare* - Ernesto "Che" Guevara	Master Page 377
Part 4 -	*One Hundred Fifty Questions to a Guerrilla* - Alberto Bayo Giroud	Master Page 455
Epilogue (for camouflage)		cdxciii

PART 1

PEOPLE'S WAR
PEOPLE'S ARMY

VO NGUYEN GIAP

1961

NOTICE TO THE READER

The author, General Vo Nhuyen Giap, was the commanding general of the Vietminh army at Dien Bien Phu, and he is its commanding general now. The original publisher is in Communist territory and may be presumed to be a Communist instrumentality. It should be obvious, therefore, that much of the content of this book is sharply colored with Communist perspectives toward the situation and events described. We trust that the reader will provide his own discounts of this version of the political situation and behavior of the Free World powers involved, the degree of US responsibility for the French strategy, the alleged universality of the support of the Vietminh by the people of Indo-China, and the many other propaganda distortions and falsities.

This book is relevant and useful for those portions which describe General Giap's concepts of the strategy and tactics of guerrilla and revolutionary warfare, and his experience in an unquestionably effective application of those concepts. It does not necessarily follow that the US and its allies should copy all of his strategy and tactics in all comparable circumstances, but, at the minimum, they need to be understood. These concepts continue to appear in the same area and in other battlefields where the US has the problem of coping with Communist guerrilla and revolutionary warfare.

U.S. GOVERNMENT PRINTING OFFICE : 1962   O—647570

*PEOPLE'S WAR
PEOPLE'S ARMY*

General VO NGUYEN GIAP

PEOPLE'S WAR
PEOPLE'S ARMY

CONTENTS

	Page
Publisher's note	7
The Vietnamese people's war of liberation against the French imperialists and the American interventionists (1945-1954).	9
People's war, people's army	39
The great experiences gained by our Party in leading the armed struggle and building revolutionary armed forces	65
Dien Bien Phu	151

HO CHI MINH
President of the Democratic Republic of Viet Nam

PUBLISHER'S NOTE

We are very pleased to publish the English translation of a series of articles by General Vo Nguyen Giap, member of the Political Bureau of the Central Committee of the Viet Nam Workers' Party, Vice-Premier and Minister for National Defence of the Democratic Republic of Viet Nam, Commander-in-Chief of the Viet Nam People's Army.

In these articles, the author introduces the liberation war waged by the Vietnamese people, the features of that war, and deals with the reasons for victory: mobilisation of the entire people, setting up of a people's army, merging of all patriotic organisations and people into a united national front, clearsighted leadership of the Party of the working class. It lays particular stress on the problem of organisation and direction of the revolutionary armed forces of Viet Nam. In a word, it is the combination of experiences gained by the Vietnamese people in the course of a long struggle against colonialism, for national independence, struggle which ended in 1954 with the brilliant Dien Bien Phu victory and the signing of the Geneva Agreements.

Publication of this book is most timely.

It is true that since the end of World War II, the maps of Asia, Africa and Latin America have been subject to radical changes, other countries will soon be independent and colonialism is unquestionably doomed to collapse. It is no less true that great obstacles still stand in the way of the peoples struggling for their liberation. The Algerian war has just entered its seventh year. The so-called "U.N. action" in the Congo has turned out to be an imperialist plot against Lumumba's motherland. Cuba is subject to daily provocations by the U.S.A. Half of Viet Nam's territory is still under the heel of a new type colonialism "made in U.S.A."

We hope that all our friends who, like us, are still suffering from imperialist designs and threats will find in the "People's War, People's Army" what we ourselves have found: further reasons for confidence and hope.

THE VIETNAMESE PEOPLE'S WAR OF LIBERATION AGAINST THE FRENCH IMPERIALISTS AND THE AMERICAN INTERVENTIONISTS (1945-1954)

I

A FEW HISTORICAL AND GEOGRAPHICAL CONSIDERATIONS

Viet Nam is one of the oldest countries in South-east Asia.

Stretching like an immense S along the edge of the Pacific, it includes Bac Bo or North Viet Nam which, with the Red River delta, is a region rich in agricultural and industrial possibilities, Nam Bo or South Viet Nam, a vast alluvial plain furrowed by the arms of the Mekong and especially favourable to agriculture, and Trung Bo or Central Viet Nam, a long, narrow belt of land joining them. To describe the shape of their country, the Vietnamese like to recall an image familiar to them: that of a shoulder pole carrying a basket of paddy at each end.

Viet Nam extends over nearly 330.000 square kilometres on which lives a population of approximately 30 million inhabitants. During its many thousands of years old history, the Vietnamese people have always been able to maintain an heroic tradition of struggle against foreign aggression. During the 13th century in particular, they succeeded in thwarting attempts at invasion by the Mongols who had extended their domination over the whole of feudal China.

From the middle of the 19th century, the French imperialists began undertaking the conquest of the country.

Despite resistance lasting dozens of years, **Viet Nam** was progressively reduced to the state of a colony, thereafter to be integrated in 'French Indo-China' with Cambodia and Laos. But from the first day of French aggression, the national liberation movement of the Vietnamese people unceasingly developed. The repression which attempted to stifle this movement only stirred it up the more; so much so, that after the First World War, it began to take on a powerful mass character and had already won over wide circles of the intellectual and petty bourgeois levels, while penetrating deeply into the peasant masses as well as into the working class which was then beginning to form. The year 1930 saw another step forward with the founding of the Indochinese Communist Party, now the Viet Nam Workers' Party which took upon itself the mission of leading the national democratic revolution of the Vietnamese people against the imperialists and the feudal landlord class.

Just after the launching of the Second World War in 1939, France was occupied by the Nazis, while Viet Nam was progressively becoming a colony of the Japanese fascists. The Party was able in good time to appreciate the situation created by this new development. Estimating that a new cycle of war and revolution had begun, it set as a task for the whole nation the widening of the anti-imperialist national united front, the preparation of armed insurrection and the overthrow of the French and Japanese imperialists in order to reconquer national independence. The Viet Nam Doc Lap Dong Minh (League for the Independence of Viet Nam, abbreviated to Viet Minh) was founded and drew in all patriotic classes and social strata. Guerilla warfare was launched in the High Region of Bac Bo. A free zone was formed.

In August 1945, the capitulation of the Japanese forces before the Soviet Army and the Allied forces, put an end to the world war. The defeat of the German and Nippon fascists was the beginning of a great weakening of the capitalist system. After the great victory of the Soviet Union, many people's democracies saw the light of day. The socialist system was no longer confined within the frontiers of a single country. A new historic era was beginning in the world.

In view of these changes, in Viet Nam, the Indochinese Communist Party and the Viet Minh called the whole Vietnamese nation to general insurrection. Everywhere, the people rose in a body. Demonstrations and displays of force followed each other uninterruptedly. In August, the Revolution broke out, neutralising the bewildered Nippon troops, overthrowing the pro-Japanese feudal authorities, and installing people's power in Hanoi and throughout the country, in the towns as well as in the countryside, in Bac Bo as well as in Nam Bo. In Hanoi, the capital, in September 2nd, the provisional gouvernement was formed around President Ho Chi Minh ; it presented itself to the nation, proclaimed the independence of Viet Nam, and called on the nation to unite, to hold itself in readiness to defend the country and to oppose all attempts at imperialist aggression. The Democratic Republic of Viet Nam was born, the first people's democracy in South-east Asia.

But the imperialists intended to nip the republican regime in the bud and once again transform Viet Nam into a colony. Three weeks had hardly gone by when, on September 23rd, 1945, the French Expeditionary Corps opened fire in Saigon. The whole Vietnamese nation then rose to resist foreign aggression. From that day, began a war of national liberation which was to be carried on for nine years at the cost of unprecedented heroism and amidst unimaginable difficulties, to end by the shining victory of our people and the crushing defeat of the aggressive imperialists at Dien Bien Phu.

But at a time when, in the amazing enthusiasm aroused by the August Revolution, the Vietnamese people were closing their ranks around the provisional government, a new factor intervened which was to make the political situation more difficult and more complex. According to the terms of an agreement between the Allies, in order to receive the Japanese surrender, the Chinese Kuomintang forces entered in a body in the part of Viet Nam situated north of the 16th parallel, while the British forces landed in the South. The Chiang Kai-shek troop took advantage of the opportunity to pillage the population and sack the country, while using every means to help the most reactionary elements among the Vietnamese bourgeois and landlords — the members of the Viet Nam Quoc Dan Dang (the Vietnamese Kuomintang) and the pro-Japanese Phuc Quoc (Vietnamese National Restoration Party) — to stir up trouble throughout the country. After occupying the five frontier provinces, they provoked incidents even in the capital, and feverishly prepared to overthrow people's power. In the South, the British actively exerted themselves to hasten the return of the French imperialists. Never before had there been so many foreign troops on the soil of Viet Nam. But never before either, had the Vietnamese people been so determined to rise up in combat to defend their country.

These are the broad outlines of the historical and geographical conditions indispensable to an understanding of the unfolding of the war of national liberation of the Vietnamese people.

II

SUMMARY OF THE PROGRESS OF THE WAR OF NATIONAL LIBERATION

At the outset of the war, the French imperialists' scheme was to rely upon the British troops to reconquer Nam Bo and afterwards to use it as a springboard for preparing their return to the North. They had shamefully capitulated before the Japanese fascists, but after the ending of the world war, they considered the resumption of their place at the head of their former colony as an indisputable right. They refused to admit that in the meantime the situation had radically changed.

In September 1945, French colonial troops armed by the British and soon strengthened by the French Expeditionary Corps under the command of General Leclerc, launched aggression in Saigon, with the direct support of the British army. The population of Nam Bo immediately rose up to fight. In view of the extreme weakness of its forces at the beginning, people's power had to withdraw to the countryside after waging heroic street fights in Saigon and in the large towns. Almost the whole of the towns and important lines of communication in Nam Bo and the South of Trung Bo gradually fell into the hands of the adversary.

The colonialists throught they were on the point of achieving the reconquest of Nam Bo, and General Leclerc declared that occupation and pacification would be completed in ten weeks. But events took quite a different turn. Confident of the support of the whole country, the southern population continued the fight. In all the campaigns in Nam Bo the guerilla forces were going from strength to strength, their bases were being consolidated and extended, and people's power was maintained and strengthened during the nine years of the Resistance, until the reestablishment of peace.

Knowing that the invasion of Nam Bo was only the prelude to a plan of aggression by the French imperialists, our Party guided the whole nation toward preparing a long-term resistance. In order to assemble all the forces against French imperialism, the Party advocated uniting all the elements that could be united, neutralising all those that could be neutralised, and widening the National United Front by the formation of the Lien Viet (Viet Nam People's Front) urgently organising general elections with universal suffrage in order to form the first National Assembly of the Democratic Republic of Viet Nam responsible for passing the Constitution and forming a widely representative resistance government grouping the most diverse elements including even those of the Viet Nam Quoc Dan Dang (the Vietnamese Kuomintang). At that time, we avoided all incidents with the Chiang Kai-shed troops.

The problem then before the French Expeditionary Corps was to know whether it would be easy for them to return to North Viet Nam by force. It was certainly not so, because our forces were more powerful there than in the

South. For its part, our Government intended doing everything in its power to preserve peace so as to enable the newly created people's power to consolidate itself and to rebuild the country devastated by long years of war. It was thus that negotiations which ended in the Preliminary Agreement of March 6 th, 1946, took place between the French colonialists and our Government. According to the terms of this convention, limited contingents of French troops were allowed to station in a certain number of localities in North Viet Nam in order to co-operate with the Vietnamese troops in taking over from the repatriated Chiang Kai-shek forces. In exchange, the French Government recognised Viet Nam as a free state, having its own government, its own national assembly, its own army and finances, and promised to withdraw its troops from Viet Nam within the space of five years. The political status of Nam Bo was to be decided by a referendum.

Relations between the Democratic Republic of Viet Nam and France were then at a crossroads. Would there be a move towards consolidation of peace or a resumption of hostilities ? The colonialists considered the Preliminary Agreement as a provisional expedient enabling them to introduce part of their troops into the North of Viet Nam, a delaying stratagem for preparing the war they intended to continue. Therefore, the talks at the Dalat Conference led to no result and those at the Fontainebleau Conference resulted only in the signing of an unstable modus vivendi. During the whole of this time, the colonialists partisans of war were steadily pursuing their tactics of local encroachments. Instead of observing the armistice, they continued their mopping-up operations in Nam Bo, and set up a local puppet government there ; in Bac Bo they increased provocations and attacked a certain number of provinces, pillaging and massacring the population of the Hongai mining area, and everywhere creating an atmosphere of tension preparatory to attacks by force.

Loyal to its policy of peace and independence, our Government vainly endeavoured to settle conflicts in a friendly manner, many times appealing to the French Government then presided over by the S.F.I.O. (Socialist Party) to change their policy in order to avoid a war detrimental to both sides. At the same time we busied ourselves with strengthening our rear with a view to resistance. We obtained good results in intensifying production. We paid much attention to strengthening national defence. The liquidating of the reactionaries of the Viet Nam Quoc Dan Dang was crowned with success and we were able to liberate all the areas which had fallen into their hands.

In November 1946, the situation worsened. The colonialists by a *coup de force* in Haiphong seized the town. After engaging in street fights, our troops withdrew to the suburbs. In December, the colonialists provoked tension in Hanoi, massacred civilians, seized a number of public offices, sent an ultimatum demanding the disarming of our self-defence groups and the right to ensure order in the town, and finally provoked armed conflict. Obstinately the colonialists chose war which led to their ruin.

On December 19th, resistance broke out throughout the country. The next day, in the name of the Party and of the Government, President Ho Chi Minh called on the whole people to rise up to exterminate the enemy and save the country, to fight to the last drop of blood, and whatever the cost, to refuse re-enslavement.

* * *

At the time when hostilities became generalised throughout the country, what was the balance of forces? From the point of view of material, the enemy was stronger than us. Our troops were thus ordered to fight the enemy wherever they were garrisoned so as to weaken them and prevent them spreading out too rapidly, and thereafter, when conditions became unfavourable to us, to make the bulk of our forces fall back towards our rear in order to keep our forces intact with a view to a long-term resistance. The most glorious and most remarkable combats took place in Hanoi, where our troops succeeded in firmly holding a huge sector for two months before withdrawing from the capital unhurt.

The whole Vietnamese people remained indissolubly united in a fight to the death in those days when the country was in danger. Replying to the appeal by the Party, they resolutely chose the path of Freedom and Independence. The central government, having withdrawn to bases in the mountainous region of Viet Bac, military zones — soon united in interzones — were formed, and the power of local authorities was strengthened for mobilising the whole people and organising the resistance. Our government continued appealing to the Frenh government not to persist in their error and to reopen peaceful negotiations. But the latter under the pretext of negotiation demanded the disarming of our troops. We replied to the colonialists' obstinacy by intensifying the resistance.

In fact, the French High Command began regrouping forces to prepare a fairly big lightning offensive in the hope of ending the war. In October 1947, they launched a big campaign against our principal base, Viet Bac, in order to annihilate the nerve centre of the resistance and destroy our regular forces. But this large-scale operation ended in a crushing defeat. The forces of the Expeditionary Corps suffered heavy losses without succeeding in causing anxiety to our leading organisations or impairing our regular units. It was a blow to the enemy's strategy of a lightning war and a rapid solution. Our people were all the more determined to persevere along the path of a long-term resistance.

***

From 1948, realising that the war was prolonging itself, the enemy changed their strategy. They used the main part of their forces for "pacification" and for consolidating the already occupied areas, in Nam Bo especially, applying the principle: fight Vietnamese with Vietnamese, feed war with war. They set up a puppet central government, actively organised supplementary local units, and indulged in economic pillage. They gradually extended their zone of occupation in the North and placed under their control the major part of the Red River delta. During all these years, the French Expeditionary Corps followed a procedure of great dispersion, scattering their forces in thousands of military posts to occupy territory and control the localities. But ever-growing military and financial difficulties gradually led the French imperialists to let the American imperialists interfere in the conflict.

***

The enemy having altered their strategy, we then advocated the wide development of guerilla warfare, transforming the former's rear into our front line. Our units operated in small pockets, with independent companies penetrating deeply into the enemy-controlled zone to launch guerilla warfare, establish bases and protect local people's power. It was an extremely hard war generalised in all domains: military, economic and political. The enemy mopped-up, we fought against mopping-up. They organised supplementary local Vietnamese troops and installed puppet authorities; we firmly upheld local people's power, overthrew men of straw, eliminated traitors and carried out active propaganda to bring about the disintegration of the supplementary forces. We gradually formed a network of guerilla bases. On the map showing the theatre of operations besides the free zone, "red zones", which ceaselessly spread and multiplied, began to appear right in the heart of the occupied areas. The soil of the fatherland was being freed inch by inch right in the enemy's rear lines. There was no clearly-defined front in this war. It was there where the enemy was. The front was nowhere, it was everywhere. Our new strategy created serious difficulties for the enemy's plan to feed war with war and to fight Vietnamese with Vietnamese and finally brought about their defeat.

The centre of gravity of the front was gradually moving towards the enemy's rear. During this time, the free zone was continually being consolidated. Our army was growing in the struggle. The more our guerillas developed and the more our local units grew, the more we found ourselves able to regroup our forces. At the end of 1948 and the beginning of 1949, for the first time we launched small campaigns which inflicted considerable losses on our adversary. The imperialists were beginning to feel great anxiety. The commission of enquiry presided over by General Revers made a fairly pessimistic report which came to the conclusion that it was necessary to ask the United States for more aid.

* * *

1949 saw the brilliant triumph of the Chinese Revolution and the birth of the People's Republic of China. This great historic event which altered events in Asia and the world, exerted a considerable influence on the war of liberation of the Vietnamese people. Viet Nam was no longer in the grip of enemy encirclement, and was henceforth geographically linked to the socialist bloc.

At the begining of 1950, the Democratic Republic of Viet Nam was officially recognised by the People's Republic of China, the Soviet Union and the brother countries. The following year, the second Congress of the Indochinese Communist Party decided to alter the name of the Party and founded the Viet Nam Workers' Party. The Viet Minh and the Lien Viet were amalgamated. In 1953, the Party and the Government decided to carry out agrarian reform in order to liberate productive forces and give a more vigorous impulse to the Resistance. All these facts contributed to shaping to our advantage the course of our struggle.

In effect, 1950 marked the opening of a new phase in the evolution of our long Resistance. During the winter, in the frontier campaign, for the first time, we opened a relatively big counter-attack which resulted in the liberation of the provinces of Cao Bang, Lang Son and Lao Cai. Immediately after, we began a series of offensive operations on the delta front.

The enemy, routed, sent General De Lattre de Tassigny to Indo-China. The military aid granted by the United States following an agreement signed in 1950, was on the increase. The aggressive war waged by the French colonialists gradually became a war carried out with "U.S. dollars" and "French blood". It was really a 'dirty war'.

De Lattre's plan, approved by Washington, provided for a strong line of bunkers in the Red River delta to stop our progress, and for a regrouping of forces in order to launch violent mopping-up operations so as at all costs to 'pacify' the rear and create the right conditions for an offensive which would enable the French forces to recapture the initiative while attacking our free zone. In October 1951, the enemy occupied Hoa Binh. We replied by immediately launching the Hoa Binh campaign. On the one hand we contained and overwhelmed the adversary's forces on the "opposite" front, on the other hand, we took advantage of their exposed disposition of troops to get our divisions to strike direct blows at their rear in the Red River delta. Our large guerilla bases were extending further still, freeing nearly two million inhabitants. Hoa Binh was released. De Lattre's plan was checked.

In 1952, we launched a campaign in the North-Western zone and freed vast territories as far as Dien Bien Phu. At the beginning of 1953, units of Vietnamese volunteers, co-operating with the Pathet Lao liberation army, began the campaign in Higher Laos which brought about the liberation of Sam Neua.

In short, the face of the various theatres of operations was as follows:

The main front was that of North Viet Nam where most of the big battles were taking place. At the beginning of 1953, almost the whole of the mountainous region, say, more than two thirds of the territory of North Viet Nam, had been liberated. The enemy still occupied Hanoi and the

Red River delta, but outside the large towns and the important lines of communication, our enlarged guerilla bases—our free zone — already embraced nearly two thirds of the villages and localities situated in the enemy rear. In Central and South Viet Nam, we still firmly held vast free zones while continuing powerfully to develop our guerilla bases in the occupied zone.

The face of the theatres of operations had greatly altered: the zone of enemy occupation had been gradually reduced, whereas the main base of the Resistance — the free zone of North Viet Nam, had gone on extending and being consolidated day by day. Our forces constantly maintained the initiative in operations. The enemy found themselves driven into a very dangerous impasse.

The French imperialists were getting more and more bogged down in their unjust war of aggression. American aid, which covered 15 per cent of the expenditure on this war in 1950 and 1951, rose to 35 per cent in 1952, 45 per cent in 1953, soon to reach 80 per cent in 1954. But the situation of the French Expeditionary Corps remained without much hope. In Autumn 1953, taking advantage of the armistice in Korea, the American and French imperialists plotted to increase their armed forces in Indo-China in the hope of prolonging and extending hostilities.

They decided on the Navarre plan which proposed to crush the main part of our forces, to occupy the whole of Viet nam, to transform it into a colony and a Franco-American military base and to end the war victoriously within 18 months. It was, in fact, the plan of the "war-to-the-end" men, Laniel and Dulles. In order to realise the first phase of this plan, General Navarre assembled in the North more than half the entire mobile forces of the Indochinese theatre, including reinforcements newly arrived from France, launched attacks against our free zone, and parachuted troops into Dien Bien Phu to turn it into the springboard for a future offensive.

* * *

The enemy wanted to concentrate their forces. We compelled them to disperse. By successively launching strong offensives on the points they had left relatively unprotected, we obliged them to scatter their troops all over the place in order to ward off our blows, and thus created favourable conditions for the attack at Dien Bien Phu, the most powerful entrenched camp in Indo-China, considered invulnerable by the Franco-American general staff. We decided to take the enemy by the throat at Dien Bien Phu. The major part of our forces were concentrated there. We mobilised the entire potentiality of the population of the free zone in order to guarantee victory for our front line. After 55 days and 55 nights of fighting, the Viet Nam People's Army accomplished the greatest feat of arms of the whole war of liberation: the entire garrison at Dien Bien Phu was annihilated. This great campaign, which altered the course of the war, contributed decisively to the success of the Geneva Conference.

In July 1954, the signing of the Geneva Agreements re-established peace in Indo-China on the basis of respect for the sovereignty, independence, unity and territorial integrity of Viet Nam, Cambodia and Laos. It is following these agreements that North Viet Nam, with a population of 16 million inhabitants, is today entirely free. This success crowned nearly a century of struggle for national liberation, and especially the nine long and hard years of resistance war waged by the Vietnamese people. It was a crushing defeat for the French and American imperialists as well as for their lackeys. But at present, half of our country is still living under the yoke of the American imperialists and the Ngo Dinh Dien authorities. Our people's struggle for national liberation is not yet finished, it is continuing by peaceful means.

III

THE FUNDAMENTAL PROBLEMS OF OUR WAR OF LIBERATION

The Vietnamese people's war of liberation was a just war, aiming to win back the independence and unity of the country, to bring land to our peasants and guarantee them the right to it, and to defend the achievements of the August Revolution. That is why it was first and foremost a *people's war*. To educate, mobilise, organise and arm the whole people in order that they might take part in the Resistance was a crucial question.

The enemy of the Vietnamese nation was aggressive imperialism, which had to be overthrown. But the latter having long since joined up with the feudal landlords, the anti-imperialist struggle could definitely not be separated from anti-feudal action. On the other hand, in a backward colonial country such as ours where the peasants make up the majority of the population, a people's war is essentially *a peasant's war under the leadership of the working class*. Owing to this fact, a general mobilisation of the whole people is neither more nor less than the mobilisation of the rural masses. The problem of land is of decisive importance. From an exhaustive analysis, the Vietnamese people's war of liberation was essentially a people's national democratic revolution carried out under armed form and had twofold fundamental task: the overthrowing of imperialism and the defeat of the feudal landlord class, the anti-imperialist struggle being the primary task.

A backward colonial country which had only just risen up to proclaim its independence and install people's power, Viet Nam only recently possessed armed forces, equipped with still very mediocre arms and having no combat experience. Her enemy, on the other hand, was an imperialist power which has retained a fairly considerable economic and military potentiality despite the recent German occupation and benefited, furthermore, from the active support of the United States. The balance of forces decidedly showed up our weaknesses against the enemy's power. The Vietnamese people's war of liberation had, therefore, to be a hard and long-lasting war in order to succeed in creating conditions for victory. All the conceptions born of impatience and aimed at obtaining speedy victory could only be gross errors. It was necessary to firmly grasp the strategy of a long-term resistance, and to exalt the will to be self-supporting in order to maintain and gradually augment our forces, while nibbling at and progressively destroying those of the enemy ; it was necessary to accumulate thousands of small victories to turn them into a great success, thus gradually altering the balance of forces, in transforming our weakness into power and carrying off final victory.

At an early stage, our Party was able to discern the characteristics of this war : a people's war and a long-lasting war, and it was by proceeding from these premises that, during the whole of hostilities and in particularly difficult conditions, the Party solved all the problems of the Resistance. This judicious leadership by the Party led us to victory.

***

From the point of view of directing operations, our strategy and tactics had to be those of a people's war and of a long-term resistance.

Our strategy was, as we have stressed, to wage a long-lasting battle. A war of this nature in general entails several phases; in principle, starting from a stage of contention, it goes through a period of equilibrium before arriving at a general counter-offensive. In effect, the way in which it is carried on can be more subtle and more complex, depending on the particular conditions obtaining on both sides during the course of operations. Only a long-term war could enable us to utilise to the maximum our political trump cards, to overcome our material handicap and to transform our weakness into strength. To maintain and increase our forces, was the principle to which we adhered, contenting ourselves with attacking when success was certain, refusing to give battle likely to incur losses to us or to engage in hazardous actions. We had to apply the slogan : to build up our strength during the actual course of fighting.

The forms of fighting had to be completely adapted that is, to raise the fighting spirit to the maximum and rely on heroism of our troops to overcome the enemy's material superiority. In the main, especially at the outset of the war, we had recourse to guerilla fighting. In the Vietnamese theatre of operations, this method carried off great victories: it could be used in the mountains as well as in the delta, it could be waged with good or mediocre material and even without arms, and was to enable us eventually to equip ourselves at the cost of the enemy. Wherever the Expeditionary Corps came, the entire population took part in the fighting ; every commune had its fortified village, every district had its regional troops fighting under the command of the local branches of the Party and the people's administration, in liaison with the regular forces in order to wear down and annihilate the enemy forces.

Thereafter, with the development of our forces, guerilla warfare changed into a mobile warfare — a form of mobile warfare still strongly marked by guerilla warfare — which would afterwards become the essential form of operations on the main front, the northern front. In this process of development of guerilla warfare and of accentuation of the mobile warfare, our people's army constantly grew and passed from the stage of combats involving a section or company, to fairly large-scale campaigns bringing into action several divisions. Gradually, its equipment improved, mainly by the seizure of arms from the enemy — the material of the French and American imperialists.

From the military point of view, *the Vietnamese people's war of liberation proved that an insufficiently equipped people's army, but an army fighting for a just cause, can, with appropriate strategy and tactics, combine the conditions needed to conquer a modern army of aggressive imperialism.*

* *

Concerning the management of a war economy within the framework of an agriculturally backward country undertaking a long-term resistance as was the case in Viet Nam, the problem of the rear lines arose under the form of building resistance bases in the countryside. The raising and defence of production, and the development of agriculture, were problems of great importance for supplying the front as well as for the progressive improvement of the people's living conditions. The question of manufacturing arms was not one which could be set aside.

In the building of rural bases and the reinforcement of the rear lines for giving an impulse to the resistance, the agrarian policy of the Party played a determining role. Therein lay the anti-feudal task of the revolution. In a colony where the national question is essentially the peasant question, the consolidation of the resistance forces was possible only by a solution to the agrarian problem.

The August Revolution overthrew the feudal State. The reduction of land rents and rates of interest decreed by people's power bestowed on the peasants their first material advantages. Land monopolised by the imperialists and the traitors was confiscated and shared out. Communal land and ricefields were more equitably distributed. From 1953, deeming it necessary to promote the accomplishment of anti-feudal tasks, the Party decided to achieve agrarian reform even during the course of the resistance war. Despite the errors which blemished its accomplishment, it was a correct line crowned with success; it resulted in real material advantages for the peasants and brought to the army and the people a new breath of enthusiasm in the war of resistance.

Thanks to this just agrarian policy, the life of the people, in the hardest conditions of the resistance war, in general improved, not only in the wast free zones of the North, but even in the guerilla bases in South Viet Nam.

The Vietnamese people's war of liberation brought out the importance of building resistance bases in the countryside and the close and indissoluble relationships between the anti-imperialist revolution and the anti-feudal revolution.

From the political point of view, the question of unity among the people and the mobilisation of all energies in the war of resistance were of paramount importance. It was at the same time a question of the national united front against the imperialists and their lackeys, the Vietnamese traitors.

In Viet Nam, our Party carried off a great success in its policy of Front. As early as during the difficult days of the Second World War, it formed the League for the Independence of Viet Nam. At the time of and during the early years of the war of resistance, it postponed the application of its watchwords on the agrarian revolution, limiting its programme to the reduction of land rents and interest rates, which enabled us to neutralise part of the landlord class and to rally around us the most patriotic of them.

From the early days of the August Revolution, the policy of broad front adopted by the Party neutralised the wavering elements among the landlord class and limited the acts of sabotage by the partisans of the Viet Nam Quoc Dan Dang.

Thereafter, in the course of development of the resistance war, when agrarian reform had become an urgent necessity, our Party applied itself to making a differentiation within the bosom of the landlord class by providing in its political line for different treatment for each type of landlord according to the latter's political attitude, on the principle of liquidation of the regime of feudal appropriation of land.

The policy of unity among nationalities adopted by the National United Front also achieved great successes and the programme of unity with the [various religious circles attained good results.

The National United Front was to be a vast assembly of all the forces capable of being united, neutralising all those which could be neutralised, dividing all those it was possible to divide in order to direct the spearhead at the chief enemy of the revolution, invading imperialism. It was to be established on the basis of an alliance between workers and peasants and placed under the leadership of the working class. In Viet Nam, the question of an alliance between workers and peasants was backed by a dazzling history and firm traditions, the party of the working class having been the only political party to fight resolutely in all circumstances for national independence, and the first to put forward the watchword "land to the tillers" and to struggle determinedly for its realisation. However, in the early years of the resistance a certain under-estimation of the importance of the peasant question hindered us from giving all the necessary attention to the worker-peasant alliance. This error war subsequently put right, especially from the moment when the Party decided, by means of accomplishing agrarian reform, to make the peasants the real masters of the countryside. At present, after the victory of the resistance and of agrarian reform, when the Party has restored independence to half the country and brought land to the peasants, the bases of the worker-peasant alliance will daily go from strength to strength.

The war of liberation of the Vietnamese people proves that, in the face of an enemy as powerful as he is cruel, victory is possible only by uniting the whole people within the bosom of a firm and wide national united front based on the worker-peasant alliance.

IV

THE FACTORS OF SUCCESS

The Vietnamese people's war of liberation has won great victories. In North Viet Nam, entirely freed, the imperialist enemy has been overthrown, the landlords have been got rid of as a class, and the population is advancing with a firm tread on the path of building socialism to make of the North a firm base of action for the reunification of the country.

The Vietnamese people's war of liberation was victorious because it was a just war, waged for independence and the reunification of the country, in the legitimate interests of the national and the people and which by this fact succeeded in leading the whole people to participate enthusiastically in the resistance and to consent to make every sacrifice for its victory.

The Vietnamese people's war of liberation won this great victory because we had a revolutionary armed force of the people, the *heroic Viet Nam People's Army*. Built in accordance with the political line of the Party, this army was animated by an unflinching combative spirit, and accustomed to a style of persevering political work. It adopted the tactics and strategy of a people's war. It developed from nothing by combining the best elements among the workers, peasants and revolutionary students and intellectuals, stemming from the patriotic organisations of the popular masses. Born of the people, it fought for the people. It is an army led by the Party of the working class.

The Vietnamese people's war of liberation was victorious because we had a wide and firm *National United Front*, comprising all the revolutionary classes, all the nationalities living on Vietnamese soil, all the patriots. This Front was based on the alliance between workers and peasants, under the leadership of the Party.

The Vietnamese people's war of liberation ended in victory because of the existence of *people's power* established during the August Revolution and thereafter constantly consolidated. This power was the Government of alliance between classes, the government of the revolutionary classes and above all of the workers and peasants. It was the dictatorship of people's democracy, the dictatorship of the workers and peasants in fact, under the leadership of the Party. It devoted its efforts to mobilising and organising the whole people for the Resistance; it brought the people material advantages not only in the free zones, but also in the guerilla bases behind the enemy's back.

The Vietnamese people's war of liberation attained this great victory for the reasons we have just enumerated, but above all because it was *organised and led by the Party of the working class : the Indochinese Communist Party, now the Viet Nam Workers' Party*. In the light of the principles of Marxism-Leninism, it was this Party which proceeded to make an analysis of the social situation and of the balance of forces between the enemy and ourselves in order to determine the fundamental tasks of the people's national democratic revolution, to establish the plan for the armed struggle and decide on the guiding principle: long-term resistance and self-reliance. It was the Party which found a correct solution to the problems arising out of the setting up and leadership of a people's army, people's power and a national united front. It also inspired in the people and the army a completely revolutionary spirit which instilled into the whole people the will to overcome all difficulties, to endure all privations, the spirit of a long resistance, of resistance to the end. Our Party, under the leadership of President Ho Chi Minh, is the worthy Party of the working class and the nation. President Ho Chi Minh, leader of the Party and the nation, is the symbol of this gigantic uprising of the Vietnamese people.

If the Vietnamese people's war of liberation ended in a glorious victory, it is because we did not fight alone, but with the *support of progressive peoples the world over, and more especially the peoples of the brother countries, with the Soviet Union at the head*. The victory of the Vietnamese people cannot be divided from this support; it cannot be disassociated from the brilliant successes of the socialist countries and the movement of national liberation, neither can it be detached from the victories of the Soviet Red Army during the Second World War, nor from those of the Chinese people during the last few years. It cannot be isolated from the sympathy and support of progressive peoples throughout the world, among whom are the French people under the leadership of their Communist Party, and the peoples of Asia and Africa.

The victory of the Vietnamese people is that of a small and weak nation and possessing no regular army, which rose up to engage in an armed struggle against the aggression of an imperialist country with a modern army and benefiting from the support of the American imperialists. This colonial country has established and maintained a regime of people's democracy, which will open up to it the path to socialism. That is one of the great historic events in the national liberation movement and in the proletarian revolutionary movement, in the new international position born of the Second World War, in the period of transition from capitalism to socialism, in the time of the disintegration of imperialism. The Vietnamese people's war of liberation has contributed to making obvious this new historic truth: in the present international situation, a weak people which rises up resolutely to fight for its freedom is sure to triumph over all enemies and to achieve victory.

This great truth enlightens and encourages the Vietnamese people on the path of struggle for peace, socialism and the reunification of the country. This path will certainly lead it to new victories.

PEOPLE'S WAR
PEOPLE'S ARMY

On December 22, 1959, the Viet Nam People's Army will celebrate the fifteenth anniversary of its founding. I would like, on this occasion to have a few words with you about the struggle and the building up of the revolutionary armed forces in Viet Nam. At the same time I would like to lay emphasis on the fundamental points which bring out the characteristics of the military policy of the vanguard party of the Vietnamese working class and people — the Indochinese Communist Party—now the Viet Nam Workers' Party.

As Marxism-Leninism teaches us : " The history of all societies up till the present day, has been but the history of class struggle." These struggles can take either the form of political struggle or the form of armed struggle — the armed struggle being only the continuation of the political struggle. In a society which remains divided into classes, we can distinguish two kinds of politics : the politics of the classes and nations of exploiters and oppressors and that of the exploited and oppressed classes and nations. Hence two kinds of wars, of States and armies diametrically opposed to each other, the ones revolutionary, popular and just, and the others counter-revolutionary, anti-popular and unjust.

The Russian October Revolution marked a new era in the history of mankind. A state of a new type appeared, that of proletarian dictatorship, that of the liberated Soviet workers and peasants, toiling people and nationalities. An army of a new type came into being — the Red Army, a genuine people's army placed under the leadership of the Communist Party of the Soviet Union. Born in the October uprising, and steeled and tempered in the combats that followed it, the Red Army was to become, in a short time, the most powerful army in the world, always ready to defend the Soviet Motherland, the first State of workers and peasants.

In Asia, after World War One, the national democratic revolution of the Chinese people made tremendous progress under the good influence of the Russian Revolution. To free themselves, the Chinese people valiantly rose to wage an armed struggle for many decades. It was in this revolutionary war full of heroism and sacrifices that was born and grew up the Chinese Liberation Army, an army equally of a new type, genuinely popular, under the leadership of the Chinese Communist Party.

Only fifteen years of age, the Viet Nam People's Army is a young revolutionary army. It developed in the course of the national liberation war of the Vietnamese people from which it comes, and is now assuming the glorious task of defending the building of socialism in the North while contributing to make it a strong base for the peaceful reunification of the country. It also constitutes an army of a new type, a truly popular army under the leadership of the working class Party of Viet Nam.

In the U.S.S.R. as well as in China and Viet Nam, the revolutionary wars and armies have common fundamental characteristics: their popular and revolutionary nature, and the just cause they serve.

The Vietnamese revolutionary war and army however have their own characteristics. Indeed, from the very start, in the Soviet Union, the revolutionary war evolved within the framework of a socialist revolution. Moreover it proceeded in an independent country possessing a fairly important modern industrial economy, which, under the socialist regime, has not ceased to develop further. As for the revolutionary war in China, it remained for a long period within the framework of a national democratic revolution proceeding in a semi-colonial country, an immensely vast country and with a population of more than 600 million people.

The revolutionary war in Viet Nam, while advancing as in China towards the objectives of a national democratic revolution, differs for the reason that it took place in a colonial country, in a much smaller country than China in both area and population.

Therefore the history of the armed struggle and the building up of the armed forces in Viet Nam is that of a small nation subject to colonial rule and having neither a vast territory nor a large population, which, though lacking a regular army at the beginning, had to rise against the aggressive forces of an imperialist power, and triumphed over them in the end, liberating half of the country and enabling it to embark on the socialist path. As for the military policy of the vanguard Party of the Vietnamese working class, it is an application of Marxism-Leninism to the concrete conditions of the war of liberation in a colonial country.

I

Viet Nam is a nation in South-east Asia with a very old history. With its 329,600 square kilometres and 30 million inhabitants and its geographical situation in the Pacific, it has now become one of the outposts of the socialist world.

In the course of its thousands of years of history, many a time, the Vietnamese nation victoriously resisted the invasions of the Chinese feudalists. It can be proud of its traditions of undaunted struggle in safeguarding national independence.

After its invasion of Viet Nam in the second half of the 19th century, French imperialism made it their colony. Since then, the struggle against French colonialism never ceased to extend, uprisings succeeded each other in spite of repression, and daily attracting wider and wider strata belonging to all social classes.

In 1930, the Indochinese Communist Party was founded. Under its firm and clear-sighted leadership, the movement for national liberation of the Vietnamese people made new progress. After ten years of heroic political struggle, at the dawn of World War Two, the Party advocated the preparation for armed struggle, and for that the launching of a guerilla war and the setting up of a free zone. The anti-Japanese movement for national salvation, in its irresistible upsurge, led to the glorious days of the August Revolution of 1945. Taking advantage of the major events in the international situation at the time — the victory of the Soviet Red Army and Allied forces over Japanese fascism — the Vietnamese people rose up as one man in the victorious insurrection and set up the people's power. The Democratic Republic of Viet Nam was born, the first people's democracy in South-east Asia.

The political situation in Viet Nam was then particularly difficult and complicated. Chiang Kai-shek's troops had entered the North, and those of Great Britain the South of the country, to disarm the Japanese who were still in possession of all their armaments in the first days of the capitulation. It was in these conditions that French imperialists, immediately after the founding of the Democratic Republic, unleashed a war of reconquest against Viet Nam hoping to impose their domination on this country.

In response to the appeal of the Party and the Government headed by President Ho Chi Minh, the Vietnamese people rose up as one man for the defence of the Fatherland. A sacred war for national liberation began. All hopes of a peaceful settlement were not lost however. A Preliminary Agreement for the cessation of hostilities was signed in March 1946 between the Government of the Democratic Republic of Viet Nam and that of France. But the French colonialists saw it only as a delaying scheme. Therefore, immediately after the signing of the Agreement, they shamelessly violated it by successively occupying various regions. In December 1946, the war spread to the whole country. It was to rage for nine years, nine years after the end of World War Two. And it ended with the brilliant victory of the Vietnamese people.

Our war of liberation was a *people's war*, a just war. It was this essential characteristic that was to determine its laws and to decide its final outcome.

At the first gun-shots of the imperialist invasion, general Leclerc, the first Commander of the French Expeditionary Corps estimated that the operation for the reoccupation of Viet Nam would be a mere military walk over. When encountering the resistance of the Vietnamese people in the South the French generals considered it as weak and temporary and stuck to their opinion that it would take them ten weeks at the most to occupy and pacify the whole of south Viet Nam. Why did French colonialists make such an estimation? Because they considered that to meet their aggression, there must be an army. The Vietnamese army had just been created. It was still numerically weak, badly organised, led by inexperienced officers and non-commissioned officers, provided with old and insufficient equipment, a very limited stock of munitions and having neither tanks, airplanes nor artillery. With such an army how could serious resistance be undertaken and the attacks of the powerful and armoured division repelled? All it could do was to use up its stock of munitions before laying down its arms. In fact, the Vietnamese army was then weak in all respects and was destitute of everything. French colonialists were right in this respect. But is was not possible for them to understand a fundamental and decisive fact: this fact was that the Vietnamese army, though very weak materially was a people's army. This fact is that the war in Viet nam was not only the opposition of two armies. In provoking hostilities, the aggressive colonialists had alienated a whole nation. And, indeed, the whole Vietnamese nation, the entire Vietnamese people rose against them. Unable to grasp this profound truth, the French generals who believed in an easy victory, went instead to certain defeat. They thought they could easily subdue the Vietnamese people, when, in fact, the latter were going to smash them.

Even to this day bourgeois strategists have not yet overcome their surprise at the outcome of the war in Indo-China. How could the Vietnamese nation have defeated an imperialist power such as France which was backed by the U.S.? They try to explain this extraordinary fact by the correctness of strategy and tactics, by the forms of combat and the heroism of the Viet Nam People's Army. Of course all these factors contributed to the happy outcome of the resistance. But if the question is put: "Why were the Vietnamese people able to win?" the most precise and most complete answer must be: "The Vietnamese people won because their war of liberation was a people's war."

When the Resistance War spread to the whole country, the Indochinese Communist Party emphasized in its instructions that our Resistance War must be the work of the entire people. Therein lies the key to victory.

Our Resistance War was a people's war, because its political aims were to smash the imperialist yoke to win back national independence, to overthrow the feudal landlord class to bring land to the peasants; in other words, to radically solve the two fundamental contradictions of Vietnamese society — contradiction between the nation and imperialism on the one hand, and contradiction between the people, especially between the peasants and the feudal landlord class on the other — and to pave the socialist path for the Vietnamese revolution.

Holding firmly to the strategy and tactics of the national democratic revolution, the Party pointed out to the people the aims of the struggle: independence and democracy. It was, however, not enough to have objectives entirely in conformity with the fundamental aspirations of the people. It was also necessary to bring everything into play to enlighten the masses of the people, educate and encourage them, organise them in fighting for national salvation. The Party devoted itself entirely to this work, to the regrouping of all the national forces, and to the broadening and strengthening of a national united front, the Viet Minh, and later the Lien Viet which was a magnificent model of the unity of the various strata of the people in the anti-imperialist struggle in a colonial country. In fact, this front united the patriotic forces of all classes and social strata, even progressive landlords; all nationalities in the country — majority as well as minority; patriotic believers of each and every religion. " Unity, the great unity, for victory, the great victory "; this slogan launched by President Ho Chi Minh became a reality, a great reality during the long and hard resistance.

We waged a people's war, and that in the framework of a long since colonised country. Therefore the national factor was of first importance. We had to rally all the forces likely to overthrow the imperialists and their lackeys. On the other hand, this war proceeded in a backward agricultural country where the peasants, making up the great majority of the population, constituted the essential force of the revolution and of the Resistance War. Consequently the relation between the national question and the peasant question had to be clearly defined, with the gradual settlement of the agrarian problem, so as to mobilise the broad peasant masses, one of the essential and decisive factors for victory. Always solicitous about the interests of the peasantry, the Party began by advocating reduction of land rent and interest. Later on, as soon as the stabilisation of the situation allowed it, the Party carried out with great firmness the mobilisation of the masses for land reform in order to bring land to the tillers, thereby to maintain and strengthen the Resistance.

During the years of war, various erroneous tendencies appeared. Either we devoted our attention only to the organisation and growth of the armed forces while neglecting the mobilisation and organisation of large strata of the people, or we mobilised the people for the war without heeding seriously their immediate everyday interests; or we thought of satisfying the immediate interests of the people as a whole, without giving due attention to those of the peasants. The Party resolutely fought all these tendencies. To lead the Resistance to victory, we had to look after the strengthening of the army, while giving thought to mobilising and educating the people, broadening and consolidating the National United Front. We had to mobilise the masses for the Resistance while trying to satisfy their immediate interests to improve their living conditions, essentially those of the peasantry. A very broad national united front was indispensable, on the basis of the worker-peasant alliance and under the leadership of the Party.

The imperatives of the people's war in Viet Nam required the adoption of appropriate strategy and tactics, on the basis of the enemy's characteristics and of our own, of the concrete conditions of the battlefields and balance of forces facing each other. In other words, the strategy and tactics of a people's war, in an economically backward, colonial country.

First of all, this strategy must be the *strategy of a long-term war*. It does not mean that all revolutionary wars, all people's wars must necessarily be long-term wars. If from the outset, the conditions are favourable to the people and the balance of forces turn in favour of the revolution, the revolutionary war can end victoriously in a short time. But the war of liberation of the Vietnamese people started in quite different conditions: We had to deal with a much stronger enemy. It was patent that this balance of forces took away from us the possibility of giving decisive battles from the opening of the hostilities and of checking the aggression from the first landing operations on our soil. In a word, it was impossible for us to defeat the enemy swiftly.

It was only by a long and hard resistance that we could wear out the enemy forces little by little while strengthening ours, progressively turn the balance of forces in our favour and finally win victory. We did not have any other way.

This strategy and slogan of long term resistance was decided upon by the Indochinese Communist Party from the first days of the war of liberation. It was in this spirit that the Viet Nam People's Army, after fierce street-combats in the big cities, beat strategical retreats to the countryside on its own initiative in order to maintain its bases and preserve its forces.

The long-term revolutionary war must include several different stages: stage of contention, stage of equilibrium and stage of counter-offensive. Practical fighting was, of course, more complicated. There had to be many years of more and more intense and generalised guerilla fighting to realise the equilibrium of forces and develop our war potentiality. When the conjunctures of events at home and abroad allowed it, we went over to counter-offensive first by a series of local operations then by others on a larger scale which were to lead to the decisive victory of Dien Bien Phu.

The application of this strategy of long-term resistance required a whole system of education, a whole ideological struggle among the people and Party members, a gigantic effort of organisation in both military and economic fields, extraordinary sacrifices and heroism from the army as well as from the people, at the front as well as in the rear. Sometimes erroneous tendencies appeared, trying either to by-pass the stages to end the war earlier, or to throw important forces into military adventures. The Party rectified them by a stubborn struggle and persevered in the line it had fixed. In the difficult hours, certain hesitations revealed themselves, the Party faced them with vigour and with determination in the struggle and faith in final victory.

The long-term people's war in Viet Nam also called for appropriate forms of fighting: appropriate to the revolutionary nature of the war as well as to the balance of forces which revealed at that time an overwhelming superiority of the enemy over the still very weak material and technical bases of the People's Army. *The adopted form of fighting was guerilla warfare.* It can be said that the war of liberation of the Vietnamese people was a long and vast guerilla war proceeding from simple to complex then to mobile war in the last years of the Resistance.

Guerilla war is the war of the broad masses of an economically backward country standing up against a powerfully equipped and well trained army of aggression. Is the enemy strong? One avoids him. Is he weak? One attacks him. To his modern armament one opposes a boundless heroism to vanquish either by harassing or by annihilating the enemy according to circumstances, and by combining military operations with political and economic action; no fixed line of demarcation, the front being wherever the enemy is found.

Concentration of troops to realize an overwhelming superiority over the enemy where he is sufficiently exposed in order to destroy his manpower; initiative, suppleness, rapidity, surprise, suddenness in attack and retreat. As long as the strategic balance of forces remains disadvantageous, resolutely to muster troops to obtain absolute superiority in combat in a given place, and at a given time. To exhaust little by little by small victories the enemy forces and at the same time to maintain and increase ours. In these concrete conditions it proves absolutely necessary not to lose sight of the main objective of the fighting that is the destruction of the enemy manpower. Therefore losses must be avoided even at the cost of losing ground. And that for the purpose of recovering, later on, the occupied territories and completely liberating the country.

In the war of liberation in Viet Nam, guerilla activities spread to all the regions temporarily occupied by the enemy. Each inhabitant was a soldier, each village a fortress, each Party cell, each village administrative committee a staff.

enemy's designs and of the characteristics of the theatre of operations. *The thorough understanding of the contradictions and general laws of the aggressive war enabled us to detect the enemy's great weakness arising from the concentration of his forces. Always convinced that the essential thing was to destroy the enemy's manpower, the Central Committee worked out its plan of action on the basis of scientific analysis: to concentrate our forces to move to the offensive against important strategic points where the enemy's forces were relatively weak in order to wipe out a part of his manpower, at the same time compelling him to scatter his forces to cope with us at vital points which he had to defend at all costs. Our strategic directives were: dynamism, initiative, mobility and rapidity of decision in face of new situations.*

The Central Committee's strategic direction proved itself correct and clear-sighted: while the enemy was concentrating big forces in the delta to threaten our free zone, instead of leaving our main forces in the delta or scattering our forces in the free zone to defend it by a defensive action, we regrouped our forces and boldly attacked in the direction of the North-West. Indeed, our divisions marched on the North-West with an irresistible impetus, swept away thousands of local bandits at Son La and Thuan Chau, liberated Lai Chau, cutting to pieces the greater part of the enemy's column which fled from Lai Chau. Simultaneously, we encircled Dien Bien Phu, thus compelling the enemy to carry out in haste a movement of forces to reinforce Dien Bien Phu in order to save it from being wiped out. Besides the Red River delta, Dien Bien Phu became a second point of concentration of enemy forces.

Concurrently with our offensive in the North-West, the Laos-Viet Nam joint forces launched a second offensive in an important direction where the enemy was relatively exposed, the Middle Laos front.

Several enemy mobile units were annihilated and the town of Thakhek was liberated. The joint forces pushed on in the direction of Seno, an important enemy air base in Savannakhet. The enemy had to rush forces in haste from the Red River delta and from all other battlefields to reinforce Seno, thus turning it into a third point of concentration of his forces.

Early in 1954, while the enemy was feverishly making preparations for his offensive against our free territory in the Fifth zone, our plan was to leave only a small part of our forces to protect our rear and to concentrate big forces to attack on the Western Highlands, which was an important strategic position where the enemy was relatively exposed. Our advance to the Western Highlands was accompanied by resounding victories: important enemy units were wiped out, the town and whole province of Kontum were liberated. Our troops made a raid on Pleiku, compelling the enemy to dispatch more troops there in reinforcement, turning Pleiku and various bases on the Western Highlands into a fourth point of concentration of French forces.

During the same period, to create a diversion in order to secure conditions for our troops to step up preparations at Dien Bien Phu, the Laos-Viet Nam joint forces had, from Dien Bien Phu, launched an offensive in Upper Laos. Several enemy units were wiped out and the vast Nam Hu basin was liberated. The enemy was compelled to rush more forces to Luang Prabang, which became the fifth point of concentration of French forces.

The people as a whole took part in the armed struggle, fighting according to the principles of guerilla warfare, in small packets, but always in pursuance of the one and same line, and the same instructions, those of the Central Committee of the Party and the Government.

At variance with numerous other countries which waged revolutionary wars, Viet Nam, in the first years of its struggle, did not and could not engage in pitched battles. It had to rest content with guerilla warfare. At the cost of thousands of difficulties and countless sacrifices, this guerilla war developed progressively into a form of mobile war that daily increased in scale. While retaining certain characteristics of guerilla war, it involved regular campaigns with greater attacks on fortified positions. Starting from small operations with the strength of a platoon or a company to annihilate a few men or a group of enemy soldiers, our army went over, later, to more important combats with a battalion or regiment to cut one or several enemy companies to pieces, finally coming to greater campaigns bringing into play many regiments, then many divisions to end at Dien Bien Phu where the French Expeditionary Corps lost 16,000 men of its crack units. It was this process of development that enabled our army to move forward steadily on the road to victory.

People's war, long-term war, guerilla warfare developing step by step into mobile warfare, such are the most valuable lessons of the war of liberation in Viet Nam. It was by following that line that the Party led the Resistance to victory. After three thousand days of fighting, difficulties and sacrifices, our people defeated the French imperialists and American interventionists. At present, in the liberated half of our country, sixteen million of our compatriots, by their creative labour, are healing the horrible wounds of war, reconstructing the country and building socialism. In the meantime the struggle is going on to achieve the democratic national revolution throughout the country and to reunify the Fatherland on the basis of independence and democracy.

II

After this account of the main lines of the war of liberation waged by the Vietnamese people against the French and American imperialists, I shall speak of the Viet Nam People's Army.

Being the armed forces of the Vietnamese people, it was born and grew up in the flames of the war of national liberation. Its embryo was the self-defence units created by the Nghe An Soviets which managed to hold power for a few months in the period of revolutionary upsurge in the years 1930-1931. But the creation of revolutionary armed forces was positively considered only at the outset of World War Two when the preparation for an armed insurrection came to the fore of our attention. Our military and paramilitary formations appeared at the Bac Son uprising and in the revolutionary bases in Cao Bang region. Following the setting up of the first platoon of National Salvation, on December 22, 1944, another platoon-strong unit was created: the Progaganda unit of the Viet Nam Liberation Army. Our war bases organised during illegality were at the time limited to a few districts in the provinces of Cao Bang, Bac Can and Lang Son in the jungle of the North. As for the revolutionary armed forces they still consisted of people's units of self-defence and a few groups and platoons completely free from production work. Their number increased quickly and there were already several thousands of guerillas at the beginning of 1945, at the coup de force by the Japanese fascists over the French colonialists. At the time of the setting up of the people's power in the rural regions of six provinces in Viet Bac which were established as a free zone, the existing armed organisations merged to form the Viet Nam Liberation Army.

During the August insurrection, side by side with the people and the self-defence units, the Liberation Army took part in the conquest of power. By incorporating the paramilitary forces regrouped in the course of the glorious days of August, it saw its strength increase rapidly. With a heterogeneous material wrested from the Japanese and their Bao An troops — rifles alone consisted of sixteen different types including old French patterns and even rifles of the czarist forces taken by the Japanese — this young and poorly equipped army soon had to face the aggression of the French Expeditionary Corps which had modern armaments. Such antiquated equipment required from the Vietnamese army and people complete self-sacrifice and superhuman heroism.

Should the enemy attack the regions where our troops were stationed, the latter would give battle. Should he ferret about in the large zones where there were no regular formations, the people would stay his advance with rudimentary weapons: sticks, spears, scimitars, bows, flintlocks. From the first days, there appeared three types of armed forces: para-military organisations or guerilla units, regional troops and regular units. These formations were, in the field of organisation, the expression of the general mobilisation of the people in arms. They co-operated closely with one another to annihilate the enemy.

Peasants, workers and intellectuals crowded into the ranks of the armed forces of the Revolution. Leading cadres of the Party and the State apparatus became officers from the first moment. The greatest difficulty to be solved was the equipment problem. Throughout Viet Nam there was no factory manufacturing war materials. Throughout nearly a century, possession and use of arms were strictly forbidden by the colonial administration. Importation was impossible, the neighbouring countries being hostile to the Democratic Republic of Viet Nam. The sole source of supply could only be the battlefront: to take the material from the enemy to turn it against him. While carrying on the aggression against Viet Nam the French Expeditionary Corps fulfilled another task: it became, unwittingly, the supplier of the Viet Nam People's Army with French, even U.S. arms. In spite of their enormous efforts, the arms factories set up later on with makeshift means were far from being able to meet all our needs. A great part of our military materials came from war-booty.

As I have stressed, the Viet Nam People's Army could at first bring into combat only small units such as platoons or companies. The regular forces were, at a given time, compelled to split up into companies operating separately to promote the extension of guerilla activities while mobile battalions were maintained for more important actions. After each victorious combat, the people's armed forces marked a new step forward.

Tempered in combat and stimulated by victories, the guerilla formations created conditions for the growth of the regional troops. And the latter, in their turn, promoted the development of the regular forces. For nine successive years, by following this heroic path bristling with difficulties, our people's army grew up with a determination to win at all costs. It became an army of hundreds of thousands strong, successively amalgamating into regiments and divisions and directing towards a progressive standardisation in organisation and equipment. This force, ever more politically conscious, and better trained militarily, succeeded in fighting and defeating the five hundred thousand men of the French Expeditionary Corps who were equipped and supplied by the United States.

***

The Vietnamese Army is indeed a *national one*. In fighting against imperialism and the traitors in its service, it has fought for national independence and the unity of the country. In its ranks are the finest sons of Viet Nam, the most sincere patriots from all revolutionary classes, from all nationalities — majority as well as minority people. It symbolises the irresistible rousing of the national conscience, the union of the entire Vietnamese people in the fight against imperialist aggression to save the country.

Our army is a *democratic army*, because it fights for the people's democratic interests, and the defence of people's democratic power. Impregnated with the principles of democracy in its internal political life, it submits to a rigorous discipline, but one freely consented to.

Our army is a *people's army*, because it defends the fundamental interests of the people, in the first place those of the toiling people, workers and peasants. As regards social composition, it comprises a great majority of picked fighters of peasant and worker origin, and intellectuals faithful to the cause of the Revolution.

It is *the true army of the people, of toilers, the army of workers and peasants, led by the Party of the working class*. Throughout the war of national liberation, its aims of struggle were the very ones followed by the Party and people: independence of the nation, and land to the tillers. Since the return of peace, as a tool of proletarian dictatorship, its mission is to defend the socialist revolution and socialist building in the North, to support the political struggle for the peaceful reunification of the country, and to contribute to the strengthening of peace in Indo-China and South-east Asia.

In the first of the ten points of his Oath of Honour, the fighter of the Viet Nam People's Army swears:

"To sacrifice himself unreservedly for the Fatherland, fight for the cause of national independence, democracy and socialism, under the leadership of the Viet Nam Workers' Party and of the Government of the Democratic Republic, to build a peaceful, reunified, independent, democratic and prosperous Viet Nam and contribute to the strengthening of peace in South-east Asia and the world."

This is precisely what makes the Viet Nam People's Army a true child of the people. The people, in return, give it unsparing affection and support. Therein lies the inexhaustible source of its power.

The Viet Nam People's Army has been created by the Party, which ceaselessly trains and educates it. It has always been and will always be under the *leadership of the Party* which, alone, has made it into a revolutionary army, a true people's army. Since its creation and in the course of its development, this leadership by the Party has been made concrete on the organisational plan. The army has always had its political commissars. In the units, the military and political chiefs assume their responsibilities under the leadership of the Party Committee at the corresponding echelon.

The People's Army is the instrument of the Party and of the revolutionary State for the accomplishment, in armed form, of the tasks of the revolution. Profound awareness of the aims of the Party, boundless loyalty to the cause of the nation and the working class, and a spirit of unreserved sacrifice are fundamental questions for the army, and questions of principle. Therefore, the political work in its ranks is of the first importance. *It is the soul of the army.* In instilling Marxist-Leninist ideology into the army, it aims at raising the army's political consciousness and ideological level, at strengthening the class position of its cadres and soldiers. During the liberation war, this work imbued the army with the policy of long-drawn-out resistance and the imperative necessity for the people and army to rely on their own strength to overcome difficulties. It instilled into the army the profound significance of mass mobilisation in order to achieve rent reduction and agrarian reform, which had a decisive effect on the morale of the troops. In the new stage entered upon since the restoration of peace, political work centres on the line of socialist revolution in the North and of struggle for the reunification of the country.

But that is not all. Political work still bears upon the correct fulfilment in the army of the programmes of the Party and Government, and the setting up of good relations with the population and between officers and men. It aims at maintaining and strengthening combativeness, uniting true patriotism with proletarian internationalism, developing revolutionary heroism and the great tradition of our army summed up in its slogan: " Resolved to fight, determined to win ". Political work is the work of propaganda among and education of the masses ; it is, furthermore, the organisational work of the Party in the army. We have always given particular attention to the strengthening of organisations of the Party in the units. From 35 to 40 per cent of officers and armymen have joined it, among the officers, the percentage even exceeds 90 per cent.

The Viet Nam People's Army has always seen to establishing and maintaining *good relations with the people*. These are based upon the identity of their aims of struggle : in fact, the people and army are together in the fight against the enemy to save the Fatherland, and ensure the full success of the task of liberating the nation and the working class. The people are to the army what water is to fish, as the saying goes. And this saying has a profound significance. Our Army fought on the front ; is has also worked to educate the people and helped them to the best of its ability. The Vietnamese fighter has always taken care to observe point 9 of its Oath of Honour :

" In contacts with the people, to follow these three recommendations :
— To respect the people
— To help the people
— To defend the people... in order to win their confidence and affection and achieve a perfect understanding between the people and the army ".

Our army has always organised days of help for peasants in production work and in the struggle against flood and drought. It has always observed a correct attitude in its relations with the people. It has never done injury to their property, not even a needle or a bit of thread. During the Resistance, especially in the enemy rear, it brought everything into play to defend ordinary people's lives and property ; in the newly liberated regions, it strictly carried out the orders of the Party and Government, which enabled it to win the unreserved support of the broadest masses, even in the minority peoples' regions and catholic villages. Since the return of peace, thousands of its officers and men have participated in the great movements for the accomplishment of agrarian reform for agricultural collectivisation and socialist transformation of handicrafts, industry and private trade. It has actively taken part in the economic recovery, and in socialist work days. It has participated in the building of lines of communication, it has built its own barracks and cleared land to found State farms.

The Viet Nam People's Army is always concerned to establish and maintain *good relations between officers and men as well as between the officers themselves*. Originating from the working strata, officers and men also serve the people's interests and unstintingly devote themselves to the cause of the nation and the working class. Of course every one of them has particular responsibilities which devolve upon him. But relations of comradeship based on political equality and fraternity of class have been established between them. The officer likes his men ; he must not only guide them in their work and studies, but take an interest in their life and take into consideration their desires and initiatives. As for the soldier, he must respect his superiors and correctly fulfil all their orders. The officer of the People's Army must set a good example from all points of view : to show himself to be resolute, brave, to ensure discipline and internal democracy, to know how to achieve perfect unity among his men. He must behave like a chief, a leader, vis-à-vis the masses in his unit. The basis of these relations between armymen and officers, like those between officers or between soldiers is solidarity in the fight, and mutual affection of brothers-in-arms, love at the same time pure and sublime, tested and forged in the battle, in the struggle for the defence of the Fatherland and the people.

The Viet Nam People's Army practises a strict *discipline*, allied to a wide internal *democracy*. As requires point 2 of its Oath of Honour : " The fighter must rigorously carry out the orders of his superiors and throw himself body and soul into the immediate and strict fulfilment of the tasks entrusted to him ". Can we say that guerilla warfare did not require severe discipline ? Of course not. It is true that it asked the commander and leader to allow each unit or each region a certain margin of initiative in order to undertake every positive action that it might think opportune. But a centralised leadership and a unified command at a given degree always proved to be necessary. He who speaks of the army, speaks of strict discipline.

Such a discipline is not in contradiction with the internal democracy of our troops. In cells, executive committees of the Party at various levels as well as in plenary meetings of fighting units, the application of principles of democratic centralism is the rule. The facts have proved that the more democracy is respected within the units, the more unity will be strengthened, discipline raised, and orders carried out. The combativeness of the army will thereby be all the greater.

The restoration of peace has created in Viet Nam a new situation. The North is entirely liberated, but the South is still under the yoke of American imperialists and the

Ngo Dinh Diem clique, their lackeys. North Viet Nam has entered the stage of socialist revolution while the struggle is going on to free the South from colonial and feudal fetters. To safeguard peace and socialist construction, to help in making the North a strong rampart for the peaceful reunification of the country, the problem of forces of national defence should not be neglected. The People's Army must face the bellicose aims of American imperialists and their lackeys and step by step become *a regular and modern army*.

First of all, it is important to stress that, in the process of its transformation into a regular and modern army, our army always remains a revolutionary army, a people's army. That is the fundamental characteristic that makes the people's regular and modern army in the North differ radically from Ngo Dinh Diem's army, a regular and modern army too, but anti-revolutionary, anti-popular and in the hands of the people's enemies. The People's Army must necessarily see to the strengthening of the leadership of Party and political work. It must work further to consolidate the solidarity between officers and men, between the troops and the people, raise the spirit of self-conscious discipline, while maintaining internal democracy. Taking steps to that end, the Party has during the last years, given a prominent place to the activities of its organisations as well as to the political work in the army. Officers, warrant officers and armymen, all of them have followed political courses to improve their understanding of the tasks of socialist revolution and the struggle for national reunification, consolidating their class standpoint and strengthening Marxist-Leninist ideology. This is a particularly important question, more especially as the People's Army has grown up in an agricultural country, and has in its ranks a great majority of toiling peasants and urban petty-bourgeois. Our fighters have gone through a dogged political education and their morale has been forged in the combat. However, the struggle against the influence of bourgeois and petty-bourgeois ideology remains necessary. Thanks to the strengthening of ideological work, the army has become an efficacious instrument in the service of proletarian dictatorship, and has been entirely faithful to the cause of socialist revolution and national reunification. The new advances realised by it in the political plan have found their full expression in the movement "with giant strides, let us overfulfil the norms of the programme," a broad mass movement which is developing among our troops, parallel with the socialist emulation movement among the working people in North Viet Nam.

It is essential actively and firmly to continue, on the basis of a constant strengthening of political consciousness, the progressive transformation of the People's Army into a regular and modern army. Thanks to the development realised during the last years of the Resistance War, our army, which was made up of infantry-men only, is now *an army composed of different arms*. If the problem of improvement of equipments and technique is important, that of cadres and soldiers capable of using them is more important. Our army has always been concerned with the training of officers and warrant officers of worker and peasant origin or revolutionary intellectuals tested under fire. It helps them raise their cultural and technical level to become competent officers and warrant officers of a regular and modern army.

To raise the fighting power of the army, to bring about a strong centralisation of command and a close cooperation between the different arms, it is indispensable to enforce *regulations fitted to a regular army*. It is not that nothing has been done in this field during the years of the Resistance War; it is a matter of perfecting the existing regulations. The main thing is not to lose sight of the principle that any new regulation must draw its inspiration from the popular character of the army and the absolute necessity of maintaining the leadership of the Party. Along with the general regulations, the statute of officers has been promulgated; a correct system of wages has taken the place of the former regime of allowances in kind; the question of rewards and decorations has been regularised. All these measures have resulted in the strengthening of discipline and solidarity within the army, and of the sense of responsibility among officers and warrant officers as well as among soldiers.

Military training, and political education, are key tasks in the building of the army in peace-time. The question of fighting regulations, and that of tactical concepts and appropriate tactical principles gain a particular importance. The question is to synthesize past experiences, and analyse well the concrete conditions of our army in organization and equipment, consider our economic structure, the terrain of the country — land of forests and jungles, of plains and fields. The question is to assimilate well the modern military science of the armies of the brother countries. Unceasing efforts are indispensable in the training of troops and the development of cadres.

For many years, the Viet Nam People's Army was based on voluntary service: all officers and soldiers voluntarily enlisted for an undetermined period. Its ranks swelled by the affluence of youth always ready to answer the appeal of the Fatherland. Since the return of peace, it has become necessary to replace voluntary service by *compulsory military service*. This substitution has met with warm response from the population. A great number of volunteers, after demobilisation returned to fields and factories; others are working in units assigned to production work, thus making an active contribution to the building of socialism. Conscription is enforced on the basis of the strengthening and development of the self-defence organisations in the communes, factories and schools. The members of these para-military organisations are ready not only to rejoin the permanent army, of which they constitute a particularly important reserve, but also to ensure the security and defence of their localities.

The People's Army was closely linked with the national liberation war, in the fire of which it was born and grew up. At present, its development should neither be disassociated from the building of socialism in the North, nor from the people's struggle for a reunified, independent and democratic Viet Nam. Confident of the people's affection and support, in these days of peace as during the war, the People's Army will achieve its tasks: to defend peace and the Fatherland.

III

"...As is said above, the history of the national liberation war of the Vietnamese people, that of the Viet Nam People's Army, is the history of the victory of a weak nation, of a colonised people who rose up against the aggressive forces of an imperialist power. This victory is also that of Marxism-Leninism applied to the armed revolutionary struggle in a colonised country, that of the Party of the working class in the leadership of the revolution that it heads, in the democratic national stage as well as in the socialist one.

The vanguard Party of the Vietnamese working people, headed by President Ho Chi Minh, the great leader of the people and the nation, is the organiser and guide that has led the Vietnamese people and their army to victory. In the light of Marxism-Leninism applied to the national democratic revolution in a colonised country, it has made a sound analysis of the contradictions of that society, and stated clearly the fundamental tasks of the revolution. On the question of the national liberation war, it has dialectically analysed the balance of opposing forces and mapped out appropriate strategy and tactics. In the light of Marxism-Leninism, it has created and led a heroic people's army. It has ceaselessly instilled revolutionary spirit and the true patriotism of the proletariat into the people and their army.

The Party has known how to learn from the valuable experiences of the October Revolution which, with the Soviet Red Army, showed the road of liberation not only to the workers of the capitalist countries, but also to colonial people ; and those of the Chinese Revolution and Liberation Army which have enriched the theories of the national democratic revolution, of revolutionary war and army in a semi-colonised country. Their wonderful examples have ceaselessly lighted the road of the struggle and successes of the Vietnamese people. In combining the invaluable experiences of the Soviet Union and People's China with its own, our Party has always taken into account the concrete reaiity of the revolutionary war in Viet Nam, thus is able in its turn to enrich the theories of revolutionary war and army.

At present, on the international plane, the forces of socialist countries, led by the Soviet Union, have become a power previously unknown; the national liberation movement has developed considerably everywhere; the possibilities for achieving lasting peace in the world are greater. However, imperialism is still pursuing its war preparations and seeking to strengthen its military alliance for aggression. While there is a certain relaxation of tension in the international situation, South-east Asia still remains one of the centres of tension in the world. American imperialism is ceaselessly strengthening its military and political hold on the South of our country. It is pursuing the same policy of interference in Laos, aimed at turning it into a colony and military base for a new war of aggression.

Profoundly peace-loving, the Vietnamese people and their army support every effort for disarmament, every effort to relax tension and establish a lasting peace. But they must at the same time heighten their vigilance, strengthen their combativeness, increase their potentiality for defence, and contribute to strengthening the fraternal bonds between the peoples and the revolutionary armed forces of the socialist countries. They are determined to fulfil their sacred duties, to defend the work of socialist revolution and the building of socialism in the North, to pursue the struggle for the peaceful reunification of the Fatherland, to be ready to break every imperialist attempt to provoke a war of aggression, and to contribute to the safeguarding of peace in South-east Asia and throughout the world.

THE GREAT EXPERIENCES GAINED BY OUR PARTY IN LEADING THE ARMED STRUGGLE AND BUILDING REVOLUTIONARY ARMED FORCES

Our Party was born at a time when the revolutionary movement was growing in our country. From the first days, the Party led the peasants to carry out armed uprising and establish Soviet power. Therefore, the Party soon gained a knowledge of the problems of revolutionary power and armed struggle. As Marxism-Leninism teaches, the question of State power is the most important question in every revolution, and the way of setting up revolutionary power, " the only way of liberation " is " the way of armed struggle of the masses ". (1)

The 1930-1931 movement was put down; our Party continued to lead the masses to carry on political struggle, sometimes illegally, sometimes semi-legally, endeavouring to restore the revolutionary bases to push the movement forward. In 1939, when World War Two broke out, the situation in the world and at home underwent changes, and the question of preparation for armed uprising aimed at liberating the nation again arose. From that time, our Party led our people to prepare for armed uprising and successfully carry out the general insurrection in August 1945; then for 9 years on end, our Party led the long Resistance War of our people to victory.

During the first 15 years of the 30 years of revolutionary mobilisation, the Party was in an illegal position; afterwards, revolutionary power was founded all over the

---
(1) The Program of Action of the Indochinese Communist Party, 1932

country and the Party became a Party in power. Since the State has been under the leadership of the Party, in the years of armed struggle as well as in recent years after the restoration of peace, the building of the revolutionary armed forces has always been regarded as one of the most important tasks of the Party, because the revolutionary armed forces are the main part of the revolutionary State.

Looking back over the path covered, it is obvious that the armed struggle has played a very important part in the process of the revolutionary mobilisation of our people under the leadership of the Party. Through the years of extremely hard and valiant armed struggle, our Party accumulated most valuable experiences. The study of these experiences is of great significance for the strengthening of the revolutionary armed forces, for the consolidation of national defence in the North, as well as for the completion of the national democratic revolution throughout the country.

Revolutionary armed struggle in any country has common fundamental laws. Revolutionary armed struggle in each country has characteristics and laws of its own too.

Russia was originally an imperialist country, with capitalist economy developed to a certain extent. The Russian October Revolution was an uprising of the working class and the urban labouring people to overthrow capitalism, and to establish the worker-peasant Soviet power. The revolutionary war that followed it was the revolutionary civil war of the Soviet labouring people against the white guards of the reactionary bourgeoisie and landlord class ; at the same time, it was the war to defend the socialist Fatherland against the interventionist joint armies of fourteen capitalist countries. Later, the great war for the defence of the Soviet Union against the aggressive fascist armies was a revolutionary war of the labouring people of a socialist country which had become powerful, but was still encircled by capitalism.

China was a semi-colonial and semi-feudal country, covering an immense area, with the biggest population in the world and a backward agricultural economy. For a long period, the armed struggle in China was the long-term revolutionary civil war waged by the Chinese people against the feudal forces and the bureaucratic capitalists, henchmen of the imperialists. In the resistance war, this was a long-term war of the Chinese people against the aggressive imperialists. This armed struggle aimed at achieving the political goals of the national democratic revolution and paving the way for the Chinese revolution to advance to socialism.

Viet Nam was a small, weak, colonial and semi-feudal country, covering a fairly small area, with a small population and an extremely backward agricultural economy. There was the struggle of the people throughout the country, under forms of armed uprising and long-term resistance to overthrow imperialism and the reactionary feudal forces. The aim was to realise the political goals of the national democratic revolution as in China, to recover national independence and bring land to the peasants, creating conditions for the advance of the revolution of our country to socialism.

Therefore, the revolutionary armed struggle in Viet Nam was naturally the reflection of the laws of the revolutionary armed struggle in general, but simultaneously it had its own characteristics and laws. The success of our Party in leading the revolutionary armed struggle and in building revolutionary armed forces is the success of Marxism-Leninism. This is the success of wise and creative application of Marxist-Leninist principles on revolutionary war and revolutionary armed forces to the practical situation of a small, weak, colonial and semi-feudal country which had to fight for a long period against a powerful enemy while encircled by imperialism.

I

OUR PARTY SUCCESSFULLY LED THE PREPARATION FOR THE ARMED UPRISING AND THE AUGUST 1945 GENERAL INSURRECTION

In 1939, immediately after the outbreak of World War Two in Europe, the Party Central Committee realised in time that a new cycle of war and revolution had begun, "the situation in Indo-China has advanced to the stage of national liberation". In 1940 and early 1941, insurrections broke out successively in Bac Son, Nam Ky and Do Luong. Although these uprisings were ferociously repressed by the enemy, they were, however, "the signs of the national insurrection and the first steps of armed struggle of the Indochinese peoples." (1)

In fact, in the very difficult conditions prevailing at that time, our people lived very miserably under the double yoke of the French and Japanese imperialists, and the revolutionary movement was mercilessly repressed. But our Party continued to do its utmost to step up propaganda and agitation among the people, to gather all patriotic forces into the Viet Minh, to build guerilla bases, set up revolutionary armed forces and make preparations for armed insurrection.

---

(1) Resolution of the 8th Session of the Central Committee, 1941.

In August 1945, the armies of the Soviet Union and the Allied powers were completely victorious. The Japanese fascists surrendered. An atmosphere of insurrection was seething all over Viet Nam. Millions of people in the towns and countryside demonstrated and displayed their forces. The general insurrection broke out. The August Revolution was victorious. On September 2nd, President Ho Chi Minh, on behalf of the provisional government, read the Declaration of Independence before the Vietnamese people and the world. The Democratic Republic of Viet Nam came into being, the first people's democratic country in Southeast Asia.

The August General Insurrection ended in a great victory: the high tide of the struggle against the French and Japanese fascists waged by the Vietnamese people during World War Two overthrew the near century-old imperialist domination and the thousand-years-old monarchy, to set up the democratic republican regime. The August General Insurrection opened a new era in the history of the Vietnamese people, an era in which they took their destiny into their own hands.

The preparations for the insurrection during World War Two, and the August insurrection taught us many valuable lessons. In his book, *The August Revolution*, comrade Truong Chinh has analysed the strong and the weak points of this revolution and made many apt observations. In this article, we shall deal with the whole process of the preparation for armed insurrection to the victory of the general insurrection, pointing out some of the main experiences which are also a great success for the leadership of our Party.

1. The August general insurrection was successful, first of all, because of the correct strategic guidance of the Party Central Committee in the question of national liberation. The Central Committee regarded it as the main task for the whole Party and for the entire people and deemed it necessary to rally all patriotic forces in order to carry it out successfully and by every means

Our country was a colonial and semi-feudal one. The two basic contradictions in our society were the contradiction between imperialism and our nation and the contradiction between the feudal landlord class and our people, chiefly the peasantry; of these two contradictions, that between imperialism and our nation had to be considered as the most essential. That is why the revolution in Viet Nam, which was a national democratic revolution, had two fundamental tasks: the anti-imperialist and the anti-feudal task. Among these two tasks, the anti-imperialists task, the task of wiping out imperialism to liberate the people, had to be regarded as the most essential. After 1930, the Party analysed the two contradictions in our society and laid down the two tasks of our revolution, thereby mobilising a wide and deep revolutionary movement. But it was not until 1939-1941 that the anti-imperialist task, the task of national liberation was clearly conceived as the most essential. Moreover, due to the appreciation of the great events in the world and at home at that time, the Party set the national liberation task as an urgent one for the whole people.

The 6th session of the Central Committee held late in 1939, clearly pointed out: " Nowadays, the situation has changed. At the present time, French imperialism has taken part in launching a world war. The domination imposed on colonies such as Indo-China, which is clearly a fascist militarist regime, and the scheme of compromising and surrendering to Japan have set a vital problem before the Indochinese people. *In the life and death struggle of the Indochinese nations there is no way out other than that of overthrowing French imperialism, opposing all foreign invaders, either white or yellow-skinned, in order to regain liberation and independence.*"

The 8th enlarged session of the Central Committee held in 1941 perfected the change of direction in the Party's leadership of the Revolution and mapped out the concrete political program for the national liberation revolution. It was estimated by the Session that "at the present time, the watchword of the Party is first *to liberate the Indochinese people from the Japanese and French yoke.* For the fulfilment of these tasks, the forces of the whole of Indo-China must be concentrated, those who love their country will unite together into a Front and rally all their forces to fight for national independence and freedom, to smash the French and the Japanese invaders. The alliance of all forces of the classes and parties, the patriotic revolutionary groups, religious groups and all people fighting the Japanese, is the essential work of our Party."

The Central Committee also put out a *New policy for the Party*, temporarily put aside its slogan for agrarian reform and replaced it by the slogan of reduction of land rents and interest charges, and confiscation of land belonging to imperialists and Vietnamese traitors and its distribution to the peasants. Simultaneously, it decided to found the League for Independence of Viet Nam (called

Viet Minh for short) and various organisations for national salvation.

Stressing the task of national liberation, the resolution of the 8th session of the Central Committee was extremely precise and clear, and conformed to the conditions prevailing at that time, to the deep and fundamental aspirations of each class and each patriotic stratum. It was precisely for this reason that within a short period, the Viet Minh gathered together the great forces of the people and became the most powerful political organisation of the broad revolutionary masses. The program of the Viet Minh was warmly welcomed by all sections of the people. The Viet Minh was well known throughout the country. The resolution of the 8th session of the Central Committee was a concrete platform which had a decisive effect on the victory of the August Revolution.

It is necessary to add that at that time, from the strategic point of view, the feudal landlord class was not definitely assessed as the object of the revolution. From the theoretical point of view, national liberation was somewhat seen as being apart from the bourgeois democratic revolution; as far as the immediate task was concerned, the landlord class was too highly evaluated and there was lack of consideration for the worker-peasant alliance in the National United Front. These shortcomings were to have an influence on the thinking and work of our Party in the future; for instance, they led partly to the slighting of the anti-feudal task in the first years of seizure of power and of the Resistance War.

2. The August general insurrection triumphed because, while pointing out the above-mentioned change of direction for the revolutionary task, the Party Central Committee timely laid down the change of the forms of struggle and put forward the question of preparing for armed insurrection

The shifting from political struggle to armed struggle was a very great change that required a long period of preparation. If insurrection is said to be an art, the main content of this art is to know how to give to the struggle forms appropriate to the political situation at each stage, how to maintain the correct relation between the forms of political struggle and those of armed struggle in each period. At the beginning, the political struggle was the main task, the armed struggle a secondary one. Gradually, both the political struggle and armed struggle became equally important. Later, we went forward to the stage when the armed struggle occupied the key role. But even in this period, we had to define clearly when it occupied the key role within only a certain region and when throughout the nation. We had to base ourselves on the guiding principle with regard to forms of struggle to clearly lay down the guiding principles for our work and for the forms of organisation. In the then situation, the struggle between the enemy and us was extremely hard and fierce. If guidance in struggle and organisation was not precise, that is to say did not correctly follow the guiding principle of both determination and carefulness, and of knowing how to estimate the subjective conditions and compare the revolutionary forces with the counter-revolutionary forces, we would certainly have met with difficulty and failure. The correct leadership in the preparation for armed insurrection had to secure steady and timely development of the revolutionary forces until the time for launching the insurrection was ripe.

It was clearly pointed out at the 8th session of the Central Committee: "To prepare forces for an insurrection, our Party has to:

"1 — Develop and consolidate the organisations for national salvation.

"2 — Expand the organisations to the cities, enterprises, mines and plantations.

"3 — Expand the organisations to the provinces where the revolutionary movement is still weak and to the minority areas.

"4 — Steel the Party members' spirit of determination and sacrifice.

"5 — Steel the Party members so that they may have capacity and experience to enable them to lead and cope with the situation.

"6 — Form small guerilla groups and soldiers' organisations..."

When speaking of insurrection, V. Lenin stressed that "uprising must rely upon the high tide of the revolutionary movement of the masses", and "not upon a conspiracy". To speak of preparations for armed insurrection and of insurrection does not mean that we will pay no more attention to the political movement of the masses; on the contrary, insurrection could not be victorious without a deep and wide political movement waged by the revolutionary masses. Therefore, to make good preparations for armed insurrection, the most essential and important task was to make propaganda among the masses and organise them, to "develop and consolidate the organisations for national salvation". Only on the basis of strong political organisations could semi-armed organisations be set up firmly, guerilla groups and guerilla units organised which have close connection with the revolutionary masses, eventually to further their activities and development.

In the early years, as the political movement of the masses was not strong enough and the enemy's forces still stable, the political mobilisation among the masses had all the more to be considered as the main task for the preparation of armed insurrection. The propaganda and organisation of the masses carried out everywhere in the country, particularly at the key points was of decisive importance. Viet Bac mountain regions were soon chosen by the Party Central Committee as the armed bases, the two central areas being Bac Son - Vu Nhai and Cao Bang. In the then conditions, the armed bases must be held secret, must be localities where the revolutionary movement was firm and the mass organisations strong; on the basis of the political organisations of the masses, self-defence groups and fighting self-defence groups (1) were set up which swelled afterwards to local armed groups, or armed platoons freed or partially freed from production, and eventually to bigger guerilla units. Underground operating cadres' teams, underground militarized teams, armed shock teams and local armed groups and platoons gradually appeared. The most

---

(1) Self-defence force ensures public security and order in the village and take part in the fighting in the last extremity only. Fighting self-defence force has the task to fight the enemy as soon as he arrives at the village.

appropriate guiding principle for activities was *armed propaganda (1); political activities were more important than military activities, and fighting less important than propaganda ;* armed activity was used to safeguard, consolidate and develop the political bases. Once the political bases were consolidated and developed, we proceeded one step further to the consolidation and development of the semi-armed and armed forces. These had to be in strict secrecy with central points for propaganda activity or for dealing with traitors. Their military attacks were strictly secret and carried out with rapidity. Their movements had to be phantom-like. A position of legal struggle was maintained for the broad masses. The setting up of revolutionary power was not then opportune. There were regions in which the whole masses took part in organisations of national salvation, and the village Viet Minh Committees had, as a matter of course, full prestige among the masses as an underground organisation of the revolutionary power. But even in these localities, we had not to attempt to overthrow the enemy, but try to win over and make use of him. It was in keeping with that direction that the Party Central Committee gave instructions to the armed units for national salvation at Bac Son-Vu Nhai. It was in line with that direction that President Ho Chi Minh pointed out the guiding principle of armed propaganda for the armed organisations at Cao Bang-Bac Can, chiefly when giving orders for the setting up of the Viet Nam Liberation Armed Propaganda Unit. Experiences have proved that in the first period of the preparation for armed insurrection, if the above-mentioned guiding principles were not thoroughly understood, the revolutionary movement would

---

(1) Propaganda carried out by armed units.

often meet with temporary difficulties and losses, thus affecting the preparation for armed insurrection.

With the Japanese Coup de force on March 9, 1945, the situation underwent great changes. The French fascists collapsed. Japanese fascism, the main immediate enemy of the Vietnamese people, had not yet time to consolidate its rule in Indo-China and was suffering successive defeats on various battefields. The Party Central Committee realised very clearly and timely the new political crisis caused by the Japanese Coup d'Etat and issued the order "to mobilise a strong high tide of anti-Japanese movement for national salvation which would be the prerequisite for the general insurrection". Preparations were made to be ready to shift to the general uprising when conditions were ripe. The Party Central Committee also worked ont the policy of "promoting guerilla warfare to occupy the resistance bases", brought all armed forces under a single command and set up revolutionary power in areas where the guerillas were operating. Where mass bases were fairly strong, power was exercised secretly.

From Cao Bang - Bac Can to Thai Nguyen - Tuyen Quang and in serveral localities in the Midland, the Liberation Armed Propaganda units and National Salvation units attacked the district towns and set up revolutionary power. The attacks on Japanese paddy stores developed strongly everywhere, the picked-units of the Viet Minh operated right in the cities. The network of self-defence and fighting self-defence units, People's committees and Liberation Committees spread to all localities. In Quang Ngai province, the Ba To guerilla unit was born. A seething atmosphere prevailed in the whole country.

In April 1945, the Military Conference of north Viet Nam decided to unify the revolutionary armed forces under the name of Viet Nam Liberation Army; resistance zones were set up, and the North Viet Nam Revolutionary Military Committee organised. In June, the free zone was founded, the ten-point policy of the Viet Minh being implemented everywhere in the 6 provinces of the free zone. The Viet Minh was gaining increasing strength. The influence of the free zone and that of the Liberation Army which spread widely and rapidly, further encouraged our people to prepare to take advantage of the good opportunity, winning the wavering elements to side with the revolutionary forces, and throwing the enemy's ranks into confusion.

After May 1945, the preparation for armed insurrection entered a new stage with the following features: *the upsurge of the anti-Japanese movement for national salvation strongly increased everywhere, regional guerilla war was launched; local revolutionary power was established, and anti-Japanese bases set up.* Under the leadership of our Party, the movement advanced boldly and firmly.

3. **The August General Insurrection was successful because the Party Central Committee accurately and clearsightedly laid down what the conditions should be for the General Insurrection to break out and gain success; this therefore, would help mobilise the whole Party and the entire people, promote the spirit of determination, courage, positiveness and creativeness of the masses.**

When speaking of insurrection, J. Stalin pointed out that selecting the right opportunity is one of the essential conditions to lead the uprising to victory : " The selection of the moment for the decisive blow, of the moment for starting the insurrection, so timed as to coincide with the moment when the crisis has reached its climax, when it is already the case that the vanguard is prepared to fight to the end, the reserves are prepared to support the vanguard, and maximum consternation reigns in the ranks of the enemy "

Our Party had drawn bloody experiences from the Nghe - Tinh and Nam Ky insurrections. All these spoke out the decisive importance of the right opportunity for uprising. Therefore, as early as 1941, the 8th session of the Central Committee took care to clearly set out what the conditions should be when people could be led to carry out the insurrection : " The revolution in Indo-China must be ended with an armed uprising ; in order to wage an armed insurrection, conditions must be as follows :

" the National Salvation Front is already unified all over the country ;

" the masses can no longer live under the French-Japanese yoke, and are ready to sacrifice themselves in launching the insurrection ;

" the ruling circles in Indo-China are driven to an economic, political and military crisis ;

" the objective conditions are favourable to the uprising such as the Chinese army's great triumph over the Japanese army, the outbreak of the French or Japanese revolution, the total victory of the democratic camp in the Pacific and in the Soviet Union, the revolutionary fermentation in the French and Japanese colonies, and particularly the landing of Chinese or British-American armies in Indo-China "

The instruction on the *preparations for the Insurrection* issued by the Viet Minh Central Committee in May, 1944, also pointed out clearly the moment the people should rise up :

" 1. The enemy's ranks at that moment are divided and dismayed to the extreme.

" 2. The organisations for national salvation and the revolutionaries are resolved to rise up and kill the enemy.

" 3. Broad masses wholeheartedly support the uprising and determinedly help the vanguard.

" If we launch the insurrection at the right time, our revolution for national liberation will certainly triumph. We must always be on the alert to feel the pulse of the movement and know the mood of the masses, estimate clearly the world situation and the situation in each period in order to seize the right opportunity and lead the masses of people to rise up in time ".

After the Japanese Coup d'Etat, in the historical instruction on March 12, 1945, the Party Central Committee clearsightedly estimated that the situation was offering many new favourable conditions, but " conditions for the insurrection are not yet ripe ". At the same time, it pointed out that when the Allied troops landed in Indo-China and the Japanese sent their troops to intercept them, thus exposing their rear, then would be the most favourable opportunity for launching the insurrection. The instruction added : " *If revolution breaks out and people's revolutionary power is set up in Japan, or if Japan is occupied as was France in 1940, and the Japanese expeditionary army demoralised, then, even if the Allied forces have not yet arrived in our country, our general insurrection can be launched and win victory*". The high tide of the revolution was sweeping throughout the country. The world situation changed rapidly. On August 8, 1945 the Soviet Red Army attacked in North-east China, and within a few days, the crack Japanese Kwantung Army was entirely routed. Japanese fascism was in an extremely critical situation and sought unconditional surrender. At that moment, the National Conference of the Party, which was being held at Tan Trao village, together with the Viet Minh General Committee, took a decision to order the launching of the General Insurrection and the setting up of people's power throughout the country. The National Insurrection Committee was established. Afterwards, the National People's Congress held at Tan Trao set up the Viet Nam National Liberation Committee that was tantamount to the Provisional Government of the Democratic Republic of Viet Nam, headed by President Ho Chi Minh.

News of the Japanese capitulation spread rapidly. Having thoroughly understood the Party's instructions, and taking advantage of the extreme demoralisation of the Japanese forces, the consternation of the puppet government and the vacillation of the security troops, the local Party organisations and Viet Minh organisation immediately took the initiative to lead the people to seize power even before receiving the insurrection order. On August 11, insurrection broke out in Ha Tinh province ; on August 12, the insurrection order was proclaimed in the free zone, the Liberation army stormed many enemy's posts and a few days later marched to liberate Thai Nguyen. On August 13, the people in Quang Ngai province arose . On August 19, a splendid victory was scored by the insurrection in Hanoi Capital ; on August 23, the insurrection was successful in Hue, and on August 25, in Saigon. On August 29, the first unit of the Viet Nam Liberation Army entered the Capital city of Hanoi. Throughout the country, in town and countryside, millions of people rose up to wrest back power from the hands of the Japanese fascists and pro-Japanese puppets, shattering the fetters of the imperialists and feudalists. Basing ourselves on the powerful political forces of the people, backed by military and para-military forces, and on our skill in neutralising the Japanese army then in dismay, the insurrection cost little blood and rapidly gained success from North to South. In face of the people's powerful strength, Bao Dai abdicated, the Tran Trong Kim puppet government surrendered. On September 2, the Provisional Government appeared before the people. At historic Ba Dinh square, President Ho read the Declaration of Independence. The Democratic Republic of Viet Nam came into being, a great historical event of South-east Asia.

Availing itself of the right opportunity, our Party led the August General Insurrection to victory. Had the insurrection broken out sooner, it would have certainly met with numerous difficuties. It would have been in a dangerous situation had it broken out later, when the Chiang Kaishek and British armies had arrived in our country. The Party led the people to seize power immediately after the Japanese capitulation and before the Allied forces arrived in Indo-China. The splendid victory of the insurrection was due to its timely launching.

The above-mentioned lessons were victories for the leadership of our Party, the subjective conditions leading the August General Insurrection to victory. Of course, the triumph of the August General Insurrection was also due to very important objective conditions. The great victory of the Soviet Red army and Allied forces over the German-Italian-Japanese fascists created favourable conditions for the liberation of the oppressed peoples in the world In our country, during the days of the August Insurrection, the enemy of the revolution was brought to an utter crisis. After France was occupied by German troops, the French colonialists' forces and influence in our country grew markedly weaker. Then, due to acute contradictions between the competing invaders, the enemy who had ruled for long over our country was defeated by the Japanese. Consequently, the feudal administration set up by the French also disintegrated. After March 9, the Japanese fascists, the main enemy of our people, were in a critical stage due to successive failures. The pro-Japanese puppet government having no solid foundation, was not able to cope with the ever growing revolutionary movement. Following the great victory of the Soviet Red Army and the Japanese fascists' capitulation, the spirit of the Japanese army in Indo-China was brought to utter crisis, their dream of enslaving our country had shattered. In view of this situation, the revolutionary masses were able to neutralise the Japanese. Thanks to this, no fierce resistance was met with and the General Insurrection triumphed rapidly. The above-mentioned objective factors proved the great influence and assistance of the socialist, democratic and peaceful forces, particularly that of the Soviet Red Army's great victory, towards our revolution. It has proved that the Vietnamese revolution is part of the world socialist revolution, our era is the era of socialist revolution and national liberation revolution, an era of decay and disintegration of imperialism and colonialism. This important objective condition does not in any way belittle the Party's precise and clearsighted leadership and its effect on the success of the August General Insurrection. To realise the decisive role of correct leadership, suffice it to compare the situation of our country with that of a number of countries in South-east Asia at that time. In the same month of August and under the same favourable objective conditions, not only did the revolution in these countries not win great success but it even met with setbacks.

The August General Insurrection was a great victory for our people and our Party. *This was a successful uprising of the people of a colonial and semi-feudal country, under the leadership of the Communist Party. Through a long political struggle, it developed into a regional armed struggle in the pre-insurrectionary period. In the end, seizing the right opportunity, when the enemy was in utter crisis, and making use mainly of the masses' political forces with the support of the armed and semi-armed forces, we heroically rose in the cities and the countryside, smashed the rule of the imperialists and feudalists and set up people's democratic power. The success of the August General Insurrection proves that the liberation movement of the oppressed nations, in given historical conditions, can be victorious through insurrection.*

This was the first time the people in a weak and small colony gallantly rose up under the leadership of the Communist Party and freed themselves from the chains of the imperialists and their stooges. The August General Insurrection was a good contribution to the upsurge of the national liberation movement in the world which rose like a high tide in and after World War Two, foretelling the coming disintegration of colonialism.

II

OUR PARTY VICTORIOUSLY LED THE LONG-TERM RESISTANCE WAR AGAINST THE FRENCH IMPERIALISTS AND U.S. INTERVENTIONISTS

Soon after the victory of the August Revolution and the founding of the Democratic Republic of Viet Nam, assisted by the British forces, the French colonialists launched war and occupied Saigon on September 23, 1945, thus invading our country once more. Our Southern compatriots resolutely resisted. On December 19, 1946, the nation-wide Resistance War broke out. This was a long, hard and extremely heroic liberation war of our people against the French imperialists and the U.S. interventionists. This resistance was carried on for 9 years and ended with our great victory at Dien Bien Phu and at the Geneva Conference. Peace has been restored in Indo-China on the basis of respect of the sovereignty, independence, unity and territorial integrity of our country, Cambodia and Laos. The North of our country has been entirely liberated.

The successful nine years of Resistance are the most glorious pages in the history of our people in the process of mobilisation for national liberation. Under the Party leadership, our people, from the days they rose against the enemy in Nam Bo until the Dien Bien Phu victory, fought and defeated an aggressive army of a powerful imperialist country. In this article, we try to point out some essential experiences of our Party in leading the revolutionary war.

1. First of all, our resistance won victory because the Party's policy of uniting the entire people to resolutely wage the Resistance war was a correct policy.

As at the beginning of World War Two, the policy advanced by the Party was to prepare for armed uprising to liberate our country. In the years 1945-1946, immediately after the establishment of the democratic republican regime, the Party put forward the policy of uniting the entire people resolutely to wage the Resistance War to safeguard the achievements of the August Revolution and newly-recovered independence.

Following the success of the Revolution, the Party clearly realised the danger of aggression from the French colonialists. Even in the Declaration of Independence and in the Oath of Independence, the Party called for a heightening of vigilance, and mobilised the people to be prepared to defend the Fatherland.

The French colonialists' aggressive war broke out in Saigon when people's power had not yet been consolidated and great difficulties in all fields lay ahead of us. Never had our country borne the yoke of so many foreign armies. The Japanese had capitulated but were still in possession of their arms. The Chiang Kai-shek army which landed in the North, did its best to assist the Viet Nam Quoc Dan Dang* to overthrow people's power. In the South, British forces occupied the country up to the 16th parallel and tried to help the French colonialists expand their aggressive war.

Our Party led the people in Nam Bo to wage a resistance war against the French colonialists. To spearhead all its forces at the principal enemy, the Party carried out the line of winning more friends and less enemies, endeavoured to widen the national united front, founded the Viet Nam National United Front (called Lien Viet for short), united all forces which could be united, neutralised all those which could be neutralised and differentiated between forces which could be differentiated. At the same time, it consolidated power, developed and consolidated the armed forces, elected the National Assembly and formed the Government of coalition for the Resistance.

In its foreign policy, our Party tried by every means to realise a cordial policy with the Chiang Kai-shek army and avoid all conflicts. Dealing with the main enemy, the French aggressive colonialists, on the one hand, our Party led the people and army in Nam Bo resolutely to resist against their aggressive army, mobilised the entire people throughout the country to do their best in supporting the South, sent troops there and at the same time actively prepared for the resistance in case the war spread out. On the other hand, it did not miss any opportunity to take advantage of the contradictions between the French and Chiang Kai-shek forces and to negotiate with the French Government to secure a detente and preserve peace.

---

* The Vietnamese Kuomintang

The signing of the Preliminary Convention on March 6, 1946 between the French and our forces was the result of this correct policy and strategy. Due to the concession granted by us, part of the French army could land at certain localities in north Viet Nam to relieve the Chiang troops. On the French Government's side, it recognised that the Democratic Republic of Viet Nam was a free country within the framework of the French Union, having its own government, army, parliament, finance, etc.. Thus, we succeeded in driving 200,000 Chiang Kai-shek troops out of our country. Following this, the counter-revolutionary army of the Viet Nam Quoc Dan Dang which still occupied 5 provinces along the frontier and the Midland of north Viet Nam was also annihilated. The democratic republican regime grew stronger.

With the Preliminary Convention, we had carried out the policy of " making peace to go forward ". Immediately after the signing of the Convention, there was a time when illusions of peace partly influenced our vigilance towards the colonialists' reactionary schemes. But in general, the Party kept on making efforts to consolidate peace, while increasing our forces, ready to cope with all the enemy's plots. On the one hand, it adhered the convention which had been signed, on the other hand, it resolutely carried out a self-defence struggle against all enemy acts sabotaging this convention. The French colonialists' schemes were revealed with every passing day. The more concessions we made, the further they trespassed. They openly tore the convention they had signed, carried on mopping-up operations in the provisionally occupied areas in the South, indulged in provocative acts, step by step encroached upon our rights in numerous localities including Haiphong, and

Hanoi Capital. They did their utmost to occupy our country. Therefore, when realising that all possibilities for maintaining peace no longer existed, the Party called on the entire people to wage the Resistance War.

Realities clearly showed the people that our Party and Government had done their utmost to maintain the policy of peace, but the French colonialists were determined to invade our country once more. It was obvious that there was no other way for our nation but to take up arms resolutely to safeguard the Fatherland. Practical deeds had clearly shown to the French people and peace-loving peoples all over the world that we wished to live in peace, but the French colonialists determinedly provoked war. That is why our people's War of Resistance won ever greater sympathy and support from the broad masses in France and the world over.

Our Party's policy of resistance was a precise one, in conformity with the masses' requirements, whose wrath towards the aggressors had reached a climax. For this very reason, in response to President Ho's appeal for carrying out a resistance war, our army and people did not shun hardships and sacrifices. Like one man, they were determined to wage the War of Resistance to final victory and annihilate the aggressors.

2. Throughout the resistance war, our Party fundamentally stuck firmly to the national democratic revolution line, therefore succeeded in launching the people's war and defeating the enemy

The Resistance War waged by our people was the continuation of the national democratic revolution by armed struggle. Therefore, to hold firm the national democratic revolution line in leading the Resistance War was a nodal, decisive question.

As is said above, Viet Nam was originally a colonial and semi-feudal country. Our society underwent great changes as a result of the August Revolution. Imperialist rule had been overthrown. The power of the king and mandarins, the imperialists' stooges representing the most reactionary section of the feudal landlord class had been overthrown. However, this class still existed in our society and the land question was only partly solved.

French colonial troops re-kindled the aggressive war. The basic contradiction between our people and imperialism reappeared in the most acute form. Who was the aggressive enemy ? Obviously the French imperialists. At the beginning, owing to the fact that there were progressive elements in the French government and due to tactical necessity, we named as our enemy French reactionary colonialists. But later, especially from 1947 on, the French government definitely became reactionary, the aggressors were unmistakably the French imperialists who were the enemy of our entire people and were invading our country. In this situation, the national factor was of utmost importance. To fight French imperialism, it was necessary to unite the whole nation, all revolutionary classes, patriotic elements, to strengthen and widen the National United Front. Our Party obtained great success in its policy of uniting the people. The slogan: "Unity, unity and broad unity — success, success and great success," put forth by President Ho Chi Minh became a great reality. The anti-imperialist National United Front in our country was a model of the broadest national front in a colonial country.

The revolution for national liberation under the leadership of the Communist Party never deviated from the democratic revolution. The anti-imperialist task always went side by side with the anti-feudal task, although the former was the more urgent; Viet Nam was a backward agricultural country and the great majority of the population were peasants. While the working class is the class leading the revolution, the peasantry is the main force of the revolution, full of anti-imperialist and anti-feudal spirit. Moreover, in waging the Resistance War, we relied on the countryside to build our bases to launch guerilla warfare in order to encircle the enemy in the towns and eventually arrive at liberating the towns. Therefore, it was of particularly importance to pay due attention to the peasant question and the anti-feudal question to step up the long Resistance War to victory.

How did our Party solve the anti-feudal question with a view to mobilising the peasant force during the Resistance War? In the August Revolution, after we had overthrown the power of the king and mandarins, a number of traitors were punished, their land allotted to the peasants. Colonialists' land was also temporarily given to the peasants. After the French imperialists re-invaded our country, the collusion between the imperialists and the most reactionary section of the feudal landlord class gradually took shape.

The essential contradiction in our society at that time was the contradiction between, on the one side, our nation, our people, and on the other, the French imperialists and their henchmen, the reactionary feudalists. We accordingly put forth the slogan " To exterminate the reactionary colonialists and the traitors." As a result, as early as the first years of the Resistance War, a number of the most reactionary of the landlord class were repressed in the course of the operations against local puppet administration and traitors. Their land and that belonging to absent landlords were allotted outright or given to the trusteeship of the peasants. Thus, in practice, the anti-feudal task was carried on.

However, due to a vague conception of the content of the revolution for national liberation as early as 1941, in the first years of the Resistance War, in our mind as well as in our policies, the anti-feudal task was somewhat neglected and the peasant question underestimated in importance. Only by 1949 - 1950 was this question put in a more definite way. In 1952 - 1953, our Party decided to mobilise the masses for a drastic reduction of land rent and to carry out land reform, implementing the slogan "land to the tiller". Hence, the resistance spirit of millions of peasants was strongly roused, the peasant-worker alliance strengthened, the National United Front made firmer, the administration and army consolidated and resistance activities intensified. There were errors in land reform but they were, in the main, committed after the restoration of peace and thus did not have any effect on the Resistance War. It should be added that not only was land reform carried out in the North, but in south Viet Nam, land was also distributed to the peasants after 1951.

The carrying out of land reform during the Resistance War was an accurate policy of a creative character of our Party.

Looking back, on the whole, our Party stuck to the national democratic revolution line throughout the Resistance War. Thanks to this, we succeeded in mobilizing our people *to launch the people's war*, using the enormous strength of the people to vanquish the aggressors.

Right at the outbreak of the nation-wide Resistance War, the Party issued instructions for a *"whole nation Resistance War, all-out Resistance War"*. This was the fundamental content of the people's war. Through the struggles of the Resistance years, this content became richer and more concrete, especially after the launching of the guerilla war and after the peasant question was duly considered concurrently with the national question.

As the political objective of the Resistance War was national independence and land, and as this was in conformity with their deep and fundamental aspirations, the people arose to exterminate the aggressors and save the country. President Ho Chi Minh appealed: "All Vietnamese regardless of sex, age, creed, political tendency and nationality must stand up to fight the French colonialists and save the country. Those who have guns use guns, those who have swords use swords, and those who have no swords use picks, shovels or sticks. Everyone has to do his utmost against the colonialists to save the country." The Vietnamese people responded to President Ho Chi Minh's call and millions rose as one man to wage the Resistance War, annihilate the enemy and save the country. This people's war was, from the point of view of forces, mainly a peasants' war. The peasants had for a long time fought under the Party's banner, rosed up to seize the power in the August Revolution, and throughout the long and hard Resistance War also played a great and important part.

In fact, our Resistance War was a people's war.[1] On the battlefronts the armymen rushed forward to annihilate the enemy, while in the rear the people were striving to increase production — the peasants in the fields and the workers in the arms-factories — in order to supply the troops, to serve the front line. The people's armed forces were the regular army and the regional troops and guerilla units. With the slogan " the whole nation in arms " each person was a soldier, each village a fortress, each Party branch and Resistance committee a staff. It was so in the free zones and all the more so in the enemy-occupied zones.

Our people's Resistance War was an all-out Resistance War. Not only did we fight in the military field but also in the political, economic and cultural fields. In the political field, we had, at home, to increase the education and mobilisation of the people, unremittingly strengthen national solidarity and endeavour to smash all the enemy's schemes to divide and deceive our people, while in its foreign policy, efforts had to be made to win over the support of progressive people throughout the world, particularly to closely co-ordinate with the struggle of the French people and those in the French colonies against this dirty war. In the economic field, great efforts had to be made to build a Resistance War economy, increase production, realise self-reliance and self-sufficiency in order to perseveringly wage the long Resistance War; at the same time we had to do our utmost to sabotage the enemy's economy, to frustrate his plots to grab our manpower and wealth, to " use war to feed war ". In the cultural field, we had to develop the culture of the Resistance imbued with a mass character and to heighten patriotism and hatred. Simultaneously, we had to actively struggle to wipe out the influence of obscurantist culture in the free zones, to fight against the enemy's debased culture in the occupied zones, to break to pieces the enemy's counter-propaganda, to maintain and raise the confidence and determination to carry on the Resistance War of the whole people.

Under the Party's leadership, the people's administration played an important part in the mobilisation of manpower and wealth for the Resistance. "All for the front, all for victory" was the slogan for our nation and it showed the determination of our people to concentrate all their forces to fight to the bitter end to overthrow the French imperialists and their henchmen, liberate the country and wrest back independence and land. This was the slogan of the people's war.

3. Our Party put forth the correct strategic guiding principle: long-term resistance war, self-reliance, and the appropriate fighting principle: guerilla warfare and eventually advancing to mobile warfare.

Launching the Resistance War, our Party accurately assessed the strong and weak points of the enemy and ours and clearly saw the balance of forces and the enemy strategic schemes in order to define our strategic principle.

The *enemy*, an imperialist power much weakened after World War Two, was still strong as compared with us. Moreover, he possessed a seasoned professional army equipped with up-to-date arms, well supplied and experienced in aggressive wars. His weak point lay in the unjust character of his war. As a result, he was internally divided, not supported by the people of his own country and did not enjoy the sympathy of world opinion. His army was strong at the beginning but its fighting spirit was deteriorating. French imperialism had other weak points and difficulties, namely: limited manpower and wealth, their dirty war was strongly condemned by their countrymen, etc.

On *our side*, our country was originally a colonial and semi-feudal country whose independence was newly won back. Thus, our forces in all fields were not yet consolidated, our economy was a backward agrarian one, our army untried guerilla troops with few and obsolete arms, our supplies insufficient and our cadres lacking experiences. Our strong point lay in the just nature of our Resistance War. Hence, we succeeded in uniting our entire people. Our people and troops were always imbued with the spirit of sacrificing themselves in fighting the enemy, and enjoyed the sympathy and support of people throughout the world.

These were the main features of the two sides in the last Resistance War. They clearly pointed out that the enemy's strong points were his weak ones and our strong points were his weak ones, but the enemy's strong points were temporary ones while ours were basic ones.

Owing to the above-mentioned characteristics, *the enemy's strategic principle was to attack swiftly and win swiftly*. The more the war was protracted the lesser would be his strong points, and their weak points would grow weaker. This strategic principle was in contradiction with the French imperialists' limited forces which had grown much weaker after World War Two. Consequently, in their schemes of invading our country, they were compelled to combine their plan of attacking swiftly and win swiftly with that of invading step-by-step, and even of negotiating with us in their time-serving policy to muster additional forces. Despite the difficulties and obstacles caused by their weak points, whenever they had the possibility, they would immediately carry out their plan of attacking swiftly and winning swiftly, hoping to end the war by a quick victory. From the very beginning of the war, French colonialists had the ambition to complete their occupation and "pacification" of south Viet Nam within a few weeks. The nation-wide Resistance War broke out. On the failure of their attempt to wipe out our main forces in the cities, they did their utmost to regroup their troops and launched a big offensive in Viet Bac, expecting to annihilate our leading organs and main forces in order to score a decisive success. The offensive in Viet Bac was brought to failure, the enemy was forced to protract the war and switch over to "pacifying" the areas in his rear but he had not as yet given up his strategic plan of attacking swiftly and winning swiftly. The reshuffling of generals time and again, especially the sending of General Navarre to Indo-China were all aimed at striking decisive blows in order to quickly end the aggressive war.

Realising clearly the enemy's strong and weak points and ours, to cope with the enemy's strategic scheme, *our Party set forth the guiding principle of a long-term Resistance War.* Facing an enemy who temporarily had the upper hand, our people was not able to strike swiftly and win swiftly but needed time to overcome its shortcomings and increase the enemy's weak points. Time was needed to mobilise, organise and foster the forces of the Resistance, to wear out the enemy forces, gradually reverse the balance of forces, turning our weakness into strength and concurrently availing ourselves of the changes in the international situation which was growing more and more avantageous to our Resistance, eventually to triumph over the enemy.

The general law of a long revolutionary war is usually to go through three stages: defensive, equilibrium and offensive. Fundamentally, in the main directions, our Resistance War also followed this general law. Of course, the reality on the battlefields unfolded in a more lively and more complicated manner. Implementing the guiding principle of a long war, after a period of fighting to wear out and check the enemy troops, we carried out a strategic withdrawal from the cities to the countryside in order to preserve our forces and defend our rural bases. Following the failure of the enemy offensive in Viet Bac, equilibrium gradually came into being. We decided to launch an extensive guerilla war. From 1950 onwards, campaigns of local counter-offensives were successively opened and we won the initiative on the northern battlefront. The Dien Bien Phu campaign in early 1954 was a big counter-offensive which ended the Resistance War with a great victory.

To make everyone thoroughly understand the strategic guiding principle of long-term war was not only a big work of organisation militarily and economically but also a process of ideological education and struggle within the Party and among the people against erroneous tendencies which appeared many a time in the years of the Resistance War. These were pessimistic defeatism which presumed that our country being small, our population thin, our economy backward and our armed forces young and weak, we would be unable to face the enemy, let alone perseveringly to wage a long Resistance War. These were subjectivism, loss of patience, eagerness to win swiftly which came out in the plans of operations of a number of localities at the start of the Resistance War which were unwilling to withdraw their force to preserve our main force, and in their plan of general counter-offensive put forth in 1950 when this was not yet permitted by objective and subjective conditions.

Utmost efforts were made by the Party to correct these erroneous tendencies, to educate the people, enabling them to see clearly our difficulties and advantages and stimulate the entire people to keep firm their determination to fight. The booklet *The Resistance War Will Win* written by comrade Truong Chinh was an important contribution to the thorough understanding of the Resistance War line and policies of the Party. Here, emphasis should be laid upon the great effect of the resolutions of the First Session of the Central Committee in 1951 which reminded the whole Party that "our Resistance War is a long and hard struggle" and "we have mainly to rely on our own forces". The ideological remoulding drives in the Party and the army and the propaganda work among the people carried out on the Central Committee's instructions, basically consolidated the people's determination to wage the long Resistance War, heightened their confidence in final victory and enabled the guiding principles of long-term and self-relying Resistance War to penetrate more deeply into the masses' consciousness.

To wage a long Resistance War, we had to highlight the spirit of *self-reliance*. During the first years of the Resistance, our people had to struggle when encircled on all sides, self-reliance was then a vital question. Our people had no other way than relying on their own forces to cope with the enemy. Highlighting the spirit of self-reliance, our troops looked for their supplies on the battlefields, capturing the enemy weapons to arm themselves, economising in munitions, developing their endurance, overcoming difficulties, striving to take part in production, supplying themselves with a part of their requirements, in order to lighten the people's contributions. Our people endeavoured to build our rear, develop the economy of the Resistance to supply themselves and meet the demand of the front. We stepped up production in every aspect to supply the people with staple commodities and fought against the enemy's economic blockade. Large areas of virgin land were broken to increase the output of foodstuffs. Many arms factories were built to produce weapons for the troops. In particular, the people and troops in the Fifth zone and Nam Bo raised to great heights the spirit of self-reliance, scored many achievements in self-supply to perseveringly wage the Resistance War in extremely difficult and hard circumstances.

When the international situation changed to our advantage, but we were still meeting with many difficulties, there began to appear in the Party and among the people the psychology of waiting and relying on foreign aid. Therefore, while continuing to ideologically prepare for a long Resistance War, attention was given by our Party to rousing our self-reliance and pointing out that international sympathy and support was of importance, but only by relying on our own efforts could we ensure victory for our people's struggle for liberation.

To bring the Resistance War to victory, it was not enough to have a correct strategic guiding principle but an appropriate *guiding principle of fighting* was also necessary in order successfully to carry out that strategic guiding principle. In general, our Resistance War was a *guerilla war moving gradually to regular war, from guerilla warfare to mobile warfare combined with partial entrenched camp warfare*. Basically, we had grasped that general law hence we were successful. However, we did not thoroughly grasp it from the beginning but only after a whole process of being tested and tempered in the practice of war.

In the Resistance War, *guerilla warfare* played an extremely important role. Guerilla warfare is the form of fighting of the masses of people, of the people of a weak and badly equipped country who stand up against an aggressive army which possesses better equipment and technique. This is the way of fighting the revolutionary war which relies on the heroic spirit to triumph over modern weapons, avoiding the enemy when he is the stronger and attacking him when he is the weaker, now scattering, now regrouping one's forces, now wearing out, now exterminating the enemy, determined to fight him everywhere, so that wherever the enemy goes he would be submerged in a sea of armed people who hit back at him, thus undermining his spirit and exhausting his forces. In addition to the units which have to be scattered in order to wear out the enemy, it is necessary to regroup big armed forces in favourable conditions in order to achieve supremacy in attack at a given point and at a given time to annihilate the enemy. Successes in many small fights added together gradually wear out the enemy manpower while little by little fostering our forces. The main goal of the fighting must be destruction of enemy manpower, and ours should not be exhausted from trying to keep or occupy land, thus creating final conditions to wipe out the whole enemy forces and liberate our country.

Guerilla warfare was obviously a form of fighting in full keeping with the characteristics of our Resistance War.

In the early period of the Resistance, there was not and could not be in our country regular war but only guerilla activities. When the Resistance War started in south Viet Nam, our plan was to wage guerilla warfare, and in practice guerilla war took shape. But when the nation-wide Resistance War broke out, the policy of mainly waging guerilla warfare was not clearly put forth. At the beginning of Autumn-Winter 1947, the Party Central Committee put forth the task of launching and extending guerilla activities all over the occupied areas. One part of our main force was divided into independent companies operating separately which penetrated deep into the enemy's rear-line to carry out propaganda among the people, defend our bases and intensify guerilla war. The policy of independent companies concurrently with concentrated battalions was a very successful experience in the direction of guerilla war. As guerilla activities were intensified and widely extended, many enemy's rear-lines were turned into our frontlines.

To cope with our ever-expanding guerilla activities, great efforts were made by the enemy to launch repeated mopping-up operations with ever-bigger armed forces. The aim of these operations was to annihilate our guerilla units, destroy our political bases and crops, and plunder our property, hoping to crush our resistance forces and "pacify" his rear. That is why mopping-up operation and counter mopping-up operation became the chief form of guerilla war in the enemy's rear-line. Through the counter mopping-up operations, our people brought to the utmost their endurance of hardships and heroic fighting spirit, creating extremely rich forms of fighting. To maintain and extend guerilla activities in the enemy's rear, our Party cleverly combined the co-ordination of political and economic struggle with armed struggle. The Party strove hard to avail itself of the favourable opportunities to push the people forward to the armed struggle, develop our forces, annihilate and wear out the enemy forces, turn temporarily occupied zones into guerilla zones or the latter into our bases. When meeting with a difficult situation, our Party cleverly switched the movement in good time to preserve our forces and safeguard our bases. Guerilla activities in the enemy's rear were the highest expression of the iron will and extremely courageous spirit of our people, and at the same time a proof of the talented leadership of the Party.

From the strategic point of view, guerilla warfare, causing many difficulties and losses to the enemy, wears him out. To annihilate big enemy manpower and liberate land, guerilla warfare has to move gradually to *mobile warfare*. As our Resistance War was a long revolutionary war, therefore guerilla warfare not only could but had to move to mobile warfare. Through guerilla activities, our troops were gradually formed, fighting first with small units then with bigger ones, moving from scattered fighting to more concentrated fighting. Guerilla warfare gradually developed to mobile warfare — a form of fighting in which principles of regular warfare gradually appear and increasingly develop but still bear a guerilla character. Mobile warfare is the fighting way of concentrated troops, of the regular army in which relatively big forces are regrouped and operating on a relatively vast battlefield, attacking the enemy where he is relatively exposed with a view to annihilating enemy manpower, advancing very deeply then withdrawing very swiftly, possessing to the extreme, dynamism, initiative, mobility and rapidity of decision in face of new situations. As the Resistance War went on, the strategic role of mobile warfare became more important with every passing day. Its task was to annihilate a bigger and bigger number of the enemy forces in order to develop our own, while the task of guerilla warfare was to wear out and destroy the enemy's reserves. Therefore, mobile warfare had to go side by side with annihilating warfare. Only by annihilating the enemy's manpower, could we smash the enemy's big offensives, safeguard our bases and our rear-line, move to win the initiative in the operations, wipe out more and more important enemy manpower, liberating larger and larger localities one after the other and eventually arrive at destroying the whole enemy armed forces and liberating our whole country.

Implementing the guiding principle of moving gradually from guerilla warfare to mobile warfare, from the outset, there was in our guerilla troops, besides one part operating separately, another with concentrated activity, and this was the first seeds of mobile warfare. In 1947, with the plan of independent companies operating separately and concentrated battalions, we began to move to more concentrated fighting, then to mobile warfare. In 1948, we made relatively great ambuscades and surprise attacks with one or several battalions. In 1949, we launched small campaigns not only in the North but also on other battlefronts. From 1950, we began to launch campaigns on an ever larger scale enabling mobile warfare to play the main part on the northern battlefield, while entrenched camp warfare was on the upgrade. This fact was clearly manifest in the great Dien Bien Phu campaign.

We used to say : guerilla war must multiply. To keep itself in life and develop, guerilla warfare has necessarily to develop into mobile warfare. This is a general law. In the concrete conditions of our Resistance War, there could not be mobile warfare without guerilla warfare. But if guerilla warfare did not move to mobile warfare, not only the strategic task of annihilating the enemy manpower could not be carried out but even guerilla activities could not be maintained and extended. To say that it is necessary to develop guerilla warfare into mobile warfare does not mean brushing aside guerilla warfare, but that in the widely extended guerilla activities, the units of the regular army gradually grew up and were able to wage mobile warfare and side by side with that main force there must always be numerous guerilla troops and guerilla activities.

Once mobile warfare appeared on the battlefront of guerilla war, there must be close and correct co-ordination between these forms of fighting to be able to step up the Resistance War, wear out and annihilate bigger enemy forces and win ever greater victories. This is another general law in the conduct of the war. On the one hand, guerilla warfare had to be extended to make full use of the new favourable conditions brought about by mobile warfare, in order to co-ordinate with mobile warfare to wear out and annihilate a great number of enemy manpower and through these successes continue to step up mobile warfare. On the other hand, mobile warfare had to be accelerated to annihilate big enemy manpower, and concurrently create new favourable conditions for a further extension of guerilla warfare. In the course of the development of mobile warfare, owing to the enemy's situation and ours on the battlefields, entrenched camp warfare gradually came into being. Entrenched camp warfare which became part and parcel of the mobile warfare, kept developing and occupied a more and more important position.

The conduct of the war must maintain a correct ratio between the fighting forms. At the beginning, we had to stick to guerilla warfare and extend it. Passing to a new stage, as mobile warfare made its appearance, we had to hold firm the co-ordination between the two forms, the chief one being guerilla warfare; mobile warfare was of lesser importance but was on the upgrade. Then came a new and higher stage, mobile warfare moved to the main position, at first, only on one battlefield — local counter-offensive came into being — then on an ever wider scope. During this time, guerilla warfare extended but, contrary to mobile warfare, it moved back from the main position to a lesser but still important one, first on a given battlefront then on an ever-wider scope.

In the practice of the liberation war, on some battlefronts we met with numerous difficulties because we were not determined to advance guerilla warfare to mobile warfare; on others, rashness in speeding up mobile warfare had a bad influence on guerilla warfare, and therefore mobile warfare also met with difficulties. This manifestation was relatively widespread when the slogan "To prepare for the general counter-offensive" was put forth, but it was overcome after a certain time. In general, through tests and trials, our guidance fundamentally held firm the aforesaid ratio and was therefore successful. The Hoa Binh campaign was typical of co-ordination between guerilla warfare and mobile warfare on the northern battlefront. The Dien Bien Phu campaign and Winter-Spring 1953 — 1954 campaign were most successful models of co-ordination between mobile warfare and guerilla warfare, between the face to face battlefield and the theatres of operation in the enemy's rear, between the main battlefield and the co-ordinated battlefields all over the country.

With the forms of guerilla fighting and mobile fighting and owing to the enemy's conditions and ours in strength, shaping up of force and topography, etc., there appeared on the battlefronts the situation of free zones interlacing with enemy-controlled areas, intersecting and encircling each other. In the enemy-controlled areas, there were also guerilla zones and guerilla bases, another phenomenon of interlacement, intersecting and encircling one another. The process of development of the war was that of ever-widening of our free zones and guerilla areas and ever-narrowing of the enemy-occupied areas, advancing towards liberating wast areas, then the whole North.

The strategy of long-term war and the guiding principle of fighting from guerilla war gradually moving to regular war with the forms of guerilla warfare, mobile warfare including entrenched camp warfare, were very successful experiences of our national liberation war. These were the strategy and tactics of the people's war, the art of military conduct of the people's war, of the revolutionary war in a small and backward agricultural country under the leadership of our Party.

In the course of the national liberation war, *the building of bases* for a steadfast and long resistance was an important strategic question and also a very successful experience of our Party. It is an absolute necessity for us to make a profound study of and to sum up the rich experiences of this question.

* * *

The success of the Vietnamese Resistance War was the success of the people of a country which was originally a colonial and semi-feudal country with a small area, a small population, and an extremely backward agricultural economy, which, under the leadership of the vanguard party of the working class, rose up to wage a long armed struggle against an aggressive imperialist country.

The successful Resistance War has completely liberated the North. For the first time in nearly one hundred years in modern history, the shadows of an imperialist enemy and colonial soldiers are no longer seen on a half of our country. The successful Resistance War has brought about conditions for the drastic completion of land reform. After thousands of years of feudal rule, the system of exploitation of the landlord class has been abolished once for all over a half of our country. The victorious Resistance War has created conditions for the revolution in the liberated North to move to the socialist stage. At present, economic restoration has been achieved, land reform completed, our people are striving to accelerate the socialist transformation and socialist construction turning the North into an ever-stronger base for the struggle to achieve national unification and continue to complete the national democratic revolution throughout the country.

The sacred Resistance War of our people has continued the glorious work of the August Revolution, raising aloft the banner of national liberation against colonialism and has eloquently proved that : *" In the present international al conditions, even a weak and small nation, once united to stand up under the leadership of the working class, resolutely to struggle for independence and democracy, will have full capacity to defeat all aggressive forces. This struggle for national liberation, in given historical conditions can, through the form of a long-armed struggle — a long resistance war — come to success.''*

The successful Resistance War of our people has dealt a heavy blow to the ever-disintegrating colonialist system, thus contributing a part to the smashing of the imperialists' war-provoking plots, and to the struggle of the world people for peace, democracy and socialism.

To speak of the factors of victory in a more comprehensive way, the Resistance War of our people has obtained success thanks, first, to the leadership of the Party of the working class, second, to the fact that our Party has taken into due account the peasant question and organised the broad National United Front on the firm basis of the worker-peasant alliance, third, because we have a heroic people's army, fourth, because our State power genuinely belongs to the people, fifth, thanks to the solidarity and support of the people of the brother countries and peace-loving people in the world including the French people and those of the French colonies. Within the limit of this article, we do not analyse the causes of our success in an overall way, but we speak only of our Party's leadership to point out the great experiences in leadership.

III

OUR PARTY HAS SUCCESSFULLY LED THE BUILDING OF THE PEOPLE'S REVOLUTIONARY ARMED FORCES

In the decisive struggle to liberate the nation, overthrow imperialism and its henchmen, our people — first of all the worker-peasant masses — under our Party's leadership, stood up with arms in hand to build their own armed forces. Lenin says: "An oppressed class, if not making efforts to learn to wield arms and to obtain them, only deserves to be treated as slaves." [1] Our people have learned to wield arms and have their own armed forces, therefore their work of liberation has scored success in half of the country. With the setting up of people's power, the building of the people's armed forces was all the more urgent. This was a task of utmost importance during the Resistance War and still is a very important task now, in peace time.

Our people's revolutionary armed forces were born in the revolutionary struggle of the entire people, first of all the broad worker-peasant masses. The first resolutions of our Party already posed the question of setting up worker-peasant self-defence units and the worker-peasant army. In the Nghe-Tinh Soviet movement, there appeared the Red

---
(1) V. Lenin: Military Programme of the Proletarian Revolution.

Self-defence units which were the embryo of the revolutionary armed forces of the people under our Party's leadership.

During World War Two, when preparations for armed insurrection became the urgent task of the revolution, the self-defence and fighting self-defence forces again came into being and developed, first in the revolutionary bases in the Viet Bac mountain area, then in vast areas all over the country. The predecessors of the People's Army came into existence one after the other: the National Salvation Unit, the Viet Nam Liberation Armed Propaganda Unit, and the Ba To guerilla unit. These small troops fought very heroically, kept themselves in existence and developed in extremely difficult conditions when the enemy was hundreds of times stronger. With the anti-Japanese surging high tide in 1945, guerilla war was launched, people's power set up in the liberated zone and the revolutionary armed forces grouped under the name of Viet Nam Liberation Army. In the August General Insurrection, together with the people throughout the country, the Liberation Army and self-defence forces rose up to seize power. In the glorious days of August and after the success of the Revolution, the ranks of the Liberation Army extended very quickly and became the armed forces of the Democratic Republican State, that is the present Viet Nam People's Army. The above-mentioned years can be regarded as *the period of formation of our army*.

Through nine years of the Resistance War, the People's Army unremittingly fought against French imperialists and American interventionists. These nine years of heroic fighting and glorious victory were also the *period in which our army was tempered and grew up*. Our People's Army grew stronger with every passing day, going from one victory to another, and ended the Resistance War with the great Dien Bien Phu victory, contributing its part to the restoration of peace in Indo-China and the complete liberation of half of our country.

In the last five years, our army entered for the first time *the period of building in peace time*. Our army is speeding up every aspect of its building in peace time in order to become a powerful people's army so as to make of it a regular and modern army fulfilling its task of safeguarding the socialist revolution and construction in the North and serving as a support for the struggle to achieve national unification by peaceful means.

The Viet Nam People's Army is a revolutionary army which was born in the revolutionary movement of the people of a colony who arose to liberate themselves. Our army courageously fought the French-Japanese imperialists in the pre-insurrectional period, and together with the whole nation founded the revolutionary power, and defeated the mercenary aggressive troops of the French colonialists backed by U.S. imperialists. Our army has raised to great heights the undaunted spirit of the nation, the indomitable fighting spirit of our people, and is worthy of being the army of an heroic nation.

The success of the People's Army is the great success of our people and our Party. In the process of building and growing up of the army, the Party has always pointed out the nature and task of the army, and defined the principles of building the army politically and militarily. Thanks to that, it came into being from nought, growing from a small beginning to a bigger army, from weakness to strength and has gloriously vanquished the enemy and fulfilled its revolutionary task in the historical stages.

1. Our army was successful and mature because it is a people's army led by the Party

For what reason has our army, though still young, already an extremely glorious history, scored resounding feats of arms and contributed an important part to the success of the revolution of our entire people? Because it is a people's army, led by our Party. The Party's leadership is the decisive factor of all the successes of our army.

Our army was born and has grown up in the revolutionary struggle of the entire people. It is the implement of the Party and the revolutionary State for the carrying out of the revolutionary struggle and class struggle. It embraces the elite of the revolutionary classes, of the people of all nationalities in Viet Nam, first and mainly the finest worker and peasant elements who have volunteered to fight to the bitter end for the interests of the country, of the toiling people and of the worker-peasant masses.

Therefore, *our army is a people's army, the army of the toiling people, essentially, of the workers and peasants and led by the Party of the working class*. It is the armed forces of the people's democratic State, which was formerly in essence, the worker-peasant dictatorship and is now the proletarian dictatorship. This is *the question of the revolutionary nature and class nature of our army*. This is the basic difference between the enemy's army and ours. This is the most fundamental question which must be thoroughly understood in any stage of the building of the army.

Owing to its class nature, our army, since its setting up, has always been faithful to the revolutionary cause of the Party and people. The revolutionary task of the Party and people is also the target of the army's efforts.

The working out of a correct *revolutionary line and task* has a decisive effect in the building of the armed forces. In the previous stage, our entire people was carrying out the national democratic revolution throughout the country, aimed at overthrowing imperialism and the feudal landlord class, winning national independence, bringing land to the peasants and creating conditions to advance the revolution in our country to the socialist stage. At that time, during the hard years of the armed struggle, our people's army fought very heroically to annihilate the aggressive army of imperialism and the traitors, its henchmen. However, in the first years of the Resistance War, although the anti-imperialist task was set out clearly, the anti-feudal task was not put forth in full keeping with its importance. As a result, the national spirit and consciousness of the army was heightened but its class consciousness was rather weak, thus having a bad influence on drawing a line between the enemy and us. Since the moment our Party paid attention to the anti-feudal task, especially since the mobilisation of the masses for rent reduction and land reform, not only broad peasant masses in the rear were ideologically roused but also our army — the great majority being peasants and very eager for land — also saw clearly and more fully its own fighting objective that it not only fights for national independence, but also to bring land to the peasants, and consequently its class consciousness and fighting spirit were raised markedly.

Since our people's revolutionary struggle has entered the new stage, the task of our entire people is to struggle for national reunification, to continue to complete the national democratic revolution throughout the country ; to endeavour to advance the North to socialism and build a peaceful, united, independent, democratic, prosperous and strong Viet Nam. Basing itself on this revolutionary task, our Party has set the people's army the political task of safeguarding the socialist building in the North to serve as the mainstay of the struggle for national unification and keeping itself ready to smash all aggressive plots of imperialism, mainly U.S. imperialism and its stooges. As the common revolutionary task and political task of the army are accurately set out, the political education in the army, especially the recent ideological remoulding class, having a precise and concrete direction, have raised to new heights the socialist consciousness and patriotism of all the officers and men, creating a new revolutionary mettle in the whole army that has been showing itself in the emulation movement to advance rapidly and overfulfil the plan with a view to making as great a contribution as possible to socialism. The army has seen more clearly its task with regard to the maintaining of social order in the North while the struggle between the two paths unfolds, as well as to the defence of territorial security.

Our army is made up of officers and men who are imbued with revolutionary consciousness and fighting spirit. It has a fine revolutionary nature. However, to say so does not mean that it is not necessary to strive to maintain and strengthen that class nature. On the contrary, in its leadership, the Party must pay *great attention to the question of maintaining and strengthening the revolutionary and class nature of our army.* Only by working out and thoroughly grasping the Party's revolutionary task in the army, only by strengthening without cease the Party's leadership and increasing political work, can we succeed in doing that, in enabling the army fulfil its own revolutionary task.

Since the restoration of peace, our Party has put forth the guiding principle of building a powerful people's army that will gradually become a regular and modern army. The question of maintaining the revolutionary nature of the army is still a fundamental demand of utmost importance. Only by maintaining and strengthening the revolutionary nature of the army and by raising its socialist consciousness and patriotism, can we succeed in turning it into a regular and modern army. On the way to become a regular and modern army, our army will always be a people's army. It must become a modern revolutionary army.

The Party's leadership is the decisive key question to enable the army to maintain its class nature and carry out its revolutionary task. The Party's leadership of the army is an absolute one. This leadership reveals itself politically: to imbue the army with the Party's revolutionary line in order to make of it the faithful implement of the Party in the carrying out of the revolutionary task. This leadership shows itself ideologically: to educate the army in the ideology of the working class, in Marxism-Leninism, to use Marxism-Leninism as the compass for all activities, as the one and only guiding ideology of our army. This leadership still reveals itself organisationally: to thoroughly understand the Party's class line in the building of the Party as well as in the work concerning the cadres in our army. Only by doing that can our army be always a genuine people's army which is ready to fulfil its own revolutionary task in all circumstances, thereby becoming more and more mature and winning new successes.

2. Our Party has correctly defined the fundamental principles of political building of the army

The most fundamental principle in the building of our army is to put it under the Party's leadership, *to ceeaselessly strengthen the Party's leadership* of the army. The Party is the founder, organiser and educator of the army. Only by realising the Party's absolute leadership can the army unswervingly follow the class line, the political direction and fulfil its own revolutionary task. To carry out and strengthen the Party's leadership *great attention must be given to the work of building the Party and political works, and the system of Party Committee and political commissar must be firmly maintained*. Only thanks to the firm organisation of the Party which serves as the core and leading nucleus in the army can the Party, through its own organisation — from Party committees down to Party branches — guide the implementation of its line and policies. Here stress should be laid on the important role of the grass-root Party branch; only when it is strong can the company be strong. The method of Party committee taking the lead, and the commander allotting the work coupled with the regime of political commissar, ensures the carrying out of the principle of collective leadership. It thereby succeeds in concentrating the knowledge of many people and also consolidating the solidarity based on ideological unity, closely co-ordinating the various tasks in the army, uniting the mind and deeds and increasing the army's fighting strength. Here, we have to emphasise that the method of Party committee taking the lead must always be coupled with the method of the commander allotting the work, in line with the Party's principle of collective leadership and personal responsibility. Political work is Party work and work of mass mobilisation of the Party in the army. Political work is the soul, the sinews of the army. Political work takes care of the building of the Party, guides the education of the army in the ideas of Marxism-Leninism, in the revolutionary line and task of the Party, military line and task of the Party, ensures good relations between officers and men, between the army and the people, between the army and the State, between our army and the armies and people of the brother countries and enables our army to have a high combativeness capable of defeating all enemies.

Right at the founding of our army, the first armed groups and platoons had their Party groups and branches. The platoons had their political commissars. On coming into being, the regiments had political commissars. The method of Party committee taking the lead and the commander allotting the work also took shape from the very first days. Officers were provided with hand-books *The Political Commissar's book* or *Political Work in the Army*. After the August Revolution, the traditional method of Party's leadership and political work was basically kept up, but in the first years, there appeared the tendency of not taking into due account the part played by political work and political work did not yet grasp that the main task was political education and ideological leadership. Sometimes, the Party's political agitation in the army was not closely co-ordinated with the Party work. After the Second Party Congress, the Party's leadership was strengthened in the army as in all other branches of activity. Agitation for ideological remoulding courses in the Party and the army, brought about increased education on the long-term war and self-relying war of resistance, education in policies, mobilisation of the masses for drastic reduction of land rent and land reform. Political work in the army became richer and more concrete, thus its position was markedly raised and its strength grew visibly. The Party's leadership was consolidated. Political work had a great effect on the raising of the ideological level, educating and consolidating the class stand of officers and men. It had a great effect, a lively content full of combativeness and mass character, in our big military campaigns as well as in guerilla war on all battlefronts.

Since the restoration of peace, in the new revolutionary stage, the Party's leadership and work of consolidating and developing the Party has been carried on steadily. Through congresses of Party delegates from all echelons, inner-Party democracy was implemented and the role of the Party branches highlighted. The ideological remoulding courses which set out a clear line — strengthen solidarity, raise to new heights the fighting spirit and socialist consciousness. Especially, the ideological remoulding drive last year which aimed at thoroughly grasping the resolutions of the Party Central Committee on the revolutionary task and line of the people's war and people's army, achieved good results and contributed to the strengthening of our army's revolutionary nature. Attention was also given to education on Marxism-Leninism. Entering a new stage, and faced with the urgent demand of turning our army into a regular and modern one, in some works there appeared to a certain degree the tendency to disregard political work. When dealing with the necessity to strengthen centralisation and unification, although this principle was not yet sufficiently implemented, there appeared the tendency to slight the role of the Party branches and collective leadership of the Party committees. When dealing with the necessity to strengthen the material and technical bases in the army, to master technology, although the technical level of our army is still low and needs to be raised further, there emerged the tendency to belittle the role of politics, to divorce politics from technology, from specialisation, falling into the bourgeois viewpoint of pure militarism and technology. In the last ideological remoulding drive, these erroneous tendencies were corrected in the main. From now on, we still have to continue to strengthen political education and ideological leadership in the army, continue educating and fostering socialist ideology and patriotism, energetically combating all the expressions of bourgeois and other non-proletarian ideologies, fighting individualism and liberalism, thereby to maintain and constantly raise the solidarity and combativeness of our army.

Our army is made up of combatants who consciously fight for the revolutionary cause of the people, therefore our officers and men are completely single-minded about their fighting objective and class interest. We must always take care *to strengthen the monolithic solidarity within the army.* The relation between the officers and soldiers, between the higher and lower echelons, between one branch and another, is the relation of solidarity between comrades based on political equality and class love. This relation has been built up from the very founding of the army. Through the long years of fighting in hard conditions of dangers and privations, our officers and men have loved each other like blood brothers, sharing hardships and joys together, united for life and in death. Concurrently with the raising of their class consciousness, the solidarity between officers and men has been more and more strengthened. This unity has welded all the members of our army into an unbreakable monolithic block.

Up till now, in general the question of internal unity has always been taken into consideration and has become a fine tradition of our army. However, in entering the new stage of building the army, in a number of units and organs, internal unity has not been taken into due account. To strengthen the regular management and set up organisations are of utmost necessity, but in the actual task of turning the army into a regular army, beside correct measures, there were bothersome and unnecessary provisions on certain privileges which kept the officers apart from the soldiers, the higher from the lower echelons, having some bad effect on the comradeship and solidarity in the army. These mistakes were corrected in good time.

Our army is a revolutionary army belonging to the people and fighting under the Party's leadership, therefore the interests of the army and the people are one and the same. We must always take care to *strengthen the monolithic solidarity between the army and the people*. The army and the people are of the same heart, they are like fish and water. Our army has no other interests than those of the people, of the toiling people and the worker-peasant masses. Right from its inception, the question of single-mindedness between the army and the people has been laid down clearly in the ten-point pledge of honour and 12 points of discipline in its relations with the people. During the Resistance War, not only did our army make sacrifices and struggle for the defence of national independence and the protection of the people's lives and property, doing nothing to the detriment of the people, but it also did its utmost to give the people a helpful hand in all their activities. Side by side with the people, our army made sacrifices and fought in the Resistance War to defeat the enemy of the nation, won national independence, enthusiastically struggled for land reform to overthrow the feudal landlord class and bring land to the peasants. As a result, the solidarity between the army and the people grew stronger, and the people trusted, loved and supported the army, taking care of them as of their own children. Since the restoration of peace, the traditional unity between the army and the people has been maintained and developed. After many years of fighting against the cruel enemy for national liberation, our combatants are still working tirelessly. On the one hand, they stand ready for the defence of the people's peaceful labour and on the other, they strive to intensify the work of mobilising the people and have never spared themselves when the people are in need of help. Our army has actively taken part in stepping up agricultural cooperation as it did formerly in land reform. Through the struggle against famine, drought and flood and the building of construction sites and factories, etc., it has shown itself a faithful servant of the people, as President Ho Chi Minh has always reminded it. In recent years, in response to the Party's call, tens of thousands of officers and men volunteered to go to remote areas on the frontier to break virgin land, set up army farms to accelerate the socialist construction of the Fatherland. As our armymen are the brothers and sons of the labouring people, in their relations with the people, we must hold firm the class viewpoint, and endeavour to strengthen the solidarity between the army and the people, first of all, the worker-peasant masses. The army is regarded as an integral part of the working class, therefore its good relations with the people and the peasant masses have a great political significance. These express the political nature of the officers and men of our army, showing clearly that it is not only a fighting army but also a working army. At present, in the North, the People's Army is not only the guardian of the socialist regime but also a builder of socialism. This is a glorious tradition that our army has to maintain firmly and develop constantly.

Our Party has always paid attention to the *solidarity between our army and our people and the armies and peoples of the brother countries, and the peace-loving people in the world*. Our Party has not only educated the army in genuine patriotism but also proletarian internationalism thoroughly. The units, predecessors of the army had once fought under the slogans "national liberation", "defence of the Soviet Union". In the Resistance War, the Viet Nam People's Volunteer Units raised to great heights the proletarian international spirit, did not shun from dangers and difficulties to fight French aggressive colonialists, shoulder to shoulder with the people of the friendly countries. Many of our comrades shed their blood for the independence, peace and closer friendship of the Indochinese peoples. Our army has devoted great attention to the strengthening and development of friendship with the people and armies of the countries in the socialist camp in the struggle for peace and socialism and against the common enemy — warmongering imperialism. Our officers and men attach great importance to the learning of the invaluable experiences of the armies of the brother countries, first of all the Soviet Army and the Chinese People's Liberation Army. The success of our army is also that of the application of Marxist-Leninist military theory, the creative application to the practical conditions of our country of the advanced experiences in building and fighting of the armies of the brother countries. Great attention was paid by our army to the strengthening of the solidarity between our people and the French people and peoples of the French colonies. It is for this reason that in fighting, our army differentiated between the French aggressive colonialists and the French and colonial toiling people who were deceived or coerced into becoming mercenary soldiers.

By differentiating between the colonial high-ranking officers and the soldiers and subalterns, and the enemy's unjust war from our just war, our army carried into effect the principle of *disintegrating the enemy.* Our troops were educated by our Party to give due consideration to propaganda work among the enemy soldiers, to enlighten them so that they could understand that they were not fighting for their own interests but as cannon-fodder to bring wealth to the colonialists. They acquainted them with our lenient policy towards prisoners of war and those who had gone over to our side of their own accord so that they would join our ranks and turn their arms against the enemy. In the course of the Resistance War, thanks to good propaganda work among enemy and puppet troops and strict implementation of our policy towards prisoners of war and those who had of their own accord passed over to our side, and thanks to the skilful co-ordination between armed struggle and political offensive, our army and people brought over to our side tens of thousands of enemy soldiers, thus throwing the enemy ranks into bewilderment and disintegration and making an important contribution to our military success.

In leading the building of the army, our Party has firmly stuck to the principle of democratic centralism. That is the organisational principle of our Party; thereby it has taken care to build the army with a genuine inner democracy and also a very strict conscious discipline.

Completely different from all types of armies of the exploiting class, our army put into practice the *regime of internal democracy* from its inception because the internal relations between officers and men as well as the relations between the army and the people express complete unity of mind. Owing to the demand of the revolutionary work, there are in our army differences in ranks and offices, but they have not and cannot influence the relations of political equality in the army. For this reason, internal democracy should and could be carried out in the army. To practise democracy is also to apply the mass line of the Party in leading the army.

During the Resistance War, democracy was exercised in three ways and brought about good results. Political democracy: at grass-root level, democratic meetings and army congresses were held regularly so that men as well as officers had the opportunity to speak their views on fighting, work, study and living questions. In our army, not only have the officers the right to criticise the soldiers but the latter also have the right to criticise the former. Military democracy: in fighting as well as in training, democratic meetings were called whenever circumstances permitted, to expound plans, promote initiatives and together try to overcome difficulties in order to fulfil their tasks. Economic democracy : in our army, the officers and soldiers have the right to take part in the managament of the improvement of material life. Finance is made public. Thanks to the carrying out of democracy in an extensive way, we succeeded in promoting the activity and creativeness of the masses of officers and men, and concentrating their wisdom to solve the most difficult and complicated problems; also thanks to it, internal unity was strengthened and the combativeness of our army increased.

On the basis of the democratic regime, our army still has a very strict *conscious discipline*. When we speak of conscious discipline, it means that it is built up on the basis of political consciousness of the officers and men, and the most important method for maintaining discipline is education and persuasion, thus making the armymen of their own accord, respect and remind each other to observe discipline. When we speak of strict discipline, it means that everyone in the army, regardless of rank or office must observe discipline and no infringements are allowed.

Our army has always thought highly of discipline because it has been educated by the Party and knows that discipline is one of the factors that improve the combativeness of the army. As an armed collective whose task is fighting and to ensure single-mindedness and united action for its own preservation and destruction of the enemy, our army cannot abstain from having centralisation to a high degree and strict discipline. Therefore, right from its inception, absolute obedience to orders and strict observance of discipline were written down clearly in the ten pledges of honour. Thanks to that, the tasks set by the Party were fulfilled and all fighting orders thoroughly carried out in extremely hard and arduous circumstances, and in its contact with the people, our army has firmly maintained mass discipline. Nowadays, our army has entered the period of building itself into a regular and modern army, consequently demands in discipline, centralisation and unification are all the higher.

To carry out internal democracy and strengthen conscious discipline is a process of struggle against the deviations that manifest themselves in two opposite tendencies. The first tendency puts great emphasis upon discipline while disregarding democracy. In the early stage of the building of the army, a number of officers tainted with the militarist manner and habits of previous armies, advocated absolute reliance on orders and punishment in the management of the army. In the new stage of building the army, when the question of turning it into a regular army was posed and regulations issued, there came to light in a number of units the tendency to lay too much stress on centralisation and unification, with insufficient attention to the extension of democracy and the mass line, relying solely on punishments and administrative orders and overlooking education and persuasion. The second tendency was that of breaking down of discipline. During the Resistance War, this tendency was expressed in using the difficult circumstances of the guerilla war as a pretext to neglect reporting to and asking instructions from higher echelons and to ignore co-ordination in fighting. These were symptoms of undisciplined liberalism, loose carrying out of fighting orders, non-observance of battlefront discipline, infringements on mass discipline, etc... In the new stage of building the army, this is the tendency to slight centralisation and unification, wanting a free and easy life and having their own way, showing itself careless in carrying out organisation and regulations.

The two above-mentioned erroneous tendencies are both expressions of non-proletarian ideologies. The first one is the manifestation of the influence of bourgeois ideology in the management of the army. The second is the expression of lack of discipline of the peasantry and petty-bourgeoisie who made up the majority of our officers and men. Therefore, the key to a correct implementation of the democratic regime, consolidation of strict and self-conscious discipline is constantly to educate our army in proletarian ideology in order to wipe out the remaining non-proletarian thoughts.

Democracy and discipline in the army reflect the principle of democratic centralisation of our Party. Therefore, to put into practice genuine democracy, heighten discipline, strengthen centralisation and unification, the life of the Party organisation must be consolidated. The internal democracy and iron discipline of our Party are the basis for democratic centralism and for the strict discipline of our army.

To maintain and consolidate the absolute leadership of the Party in the army, to increase political work as the sinews of our army, to intensify proletarian ideological education for officers and men, to implement the principle of internal unity, solidarity between the army and the people and international solidarity, to cause disintegration of the enemy, to put into practice the democratic regime parallel with a strict conscious discipline, are fundamental principles of the building of our army and an essential safeguard for it to maintain its people's nature, for its development and success.

3. Together with the defining of the principles of political building of the army, our Party has successfully solved the questions of organisation of the formation, equipment, supply, training, management, etc in order gradually to turn our army from a guerilla army to a regular and modern army in the particular conditions of our country.

Unlike the armies of many other countries, ours was at first only small guerilla units born in the course of the revolutionary struggle of the people of a colonial and semi-feudal country which, with bare hands, rose up to fight imperialism and its stooges. Through a long and hard struggle, our army has grown in the fighting, has won glorious victories and liberated half of the country. The small guerilla units have now grown into a large powerful people's army and is being formed into a regular modern army, when a half of the country is liberated and is building socialism. Our Party met very big difficulties in the building of the army, owing to the backward state of our economy and the need for our army to fight without cessation. Having thoroughly grasped the class viewpoint and the practical viewpoint of Marxism-Leninism, our Party has successfully solved a series of problems in building the army and has accumulated many invaluable experiences.

First of all the question of *formation* had to be solved. The army is organised in order to defeat the enemy, therefore the formation of the army must meet the demand of the realities in fighting, and be in harmony with the strategic guiding principle and the principle of fighting in each stage of the war. The organisation must be in line with our possibilities in equipment and supply, based on the national economy and in harmony with the practical conditions of the battlefields in our country.

In the early stage of the Resistance War, our army was in extremely difficult conditions, short of arms and munitions, its formation varied from one locality to another. Parallel with the gradual development from guerilla warfare to mobile warfare and with better supply and equipment, we had, from scattered units, gradually organised concentrated ones, then regiments and divisions of a regular army. In the units of the regular army, the organisation was unified step by step. The regiments and divisions were made up at first of infantrymen only. Later there were units of support and later sappers units and light artillery units, etc. To meet the mobile conditions of guerilla warfare and mobile warfare, we worked out plans of "good soldiers and reduction of organisation to its simplest form" to lighten and strengthen the command, and increase the fighting force of the unit.

In the new stage of the building of the army, to meet the requirements of modern fighting, we have, on the basis of improving and strengthening equipment, readjusted the organisation of the army, turning it from an army of infantry into an army made up of various arms. We must continue to take into consideration the infantry and concurrently strengthen the technical arms, developing them in a harmonious way, and at the same time strengthening the command at all levels with a view to increasing the combativeness of our army in the conditions of combined operations. It is also necessary, from practical training and manœuvres, to study the improvement of the formation to make it more appropriate day by day.

Being a revolutionary army under the Party's leadership, its organisation must also be imbued with the organisational principle and method of Party's leadership in the army. For this reason, we have set up, parallel with the system of command, a system of political commissars, in line with the principle that the commander and the political commissar are both heads of the unit. Corresponding with the maturity of the army and parallel with the strengthening and improvement of the staffs and logistics, due consideration has been given to the strengthening and improvement of the organs engaged in political work at all levels in order to maintain and strengthen the Party work and political work in the army.

To organise an army, the *question of equipment* must be solved because arms and equipment are the material basis of the combativeness of the army. Without arms it is of no use to speak of organising an army and of waging armed struggle. In the first stage of the building of the army, owing to our backward national economy, with almost no industry, and with the army's rear in mountain and rural areas only, the equipping of our army encountered many difficulties. The Party pointed out to the army that it had to look for its equipment on the front line, to capture the enemy weapons to arm itself and shoot at the enemy with his guns. We scored great success in implementing this principle. The great part of our regular army and guerilla units were armed with weapons captured on the battlefronts. The French Expeditionary Corps practically became carriers engaged in supplying our army with U.S.-type arms. On the other hand, our Party guided the workers in the spirit of self-reliance, and found means to manufacture a part of the arms and munitions for the army. In circumstances of extreme hardship and privation, the workers in the arms-factories raised to new heights the heroic and creative spirit of the Vietnamese working class, overcoming very great material and technical difficulties in order to turn scrap-iron into weapons for our troops to exterminate the enemy.

In these circumstances, our Party educated the army to develop the fine nature of a revolutionnary army to increase the political supremacy in order to make good our weakness in equipment. Hence our army succeeded, with inferior arms, in defeating the enemy who was many times stronger in weapons. It has become an extremely fine tradition of our army — to vanquish modern weapons with an heroic spirit. However, because of our inferior weapons, in the Resistance War our army and people had to fight in extremely hard and difficult circumstances, to make great sacrifices and shed much blood. We must always realise that inferiority in arms and equipment is a big weakness that must be overcome at all costs.

At present, the building of the army has stepped into a new stage. Our army must, step by step, grow into a modern revolutionary army, able to frustrate the aggressive plots of the U.S. imperialists and their myrmidons. For this reason, improving and increasing technical equipment for the army has become a pressing demand. The replacing of the backward material and technical basis of our army by more and more modern equipment and technique is a real revolution. This technical revolution in the army is a part of the great technical revolution that is being carried on by our Party in the society of north Viet Nam. It requires great efforts in strengthening equipment, raising the organisational and managerial level as well as the ability to master and to use new techniques. The solving of the problems of equipment and technique for the army cannot be separated from the building of the material and technical basis of socialism. Nowadays, we have favourable conditions : peace has been restored and the North completely liberated. Efforts must be made to build economy and develop culture, step-by-step to carry out the industrialisation of the country in order to put an end to our economic backwardness. This is not only a great revolutionary task aimed at carrying the North to socialism but also of utmost importance to strengthen national defence and create new conditions for the improvement of the equipment and technical basis of our army.

To enable the army to master and skilfully use the weapons, raise the technical and tactical level, we must take into account the *training of the troops*. Good training is imperative in the active preparation for the fighting. The aim of training being to defeat the enemy, training must meet this requirement. The content of the training must be imbued with the strategic guiding principle and the leading ideas of our army in fighting, must be based on the enemy's practical situation and ours, and on those of the battlefronts. Our army is still young, with a limited fighting experience. It must endeavour to learn the advanced experiences of the armies of the brother countries, first of all, the Soviet Union and China. We must thoroughly understand the practical viewpoint and proceed from the practical situation of our country in assimilating the experiences of the other countries, doing it in a critical, selective and creative way. Thus we have to combat both empiricism and dogmatism.

In the building of the army, we have, in the main, met the above-mentioned requirements. During the Resistance War, owing to constant fighting, the training of our troops could not be carried out continuously for a lengthy period but only between battles or campaigns. We actively implemented the guiding principles " To train and to learn while we fight ". After the difficult years at the beginning of the Resistance War, we succeeded in giving good training to our army. The practical viewpoint in this training deserves to be highlighted. The content of training became most practical and rich. Training was in touch with practical fighting : the troops were trained in accordance with the next day's fighting, and victory or defeat in the fighting was the best gauge for the control and assessment of the result of the training. On the basis of gradual unification of the organisation and its equipment, the content of training in the various units of the regular army was also systematised step by step. Applying in a creative way the invaluable fighting experiences of the brother armies, particularly the Chinese People's Liberation Army, we won victories in campaigns of an ever-larger scale and concurrently enriched our own fighting experiences.

At present, in peace time, we are building a regular and modern army, where training becomes a long-term central and permanent task. It is necessary to carry out regular training systematically and according to plan, proceeding from the rank and file upwards. To meet the requirements of modern war, the army must be trained to master modern technique, tactical use of arms, co-ordinated tactics and modern military science. For that, we must on the one hand, strive to learn the advanced experiences of the brother armies, and on the other, to take into good account the invaluable fighting experiences of our army. The summing up of experiences must be combined with the study of principles of modern fighting, and an appropriate content of training must proceed from the Party's military line and the enemy's practical situation and ours and of the topography.

As is said above, step-by-step modernisation of the army is virtually a technical revolution. The more strengthened are the material basis and modern technique, the more the men are required who are able to master that technique. Otherwise, modern technical equipment cannot develop its effectiveness and the army's combativeness will not be increased. This is a great responsibility in training.

In training, *training of officers* is central. The officers have been tested and tempered in actual fighting and have experience in building the army and leading the fighting. However, because they have grown up in the circumstances of guerilla war, our officers are weak in modern tactics. Therefore, while they have to ceaselessly raise their political and ideological level, consolidate their class stand and cultivate Marxist-Leninist theory, they must do their best to advance their cultural level and level of military technical science to become good military cadres of the Party, serving as the core of a modern and regular revolutionary army. This is a work of particular importance in the building of the army at the present time.

With the development and growth of the army, in the process of changing step by step from scattered to concentrated units, the necessary *rules* and *regulations* took shape. Parallel with the step-by-step implementation of the relative unification of formation and training, we gradually worked out the system of supply, regulations for order in army life, reward and punishment and care of arms, etc.. Nevertheless, as our army was formerly in the process of changing from guerilla units to regular units, the demand for centralisation and unification was still at a low degree, thus systematic unified rules and regulations for the whole army were not issued. The building of the army has now entered a new stage, that of turning into a regular and modern army. A modern army is made up of many arms, and modern fighting is combined operations of various arms, carried out on a large scale and at a high tempo. Consequently, a high degree of centralisation and unification, of organisational spirit, disciplinary spirit, spirit of planning and accuracy in all the army's activities had to be carried out. The rules and regulations become a great necessity to serve as a united basis for everybody and to meet the demand of combined operations and united command.

Our army is a people's army placed under the Party's leadership, the provisions in its rules and regulations must fully express the revolutionary nature of the army, really grasp the organisational principle and method of Party's leadership in the army. These rules and regulations must come out of the practical situation of our country and of our army, and must maintain and develop its fine traditions and habits.

In recent time, the system of military service, system of service of officers and n.c. officers (including the system of grade and rank), the system of pay and that of rewards have been carried out and have brought good results. Regulations of inner order, discipline and military police have been issued and have an important effect on all aspects of unification of the whole army and in building a regular army.

The great experiences in the building of the army from the military aspect are dealt with briefly above. Practice has shown that parallel with a thorough understanding of the principles of building the army politically, if we do not correctly solve the complicated questions in the building of the military side, it would be impossible to turn a small and weak guerilla army with a scattered organisation, rudimentary weapons, low military level, and having no rules, into a powerful people's army with many arms, ever-improving technical equipment, regular training, and unified rules and regulations without attention to military matters. These are most valuable experiences. They are principles that must be adhered to in order to build our army into a powerful people's army, that is becoming a regular, modern army.

4. Parallel with the building of a strong people's army, our Party has paid great importance to the problem of building the militia and developing the reserve; at the same time, it correctly solved the relation between the army and the rear.

During the process of its formation and development, the people's armed force not only include a regular and local armies but also a big self-defence force. Immediately after the Party had set the task of preparing for an armed uprising, on the basis of the intensified political movement of the masses, there appeared the multiformed semi-armed and armed organisations, aimed at gradually shifting the masses' political struggle to the armed struggle. These were the organisational forms of self-defence units, of fighting self-defence units, then of guerilla teams in the underground armed bases in the Viet Bac mountain region. In a number of regions, when the first units of the People's

Army came into being, around these units, considered as the main force, local armed units were formed; in addition there were the immense semi-armed forces. When he ordered the creation of the Viet Nam Liberation Armed Propaganda Unit, President Ho Chi Minh paid great attention to the formation of the armed and semi-armed forces and the maintenance of the relation of solidarity and coordination between them. In the Resistance War, the more the armed struggle developed, the clearer became the differentiation between these three armed forces. The People's Army included the regular divisions and regiments, and also the local regiments, battalions or companies. Besides the regular and local armies, there were broad guerilla forces which developed everywhere throughout the country. The regular forces had the task of waging mobile warfare on a large battlefront aimed at annihilating the enemy forces. The local army had the task of fighting locally and combining its action with the regular army or with the guerilla units. The latter had the task of defending the villages, participating in production, and combining with the local and the regular army in the preparation of the battlefront as well as in the attack. The existence of the three abovementioned armed forces fully met the aspirations of the people, was instrumental in developing the army's and people's fighting force, and trained the whole people to fight the enemy. *It concretely embodied the policy of arming the whole people, and it was the form of organisation of the armed force of the people's war and the revolutionary war.*

Our Party advocated that to launch the people's war, it was necessary to have three kinds of armed forces. It attached great importance to the building and development of self-defence units and guerilla units. In our country, the militia was set up everywhere. It is thanks to the founding of people's administration everywhere in the countryside and the existence of Party branches in every place, that the militia spread far and wide, and the people rose to fight. In the enemy's rear, the guerilla units, in co-ordination with the regular army, scattered and wore out the enemy, nailed them to their bases, so that our regular army could launch mobile fighting to annihilate them. They turned the enemy rear into our front-line, and built guerilla bases as starting points for our regular army's offensive right in the heart of the enemy; they protected the people and their property, fought the enemy and kept up production, and frustrated the enemy's schemes to use war to feed war, and of using Vietnamese to fight Vietnamese. In the free zones, the guerilla units effectively fought the enemy and kept watch on traitors; they were effective instruments for the local administration and local Party; at the same time, they were the shock force in production and in transport and supply. Through the process of combat and work, and having been educated and trained by the Party, the guerilla units became an inexhaustible and precious source of replenishment for the regular army, supplied the people's army with men and officers, politically well-educated and rich in fighting experience. This was a very great achievement, and at the same time a rich experience for our Party in leading the war and building the revolutionary armed forces.

The situation has now changed and the revolution has shifted to a new stage, and our People's Army is becoming a regular and modern army. If a new war breaks out, it will be a modern one. But on our side, this war will always be, in nature, a people's war; the strengthening of national defence and the safeguarding of the Fatherland will always be the common task of our people, consequently, instead of playing a minor part, the militia will be more important ; the militia will always be a strategic force, and the guerilla war a strategic problem. As formerly, in the future, our armed forces will not only include the regular and modern army, but also the people's armed and semi-armed forces which co-ordinate with the army in military operations. At present, in peace time, north Viet Nam is advancing to socialism, the struggle between two paths, socialism or capitalism, is being waged in town and countryside. We must consolidate and intensify proletarian dictatorship ; thus the strengthening and reinforcement of the self-defence units in the countryside, cities, offices and enterprises have all the more a significance. Parallel with the building of a permanent army, a great reserve must be built, aimed at organising and educating the masses militarily, thus preparing everybody to defend the Fatherland and shatter the enemy's aggressive scheme. The base of the reserve is the self-defence units. Their tasks are:

a) To replenish the permanent army ;

b) To maintain security and protect production ;

c) To serve the front-line and carry out guerilla activities in war time.

This is the important part played by the militia and the reserve. After the restoration of peace, chiefly when the military service was experimenting, there appeared a tendency to belittle the militia, separating the militia from the reserve and considering the latter as the only force to replenish the regular army. Since this deviation was rectified, the situation has improved. The carrying out of the military service is supported by the masses, and the organisation of self-defence units and of the reserve is strongly developed.

To consolidate and develop the self-defence units, to build a strong reserve is a most important task, especially in peace time, when a substantial reduction has been made in the strength of the permanent army in order to divert manpower to economic reconstruction. To perform this task satisfactorily, it is necessary to thoroughly grasp the theory of the people's war and people's army, to stick to the class line in organisation and education, to develop the militia's fine tradition and precious experience, and to strengthen the close relation between the permanent army and the militia and the reserve. At the same time, the leadership of Party Committees in the local military organs in particular and the militia and the reserve in general must be improved.

* * *

One cannot speak of the armed struggle and the building of the revolutionary armed forces without mentioning *the problem of the rear*. This is an important problem of strategic significance and a decisive factor to the outcome of armed struggle and in the building of the armed forces.

At the beginning of World War Two, when our Party set the task of preparing for the armed insurrection, we had no armed forces and not a single inch of free land as a spring-board for our activities. Afterwards, the underground armed bases were gradually created, and the resistance bases in the rural areas of six Viet Bac provinces were founded. The experience gained in the August Revolution clearly proved the importance of the resistance bases. It showed the correct leadership of our Party in organising the resistance bases, and in founding the Viet Bac liberated zone.

This lesson was illustrated on a larger scale in the long Resistance War. The problem of resistance bases and the rear was stressed at the beginning of the Resistance War; throughout the Resistance War, the safeguarding of resistance bases and the consolidation of the rear were considered by our Party as of the utmost importance. Because they wanted to crush our leading organ and smash our Resistance, the French colonialists used every scheme to raze our resistance bases, but they suffered defeat after defeat and finally collapsed. Our armymen and people fought heroically to protect the Viet Bac resistance base — the main one in the Resistance War — and the free zones in the fourth and fifth interzones and in Nam Bo.

Due to the war situation, guerilla war developed everywhere throughout the regions occupied by the enemy. Consequently, besides these big bases, our army and people set up many others on all battlefronts in central, north and south Viet Nam, thus creating a very serious threat to the enemy and a spring-board for our army to attack them. Parallel with the fight against the enemy, in order to safeguard the resistance bases and consolidate the rear, our Party implemented positive lines of action in every aspect, did its utmost to mobilise, educate and organise the masses, to increase production, practise economy, and build local armed and semi-armed forces. Thanks to that, our resistance bases were continually strengthened, and constantly furthered their great effect on the development of the army as well as on the work of serving the frontline. Therefore, we could carry on our long Resistance War and win glorious victory in the end.

At present, north Viet Nam is entirely liberated; it is the vast rear of our army. We know that in modern warfare the rear is all the more important. Strengthening of the rear ranks first among the permanent factors which determine the victory of the war. Modern warfare requires the highest development of all the economic, political and military potentialities. Marxism-Leninism has shown that "at present, war is an overall test of the material and spiritual forces for each country". Having seen the importance of the problem of the rear, the resolution taken by the 12th Session of the Central Committee in 1957 pointed out : " We must have a plan for building and consolidating the rear in every aspect. We must enable our rear to have full material and spiritual abilities to ensure all the needs for the building of an army in peace time, as well as for the requirements of life and fighting in time of war. In every aspect of State work, in the State's general plan as well as in the plan of each branch, it is necessary to take into consideration the building and consolidation of the rear, and to combine economic and cultural needs with those of national defence and the needs in peace time with those in war. While carrying on the task of building the army, it is necessary for the army itself to pay due attention to and actively participate in the work of consolidating the rear, particularly the implementation of the economic and financial policies, and the work of production and economy."

Proceeding from the revolutionary task in the present stage, our rear is, on a national scale, the entirely liberated north Viet Nam which is advancing to socialism. It is the revolutionary base for the whole country. Therefore, we must fully realise the importance of this rear, in order to intensify and consolidate north Viet Nam in every aspect. Parallel with the intensification and consolidation of national defence, and the building of the armed forces, we must strive to strengthen the rear in the political and economic spheres. We must actively carry out socialist transformation, strengthen the social regime and the State regime, intensify dictatorship towards the anti revolutionaries, educate the masses in patriotism and love for socialism, and raise the people's vigilance and concern in national defence, thereby ensuring the stability of the rear against all emergencies. We must do our best to build economy, develop socialist industry and agriculture in order constantly to raise the people's livelihood, at the same time to cater for the material needs of the army.

At present, peace has been restored in our country. The world situation is developing to the advantage of peace. But our country is still partitioned. American imperialism is striving to turn south Viet Nam into a new type colony and a military base. They are intervening in Laos and threatening the security of north Viet Nam. In view of this situation, it is of utmost importance to keep the correct relation between the army and the rear, between national defence and economy. On the one hand, we continue to cut down military expenditure to concentrate on economic construction; only thus can the building of socialism, consolidation of the rear and improvement of our people's livelihood be pushed forward, and concurrently good bases created for the strengthening of national defence. On the other hand, we must do everything in our power to raise the quality of the army, develop the militia and the reserve, at the same time thoroughly realising the requirements of national defence in economic construction. If we succeed in doing so, the socialist construction in north Viet Nam will win greater victories and the North will become a more stable base for the struggle for national reunification.

∴

On the occasion of the 30th anniversary of the founding of the Party, we are happy and enthusiastic on looking back on the glorious path traversed by our Party. Our people firmly believe in and are proud of the achievements of our Party, headed by President Ho Chi Minh.

Historical experience proves that *the armed struggle has occupied an important position* in the revolutionary agitation in our country, *the people's armed forces have played an important role* in winning victories for the revolution. By taking up to armed struggle at the correct time, our people had advantageous conditions for success in the August Revolution; it is thanks to the determination to wage a long armed struggle that our people succeeded in the Resistance War.

Historical experience proves that, from its founding, our Party has held the exclusive leadership in the people's revolutionary movement, the armed struggle and the building of the revolutionary armed forces; it has led our people determinedly to fight against imperialism and feudalism. There is no party, but ours, which is capable of doing so. *It was the Party's leadership which was the fundamental guarantee of the success of our people's armed struggle.* It is only our Party, embodying the working class' determination and radical revolutionary will, which had the courage to lead our unarmed people to rise up against the

French and drive out the Japanese, wage the Resistance war with primitive weapons, and score heroic achievements. Armed with Marxist-Leninist theory full of great vigour, our Party and only our Party could work out correct political and military lines appropriate to the practical conditions of our country, to bring our people's armed struggle to success. This political line was the national people's democratic line advancing to socialism. This military line was the people's war and people's army line.

The great victory of Dien Bien Phu has gloriously ended the long period of armed struggle of our people under the leadership of our Party. At present, our people have shifted to the period of political struggle to continue the national democratic revolution in the whole country and bring north Viet Nam to socialism. Our people have a great desire for peace, our policy is to do our utmost to safeguard peace. However, the shift to the political struggle and our peace policy do not mean in any way that, from now on, in the long struggle to achieve its revolutionary task, it is no longer necessary for our Party to prepare to shatter every aggressive scheme of the enemy and to build and strengthen the armed forces. That is why, at present, in north Viet Nam, while the task of building economy and culture becomes the main work, *our Party still considers the work of consolidating national defence, strengthening the revolutionary armed forces, and building the people's army into a regular and modern army* as « one of the main tasks for the whole Party and people » (Resolution of the 12th Session of the Central Committee).

The revolutionary task of our Party and people is still very heavy. The struggle to achieve the national democratic revolution in the whole country and to bring north Viet

Nam to socialism, to win complete victory for socialism and communism in our country as well as in the world, is still a long and hard process. To strive to study the rich experiences of the armed struggle and of the building of the revolutionary armed forces, and creatively to apply them to the new historical situation in order to push forward the strengthening of national defence and buiding of the army, are tasks of practical significance on the occasion of the 30th anniversary of the founding of our glorious Party.

DIEN BIEN PHU

General Vo Nguyen Giap

Dien Bien Phu was the greatest victory scored by the Viet Nam People's Army in the long war of liberation against the aggressive Franco-American imperialists. Dien Bien Phu marked an important turn in the military and political situation in Indochina. It made a decisive contribution to the great success of the Geneva Conference which restored peace in Indochina, on the basis of respect for the principles of the national sovereignty, independence, unity and territorial integrity of Viet Nam and its two friendly neighbouring countries, Cambodia and Laos.

On the anniversary of the Dien Bien Phu victory, I want to bring out in this pamphlet a number of experiences of our Party in the conduct of the war, and to recall to memory the determination of the People's Army to fight and to win, and our people's devotion in serving the front. The solidarity of our army and people in the struggle under the leadership of the Party was the decisive factor in our success. And this is the greatest lesson we have drawn from our experiences. Dien Bien Phu taught us that:

" A weak and small nation and a people's army, once resolved to stand up, to unite together and to fight for independence and peace, will have the full power to defeat all aggressive forces, even those of an imperialist power such as imperialist France aided by the United States ".

OUTLINE OF THE SITUATION OF HOSTILITIES IN WINTER 1953 — SPRING 1954

At the start of Winter 1953 the patriotic war of our people entered its eighth year.

Since the frontier campaign (1), our army had scored successive victories in many campaigns and kept the initiative on all battlefronts in north Viet Nam. After the liberation of Hoa Binh, the guerilla bases in the Red River delta were extended, and vast areas in the North-West were won back one after the other. The enemy found themselves in a daily more dangerous situation, and were driven on to the defensive. The Franco-American imperialists saw that to save the situation they had to bring in reinforcements, re-shuffle generals and map out a new plan. At that time, the war in Korea had just come to an end. The U. S. imperialists were more and more involved in plotting to protract and extend the war in Indochina. It was in these circumstances that they worked out the "Navarre plan" — a plan to continue and extend the war — which had been carefully studied and prepared in Paris and Washington.

In a word, *the "Navarre plan" was a large-scale strategic plan aimed at wiping out the greater part of our main*

---

(1) The counter-offensive in the Viet Nam-China border region in 1950.

forces within eighteen months, and occupying our whole territory, in order to turn Viet Nam permanently into a colony and military base of the American and French imperialists.

In accordance with this plan, in the first stage, fairly strong mobile forces would be regrouped in the Red River delta to attack and wear out our main forces, at the same time occupying Dien Bien Phu with a view to turning the temporarily occupied area in the North-West into a strong springboard.

Then, availing themselves of the rainy season, when our main forces might be expected to be worn out and unable to engage in any notable activity, the enemy would rush forces to the South to occupy all our free zones and guerilla-bases in the Fifth zone [1] and Nam Bo. [2]

Then, during Autumn - Winter 1955, after the "pacification" of the South, very strong mobile forces would be regrouped on the battlefront of the North for the launching of a big offensive against our rear. Starting simultaneously from the delta and Dien Bien Phu, the powerful mobile mass of the French army would annihilate our main forces, occupy our free zone and bring the war to a successful end. Had this plan succeeded, our country would have been turned into a colony of the Franco-American imperialists, a military base from which they could carry out new aggressive schemes.

In Autumn 1953, General Navarre launched this machiavellian strategic plan. With the slogans "always keep the initiative" and "always on the offensive", the

---

[1] During the war, as the French cut off all the main lines of communication, Viet Nam was divided into many zones, each embracing five or six provinces.

[2] Administrative division of Viet Nam : Bac Bo (Northern part), Trung Bo (Central part) and Nam Bo (Southern part).

High Command of the French Expeditionary Corps concentrated in the Red River delta 44 mobile battalions, launched fierce mopping-up operations in its rear, attacked Ninh Binh, Nho Quan, threatened Thanh Hoa, parachuted troops on Lang Son and threatened Phu Tho. At the same time, they armed local bandits to sow confusion in the North-West. Then, on January 20, 1954, Navarre dropped parachute troops to occupy Dien Bien Phu. His plan was to reoccupy Na San, consolidate Lai Chau and extend the occupied zone in the North-West.

About November, after wiping out a part of the enemy's forces on the Ninh Binh battlefront, our army opened the Winter-Spring campaign to smash the "Navarre plan" of the American and French imperialists.

In December 1953, our troops marched on the North-West, annihilated an important part of the enemy's manpower, liberated Lai Chau and encircled Dien Bien Phu.

Also in December, the Pathet Lao forces and the Viet Nam People's Volunteers launched an offensive in Middle Laos, wiped out important enemy forces, liberated Thakhek and reached the Mekong river.

In January 1954, in the Fifth zone, our troops launched an offensive on the Western Highlands, annihilated considerable enemy manpower, liberated the town of Kontum, and came into contact with the newly liberated Bolovene Highlands, in Lower Laos.

Also in January of that year, the Pathet Lao forces and the Viet Nam People's Volunteers launched an offensive in Upper Laos, swept away important enemy forces, liberated the Nam Hu basin and threatened Luang Prabang.

Throughout this period, in the areas behind the enemy lines in north Viet Nam, in Binh Tri Thien [1], as well as in the southernmost part of Trung Bo and in Nam Bo, guerilla warfare was greatly intensified.

In the second week of March, thinking that the period of offensive of our troops was at an end, the enemy regrouped a part of his forces to resume the " Atlanta " campaign in the South of Trung Bo and occupy Quy Nhon on March 12.

On the next day, March 13, our troops *launched the big offensive against the Dien Bien Phu entrenched camp.*

Our troops fought on the Dien Bien Phu battlefield for 55 days and nights until the complete destruction of the entrenched camp on May 7, 1954.

The Winter-Spring campaign of our army ended with an historic victory.

This is in broad outline the situation of hostilities on the various battlefronts in Autumn-Winter 1953 and Spring 1954.

---

[1] The three provinces of Central Viet Nam : Quang Binh, Quang Tri and Thua Thien.

STRATEGIC DIRECTION

The strategic direction of the Dien Bien Phu campaign and of the Winter 1953-Spring 1954 campaign in general, *was a typical success of the revolutionary military line of Marxism-Leninism applied to the actual conditions of the revolutionary war in Viet Nam.*

The enemy's strategy in the " Navarre plan " was aimed at solving the great difficulties of the aggressive war, in an attempt to save their situation and win a decisive victory.

Our strategy, applied in this Winter-Spring campaign, was the strategy of a people's war and of a revolutionary army. Starting from a thorough analysis of the enemy's contradictions, and developing to the utmost the offensive spirit of an army still weak materially but particularly heroic, it aimed at concentrating our forces in the enemy's relatively exposed sectors, at annihilating their manpower and liberating a part of the territory, compelling them to scatter their forces, thus creating favourable conditions for a decisive victory.

The war unleashed by the Franco-American imperialists was an unjust war of aggression. This colonial war had no other aim than to occupy and dominate our country. The aggressive nature and object of the war forced the enemy to scatter his forces to occupy the invaded localities. The carrying out of the war was for the French Expeditionary Corps a continuous process of dispersal of forces. The enemy divisions were split into regiments, then into battalions, companies and platoons, to be stationed at thousands of points and posts on the various battlefronts of the Indochina theatre of operations. The enemy found himself face to face with a contradiction: Without scattering his forces it was impossible for him to occupy the invaded territory; in scattering his forces, he put himself in difficulties. His scattered units would fall easy prey to our troops, his mobile forces would be more and more reduced and the shortage of troops would be all the more acute. On the other hand if he concentrated his forces to move from the defensive position and cope with us with more initiative, the occupation forces would be weakened and it would be difficult for him to hold the invaded territory. Now, if the enemy gives up occupied territory, the very aim of the war of re-conquest is defeated.

Throughout the Resistance War, while the enemy's forces were more and more scattered, *our strategic line was to extend guerilla warfare everywhere*. And in each theatre of operations, we chose the positions where the enemy was relatively weak to concentrate our forces there and annihilate his manpower. As a result, the more we fought, the stronger we became; our forces grew with every passing day. *And parallel with the process of the enemy's dispersal of forces, our people's revolutionary armed forces unceasingly intensified and extended guerilla activities, while without cease carrying on the work of concentration and building up regular units. In the fighting, in the course of the formation of our forces, we went gradually from independent companies operating separately to mobile battalions, then from battalions to regiments and divisions.* The first appearance of our divisions in the battles in the Viet Nam-China border region marked our first major victory, which drove the enemy into a disadvantageous situation.

It was after the Frontier Campaign that General de Lattre de Tassigny was dispatched to Viet Nam to save the situation. Tassigny had seen the problem. He was aware of the too great dispersal of French forces, and of the danger arising from our guerilla warfare. So he energetically regrouped his forces and launched extremely fierce and barbarous mopping-up operations to "pacify" the areas behind the enemy's lines in the Red River delta. But he found himself very soon face to face with the same insoluble contradiction. By concentrating his forces he found it impossible to extend occupied territory. Tassigny had, in the end, to resign himself to scattering his forces to launch the famous offensive on Hoa Binh. The results were not long in coming. While his crack troops suffered very heavy losses at Hoa Binh, our guerilla bases in the delta were restored and widened very considerably.

In 1953, when the "Navarre plan" was being worked out, the French imperialists also found themselves faced with the same dilemma : lack of forces to win back the initiative, to attack and annihilate our main forces. They set themselves the task of building up their fighting forces again at all costs, and, in fact, they did concentrate big forces in the Red River delta. With these forces, they hoped to wear out our main forces, compel us to scatter our army between the delta and the mountainous regions, with a view to gradually carrying out their plan and preparing for a big decisive offensive.

Faced with this situation, our Party's Central Comittee made a thorough and clear-sighted analysis of the

The Political Bureau takes the decision to launch the Dien Bien Phu campaign (from left to right: Premier Pham Van Dong — President Ho Chi Minh — Mr. Truong Chinh — General Vo Nguyen Giap)

The Viet Nam People's Army High Command works out the plan of operations

In the first phase of the Winter-Spring campaign, after three months' activity by our army, the enemy had suffered great losses on all battlefields. Many vast areas of strategic importance had been liberated and the Navarre plan of regroupment of forces was foiled. The enemy, who had made great efforts to regroup fairly strong mobile forces on a single battlefield — the Red River delta — was compelled to change his plan by concentrating his forces on a smaller scale at many different points. In other words, the Navarre plan of active regroupment of forces had in fact been turned into a forced dispersal of these same forces. The much-vaunted " Navarre mobile corps " in the delta had been reduced from 44 to 20 battalions. It was the beginning of the end of the "Navarre plan".

For us, the first phase of the Winter-Spring campaign was a series of offensives launched simultaneously on various important sectors where the enemy was relatively exposed, in which we annihilated part of the enemy's forces and liberated occupied areas, at the same time compelling the enemy to scatter his forces in many directions. We continually kept the initiative of the operations and drove the enemy on to the defensive. Also in this period, on the main battlefront, we pinned down the enemy at Dien Bien Phu, thus creating favourable conditions for our troops on other battlefields. In the national theatre of operations, there was large-scale co-ordination between the main battlefields and the theatres of operation in the enemy's rear. In each theatre, there was also close co-ordination between the main battlefield and the fronts in the enemy's rear. On the Indochinese battlefront, Dien Bien Phu became the strongest base of regroupment of the enemy forces and therefore the most important battlefield. As Dien Bien Phu had been encircled for a long time, there were new favourable conditions for the great intensification of guerilla activities and the winning of major successes in the Red River delta, in the southern part of Trung Bo as well as in Nam Bo. The enemy lacked the forces to launch mopping-up operations on any considerable scale. During this time, our free zones were no longer threatened. Moreover, our compatriots in the free zones could go to work even in the daytime without being molested by enemy aircraft.

It was also in the course of the first phase of the Winter-Spring campaign that we completed our preparations for the assault on Dien Bien Phu. During this period, the dispositions of the fortified entrenched camp had also undergone great changes. On the one hand, the enemy's forces had been increased and their defence strengthened; on the other hand, after the successive liberation of Lai Chau, Phong Saly and the Nam Hu river valley, Dien Bien Phu was completely isolated, some hundreds of kilometres from its nearest supply bases, Hanoi and the Plain of Jars.

From March 13, 1954, there began the second period of the Winter-Spring campaign. We launched the big offensive on the Dien Bien Phu fortified entrenched camp. This was a new step in the progress of the hostilities. Sticking firmly to our strategic principles — dynamism, initiative, mobility and rapidity of decision in face of new situations — and having the conditions for victory well in hand, we directed our main attack on the most powerful entrenched camp of the enemy. The task of our regular forces on the main battlefield was no longer to encircle and immobilise the enemy in their barracks, but to go over to the attack and to concentrate forces to annihilate the Dien Bien Phu fortified entrenched camp. The task of the other battle-fronts in the

North, Centre and South of Viet Nam was to intensify activities continuously in co-ordination with Dien Bien Phu, in order to annihilate more enemy manpower, scatter and pin down enemy forces, hampering the enemy in his efforts to reinforce Dien Bien Phu. On the Dien Bien Phu battlefield, our combatants fought with remarkable heroism and stubbornness. On all the co-ordinated battlefronts our troops did their utmost to overcome very great difficulties. They re-organised their force while fighting, and carried out the order of co-ordination with admirable determination and heroism.

Such was the essence of the strategic direction of the Dien Bien Phu campaign and of the Winter-Spring campaign as a whole. This direction drew its inspiration from the principles of dynamism, initiative, mobility and rapidity of decision in face of new situations. Its main object was the destruction of enemy manpower. It took full advantage of the contradictions in which the enemy was involved and developed to the utmost the spirit of active offensive of the revolutionary army. This correct, clear-sighted and bold strategy enabled us to deprive the enemy of all possibility of retrieving the initiative, and to create favourable conditions for us to fight a decisive battle on a battlefield chosen and prepared for by us. *This strategic direction ensured the success of the whole Winter - Spring campaign which was crowned by the great victory of Dien Bien Phu.*

DIRECTION OF OPERATIONS AT DIEN BIEN PHU

We have expounded the essence of the strategic direction of the 1953-1954 Winter-Spring campaign. The spirit and guiding principles of this strategic direction posed two problems to be solved for the direction of operations on the Dien Bien Phu battlefield:

1. — To attack or not to attack Dien Bien Phu?

2. — If we attack, how should we go about it?

The parachuting of enemy troops into Dien Bien Phu was not necessarily to be followed by an attack on the fortified camp. As Dien Bien Phu was a very strongly fortified entrenched camp of the enemy we could not decide to attack it without first weighing the pros and cons very carefully. The fortified entrenched camp was a new form of defence of the enemy developed in face of the growth in the strength and size of our army. At Hoa Binh and at Na San, the enemy had already entrenched his forces in fortified camps. In the Winter-Spring campaign new fortified entrenched camps appeared not only at Dien Bien Phu but also at Seno, Muong Sai and Luang Prabang in the Laotian theatre of operations, and at Pleiku, on the Western Highlands front.

With the enemy's new form of defence, should we attack the fortified entrenched camp or should we not?

While our forces were still obviously weaker than the enemy's we always stuck to the principle of concentration of forces to attack the points where the enemy was relatively weak to annihilate his manpower. *Our position was, time and again, to pin down the enemy's main forces in the fortified camps, while choosing more favourable directions for our attack.* In Spring 1952, when the enemy erected the fortified camps at Hoa Binh, we struck hard and scored many victories along the Da river and in the enemy's rear in north Viet Nam. In Spring 1953, when the enemy fortified Na San, we did not attack his position but intensified our activities in the delta and launched an offensive in the West. During the last months of 1953 and at the beginning of 1954, when the enemy set up fortified camps in various places, our troops launched many successful offensives on sectors where the enemy was relatively weak, and at the same time stepped up guerilla warfare behind the enemy's lines.

These tactics of attacking positions other than the fortified entrenched camps had recorded many successes. But these were not the only tactics. We could also *directly attack the fortified entrenched camp to annihilate the enemy's manpower in the heart of his new form of defence.* Only when we had wiped out the fortified entrenched camp, could we open up a new situation, paving the way for new victories for our army and people.

That was why, on the Dien Bien Phu battlefield, the problem of whether to attack or not had been posed, especially as Dien Bien Phu was the enemy's strongest fortified entrenched camp in the whole Indochina war theatre, while our troops had, up to that time, attacked only fortresses defended by one or two companies, or one battalion at most.

Dien Bien Phu being the keystone of the Navarre plan, we considered that it should be wiped out if the Franco-American imperialist plot of protracting and expanding the war was to be smashed. However, the importance of Dien Bien Phu could not be regarded as a decisive factor in our decision to attack it. In the relation of forces at that time, could we destroy the fortified entrenched camp of Dien Bien Phu ? Could we be certain of victory in attacking it ? Our decision had to depend on this consideration alone.

Dien Bien Phu was a very strongly fortified entrenched camp. But on the other hand, it was set up in a mountainous region, on ground which was advantageous to us, and decidedly disadvantageous to the enemy. Dien Bien Phu was, moreover, a completely isolated position, far away from all the enemy's bases. The only means of suppling Dien Bien Phu way by air. These circumstances could easily deprive the enemy of all initiative and force him on to the defensive if attacked.

On our side, we had picked units of the regular army which we could concentrate to achieve supremacy in power. We could overcome all difficulties in solving the necessary tactical problems ; we had, in addition, an immense rear, and the problem of supplying the front with food and ammunition, though very difficult, was not insoluble. Thus we had conditions for retaining the initiative in the operations.

It was on 'the basis of this analysis of the enemy's and our own strong and weak points that we solved the question as to whether we should attack Dien Bien Phu or not. *We decided to wipe out at all costs the whole enemy force at*

Dien Bien Phu, after having created favourable conditions for this battle by launching numerous offensives on various battlefields and by intensifying preparations on the Dien Bien Phu battlefield. This important decision was a new proof of the dynamism, initiative, mobility and rapidity of decision in face of new situations displayed in the conduct of the war by the Party's Central Committee. Our plan foresaw the launching of many offensives on the points where the enemy was relatively weak, availing ourselves of all opportunities to wipe out enemy's manpower in mobile warfare. But whenever it was possible and success was certain, we were resolved not to let slip an opportunity to launch powerful attacks on strong points to annihilate the more concentrated enemy forces. Our decision to make the assault on the Dien Bien Phu fortified camp clearly marked a new step forward in the development of the Winter-Spring campaign, in the annals of our army's battles and in the history of our people's resistance war.

We had pledged to wipe out the whole enemy force at Dien Bien Phu but we still had to solve this problem: How should we do it? *Strike swiftly and win swiftly, or strike surely and advance surely! This was the problem of the direction of operations in the campaign.*

In the early stage, when we began the encirclement of Dien Bien Phu, and the enemy, having been newly parachuted into the area, had not yet had time to complete his fortifications and increase his forces, the question of striking swiftly and winning swiftly had been posed. By concentrating superior forces, we could push simultaneously from many directions deep into enemy positions, cut the fortified entrenched camp into many separate parts, then swiftly annihilate the entire enemy manpower. There were many obvious advantages if we could strike swiftly to win swiftly: by launching a big offensive with fresh troops, we could shorten the duration of the campaign and avoid the wear and fatigue of a long operation. As the campaign would not last long, the supplying of the battlefront could be ensured without difficulty. However, on further examining the question, we saw that these tactics had a very great, a basic disadvantage: our troops lacked experience in attacking fortified entrenched camps. If we wanted to win swiftly, success could not be ensured. For that reason, in the process of making preparations, we continued to follow the enemy's situation and checked and re-checked our potentialities again. And we came to the conclusion that we could not secure success if we struck swiftly. In consequence, *we resolutely chose the other tactic: to strike surely and advance surely*. In taking this correct decision, *we stricly followed this fundamental principle of the conduct of a revolutionary war: strike to win, strike only when success is certain; if it is not, then don't strike.*

In the Dien Bien Phu campaign, the adoption of these tactics demanded of us firmness and a spirit of resolution. Since we wanted to strike surely and advance surely, preparations would take a longer time and the campaign would drag out. And the longer the campaign went on the more, new and greater difficulties would crop up. Difficulties in supply would increase enormously. The danger increased of our troops being worn out while the enemy consolidated defences and lined up his forces. Above all, the longer the campaign lasted, the nearer came the rainy season with all its disastrous consequences for operations carried out on the mountains and in forests. As a result, not everybody was immediately convinced of the correctness of these tactics. We patiently educated our men, pointed out that there were real difficulties, but that our task was to overcome them to create good conditions for the great victory we sought.

It was from these guiding principles that we developed our plan of progressive attack, in which the Dien Bien Phu campaign was regarded not as *a large-scale attack on fortresses carried out over a short period, but as a large-scale campaign carried out over a fairly long period, through a series of successive attacks on fortified positions until the enemy was destroyed.* In the campaign as a whole we already had numerical superiority over the enemy. But in each attack or each wave of attacks, we had the possibility of achieving absolute supremacy and ensuring the success of each operation and consequently total victory in the campaign. Such a plan was in full keeping with the tactical and technical level of our troops, creating conditions for them to accumulate experience in fighting and to ensure the annihilation of the enemy at Dien Bien Phu.

We strictly followed these guiding principles throughout the campaign. We encircled the enemy and carried out our preparations thoroughly over a period of three months. Then, after opening the offensive, our troops fought relentlessly for 55 days and nights. Careful preparation and relentless fighting led our Dien Bien Phu campaign to resounding victory.

SOME QUESTIONS OF TACTICS

Dien Bien Phu was a fortified entrenched camp defended by fairly strong forces: 17 battalions of infantry, three battalions of artillery, without counting engineer tank units, air and transport units, etc., most of them picked elements of the French Expeditionary Corps in Indochina. The fortified camp was made up of 49 strong-posts, organised into fortified resistance centres and grouped into three sectors capable of supporting each other. In the middle of the central sector, which was effectively guarded by the resistance centres on the hill-tops in the East, were mobile forces, artillery positions and tank units, as well as the enemy headquarters. The airfield of Dien Bien Phu was near here. This whole vast defence system lay within strong underground fortifications and trenches.

The French and American military authorities believed that the fortified entrenched camp of Dien Bien Phu was impregnable. They were certain that an offensive against Dien Bien Phu would be suicidal, that failure was inevitable. Therefore, during the first weeks of the campaign, the French High Command firmly believed that there was little possiblity of an offensive against Dien Bien Phu by our army. Until the last minute, the offensive launched by our men was unexpected by the enemy.

General Navarre had over-estimated the Dien Bien Phu defences. He believed that we would be unable to crush even one centre of resistance. Because, unlike the simple strong-posts at Na San or Hoa Binh, these were centres of resistance forming a much more complex and strongly fortified defence system.

The destruction of the fortified entrenched camp as a whole was, to Navarre's mind, still less feasible. In his opinion, his artillery and air forces were powerful enough to wipe out all forces coming from outside before these could be deployed in the valley and approach the fortifications. He was not in the least worried about our artillery which he thought weak and not transportable to the approaches of Dien Bien Phu. Nor was he anxious about his own supplies, because both airfields, surrounded by the defence sectors, could not be in danger. Never did it enter his head that the whole fortified camp could be annihilated by our troops.

The enemy's estimates were obviously wishful thinking but they were not totally without foundation. In fact, the Dien Bien Phu fortified entrenched camp had many strong points which had given our army new problems of tactics to solve before we could annihilate the enemy.

The fortified entrenched camp was a defence system manned by big forces. The centres of resistance, which were closely connected to one another, were effectively supported by artillery, tank units and aircraft, and could easily be reinforced by mobile forces. This was a strong point for the enemy and for us, a difficulty. *We overcame this difficulty by applying the tactics of progressive attack*, by regrouping our forces to have great local superiority, by striving to neutralise as much as possible the enemy artillery fire and mobile forces, bringing everything into play to wipe out the centres of resistance one by one, or a group of centres at one time in a wave of attacks. By concentrating forces to achieve absolute superiority at one point, we were certain to crush the enemy, especially in the first days of the campaign, when we attacked the enemy outposts.

The fortified entrenched camp had quite powerful artillery fire, tank and air forces. This was another strong point of the enemy, a very great difficulty of ours, especially since we had only very limited artillery fire and no mechanised or air forces. *We overcame this difficulty by digging a whole network of trenches that encircled and strangled the entrenched camp,* thus creating conditions for our men to deploy and move under enemy fire. Our fighters dug hundreds of kilometres of trenches These wonderful trenches enabled our forces to deploy and move in open country under the rain of enemy napalm bombs and artillery shells. But to reduce the effect of enemy fire was not enough, *we still had to strengthen our own firepower.* Our troops cut through mountains and hacked away jungles to build roads and haul our artillery pieces to the approaches of Dien Bien Phu. Where roads could not be built, artillery pieces were moved by nothing but the sweat and muscle of our soldiers. Our artillery was set up in strongly fortified firing positions, to the great surprise of the enemy. Our light artillery played a great part in the Dien Bien Phu battle.

While neutralising the enemy's strong points, we had to make the most of his weak points. His greatest weakness *lay in his supply*, which depended entirely on his air forces. Our tactics were from the very beginning to use our artillery-fire to destroy the air-strips, and our anti-aircraft guns to cope with the activities of enemy planes. Later, with the development of the waves of attacks, everything was brought into play to hinder enemy supply and gradually stop it altogether.

These are a few of the problems of tactics we solved in the Dien Bien Phu campaign. They were solved on the basis of our analysis of the enemy's strong and weak points, combining technique with the heroism and hard-working and fighting spirit of a People's Army.

To sum up, our plan of operations based upon these tactical considerations consisted in setting up a whole system of lines of attack and encirclement, permitting our forces to launch successive attacks to annihilate the enemy. This network of innumerable trenches with firing positions and command posts encircled and strangled the enemy. It was progressively extended with our victories. From the surrounding mountains and forests, it moved down into the valley. Each enemy position, once wiped out, was immediately turned into our own. As we encircled the enemy fortified entrenched camp, a real fortified camp of our own, very mobile, gradually took shape, and kept closing in, while the enemy camp was constantly narrowed down.

In the first phase of the campaign, from our newly-built network of attack and encirclement positions, we annihilated the Him Lam and Doc Lap centres of resistance, and the whole Northern sector. The enemy made desperate efforts to destroy our firing positions. Their planes poured napalm bombs on the mountains around Dien Bien Phu. Their artillery concentrated powerful fire on our firing positions. But we held on.

In the second phase, the "axis" communication trenches, with their innumerable ramifications, starting from our bases, extended down into the valley and isolated the Central sector from the Southern sector. The fierce and successful assault on the Eastern hill-tops enabled our belt of artillery fire to close in. From the captured positions, our guns of all calibres could exert pressure on the enemy. The air-strips were completely controlled by our fire.

The enemy became increasingly active, bringing reinforcements for his mobile forces, launching counter-attacks and furiously bombing our lines in an attempt to save the situation. It was a desperate positional battle. Many hilltops were captured and recaptured many times. Some were occupied half by our troops and half by the enemy. Our tactics were to encroach, harass and wrest every inch of ground from the enemy, destroy his air-strips and narrow down his free air-space.

The third phase was that of general offensive. The enemy had been driven in to an area about 1.5 to 2 kilometres square. His forces had suffered heavy losses. Once hill A-1 had been completely occupied by our troops, all hope of continued resistance vanished and the enemy's morale sank extremely low. On March 7, our troops launched an offensive from all directions, occupied the enemy headquarters and captured the whole enemy staff. That night, the Southern sector was also wiped out.

The Dien Bien Phu campaign ended in a great victory.

Hauling artillery pieces:
The pass is high! But it does not matter.
We are determined to cross it and all our guns will pass ...

Day and night, endless convoys of civilian volunteers ensure supplies for the front

Crossing the Muong Thanh bridge for the last attack

The «Determined to fight and to win» banner of the Viet Nam People's Army flutters over General de Castries' entrenched headquarters

General de Castries, Commander of the entrenched camp, taken prisoner with his whole staff

OUR ARMY'S DETERMINATION TO FIGHT AND TO WIN

The great task assigned to the whole army and people by the Party's Central Committee and the Government was: *to concentrate forces, to be thoroughly imbued with determination, "to actively develop the spirit of heroic fighting and endurance to bring the campaign to complete victory"*. For the Dien Bien Phu campaign, as had been pointed out by President Ho Chi Minh and the Political Bureau of the Central Committee of the Viet Nam Lao Dong Party, was an historic campaign of exceptional importance to the military and political situation in our country and to the full growth of our army, as well as to the struggle for the defence of peace in South-East Asia.

Our troops fought to carry out this great task with unshakable determination. Our combatants' will to fight and to defeat the enemy was one of the decisive factors which brought the Dien Bien Phu campaign and the Winter-Spring campaign in general such brilliant victories on all battlefronts.

Throughout the history of the armed struggle of our people, never had our army been entrusted with so great and heavy a task as in Winter 1953-Spring 1954. The enemy to be annihilated was rather a strong one. Our forces thrown into the battle were very large. The theatre of operations was extensive and the operations lasted half a year. On the Dien Bien Phu battlefield, as on all other co-ordinated battlefields, our combatants, with a spirit of heroism and endurance, surmounted countless difficulties and overcame many great obstacles to annihilate the enemy and fulfil their task. This heroism and endurance were tempered and enhanced by the long years of Resistance. Particularly in Winter 1953-Spring 1954, the revolutionary enthusiasm of our combatants increased greatly after their study of the policy for the mobilization of the masses for land reform. Here, stress should be laid on the considerable contribution made by the land reform policy to the victories of the Winter-Spring campaign, particularly on the Dien Bien Phu battlefield.

On the Dien Bien Phu battlefront, in the period of preparation, our armymen opened up the supply line from Tuan Giao to Dien Bien Phu ; built through mountains and forests roads practicable for trucks to move artillery pieces into position ; built artillery emplacements ; dug trenches from the mountains to the valley ; changed the terrain ; overcame enormous obstacles, and in all ways created favourable conditions for the annihilation of the enemy. Neither difficulties, fatigue nor enemy bombing and artillery fire could shake the iron will of our men.

From the first shot touching off the offensive against Dien Bien Phu, and throughout the battle, our combatants fought with extraordinary heroism. Under the deluge of bombs from the eneny air force, and under the enemy's cross-fire, our fighters valiantly stormed and captured Him Lam and Doc Lap hills, put the enemy troops entrenched on the Eastern hills out of action, expanded our bases, cut off the airfields, repulsed counter-attacks and kept tightening our encirclement. During all this time, the enemy's napalm-bombs burned down the undergrowth on the hills surrounding Dien Bien Phu, and enemy bombs and shells ploughed deep into the fields in our zones of operations. But our combatants kept moving forward to carry out their tasks. One fell, but many others rushed forward like a sweeping rising tide that no force on earth could hold back. *We witnessed a phenomenon of collective heroism in which the most admirable deeds were performed by* To Vinh Dien, *who threw himself under the wheel of an artillery piece to prevent it from slipping back;* Phan Dinh Giot, *who silenced an enemy gun nest with his own body; the shock troops who planted the banner of " Determination to Fight and to Win" on Him Lam hill, and the shock troops who captured the enemy headquarters.*

The spirit of heroism and endurance of our fighters on co-ordinated battlefields should also be mentioned. On the Western Highlands, great successes were scored at Kontum and An Khe. In the Red River delta, our troops destroyed 78 planes on Cat Bi and Gia Lam airfields, wiped out several enemy fortified positions and cut off road No. 5, the enemy's main supply line. In south Viet Nam, more than 1,000 enemy posts were annihilated or evacuated, many stocks of bombs destroyed and ships sunk. On the battlefields of our two neighbouring countries, our people's volunteers, together with the army and the people of these friendly countries, wiped out the invaders and scored many great victories.

Never had our army fought with such endurance for so long a time as in Winter 1953-Spring 1954. There were units which marched and pursued the enemy for more than 3,000 kilometres. There were others which moved secretly for more than 1,000 kilometres on the Truong Son * mountain range to take part in fighting on a far-off battlefield. The units on Dien Bien Phu battlefield moved from the delta to the mountains, and at once set passionately to work, at the same time fighting to protect their preparatory labour. Then came the battle, and our troops lived and fought for two months in trenches after having spent three months of hardship in the jungle. While the battle was going on, certain units rushed to places two or three hundred kilometres away to launch surprise attacks on the enemy, then came back to take part in the annihilation of the enemy at Dien Bien Phu. The spirit of co-operation between the various units and various arms was enhanced during the battle, and there was close co-ordination between the various battlefields.

Our combatants' *determination to fight and to win as described above came from the revolutionary nature of our army and the painstaking education of the Party. It had been enhanced in battle and in the ideological re-moulding classes.* This does not mean that, even when the Dien Bien Phu battle was at its height, negative factors never appeared. *To maintain and develop this determination to fight and to win was a whole process of unremitting and patient political and ideological education and struggle,* tireless and patient efforts in political work on the front line. This was a great achievement of the Party's organizations and branches and of its cadres. After a series of resounding victories, we found in our ranks signs of under-estimation of the enemy. By criticism, we rectified this state of mind in good time. In the long period of preparation, particularly

---

* Range of mountains running from the North to the South of Central Viet Nam, along the Viet Nam-Laos border.

after the second phase of the campaign, when attack and defence were equally fierce, negative rightist thoughts cropped up again o the detriment of the carrying out of the task. In accordance with the instructions of the Political Bureau, we opened in the heart of the battlefield an intensive and extensive struggle against rightist passivity, and for the heightening of revolutionary enthusiasm and the spirit of strict discipline, with a view to ensuring the total victory of the campaign. This ideological struggle was very successful. This was one of the greatest achievements in political work in our army's history. It led the Dien Bien Phu campaign to complete victory.

The determination to fight and to win of our army on the Dien Bien Phu and other co-ordinated battlefields was a distinctly marked manifestation of the boundless loyalty of our People's Army to the revolutionary struggle of the people and the Party. It was a collective manifestation of proletarian ideology, of the class-stand of the officers and men and Party members in the army. It maintained the Viet Nam People's Army tradition of heroic fighting, endurance and determination in the fulfilment of duty. It made of the soldier of the People's Army an iron fighter. Dien Bien Phu will forever symbolize the traditions of fighting and winning victory of our army and people. Our military banner is the banner of *" Determination to Win "*.

PEOPLE'S DEVOTION TO SERVING THE FRONT

The Party's Central Committee and the Government decided that the whole people and Party should concentrate all their forces for the service of the front, in order to ensure the victory of the Dien Bien Phu campaign. During this campaign, and generally speaking, during the whole Winter-Spring campaign, our whole people — workers, peasants, youth, intellectuals — every Vietnamese patriot, answered the appeal for national liberation and did his utmost to achieve the slogan *"All for the front, all for victory"* with an ardent and unprecedented enthusiasm, at the cost of superhuman efforts.

Throughout the long years of the Resistance War, our people never made so great a contribution as in the Winter 1953-Spring 1954 campaign, in supplying the army for the fight against the enemy. On the main Dien Bien Phu front, our people had to ensure the supply of food and munition to a big army, operating 500 to 700 kilometres from the rear, and in very difficult conditions. The roads were bad, the means of transport insufficient and the supply lines relentlessly attacked by the enemy. There was, in addition, the menace of heavy rains that could create more obstacles than bombing.

On the Dien Bien Phu front, the supply of food and munitions was a factor as important as the problem of tactics; logistics constantly posed problems as urgent as those posed by the armed struggle. These were precisely the difficulties that the enemy thought insuperable for us. The imperialists and traitors could never appreciate the strength of a nation, of a people. This strength is immense. It can overcome any difficulty, defeat any enemy.

The Vietnamese people, under the direct leadership of the committees of supply for the front, gave proof of great heroism and endurance in serving the front.

Truck convoys valiantly crossed streams, mountains and forests; drivers spent scores of sleepless nights, in defiance of difficulties and dangers, to bring food and ammunition to the front, to permit the army to annihilate the enemy.

Thousands of bicycles from the towns also carried food and munitions to the front.

Hundreds of sampans of all sizes, hundreds of thousands of bamboo-rafts crossed rapids and cascades to supply the front.

Convoys of pack-horses from the Meo highlands or the provinces headed for the front.

Day and night, hundreds of thousands of porters and young volunteers crossed passes and forded rivers in spite of enemy planes and delayed-action bombs.

Near the firing line, supply operations had to be carried out uninterruptedly and in the shortest possible time. Cooking, medical work, transport, etc., was carried on right in the trenches, under enemy bombing and cross-fire.

Such was the situation at Dien Bien Phu, but on the co-ordinated fronts, big armed forces were also active, especially on the Western Highlands and in other remote theatres of operation. On these fronts, as at Dien Bien Phu, our people fulfilled their tasks. They admirably solved the problems of supply to enable the army to defeat the enemy, always to win new victories.

Never had so large a number of Vietnamese gone to the front. Never had so many young Vietnamese travelled so far and become acquainted with so many distant regions of their country. From the plains to the mountains, on roads and paths, on rivers and streams, everywhere, there was the same animation: the rear sent its men and wealth to the front in order to annihilate the enemy and, together with the army, to liberate the country.

The rear brought to the fighter at the front its will to annihilate the enemy, its strong unity in the resistance and the revolutionary enthusiasm of the land reform. Each day, thousands of letters and telegrams from all over the country came to the Dien Bien Phu front. Never had Viet Nam been so anxious about her fighting sons, never had the relations between the rear and the front been so intimate as in this Winter Spring campaign.

Indeed, a strong rear is always the decisive factor for victory in a revolutionary war. In the Dien Bien Phu campaign and, generally speaking, in the whole Winter-Spring campaign, our people made a worthy contribution to the victory of the nation.

We cannot forget the sympathy and hearty support of the brother peoples, of the progressive peoples all over the world, including the French people. Every day, from all corners of the earth, from the Soviet Union, China, North Korea and the German Democratic Republic,

Algeria, India, Burma, Indonesia, and other countries news reached the front through broadcasts, bringing the expression of the boundless support of progressive mankind for the just struggle of the Vietnamese people and army. This was a very great encouragement for the combatants of the Viet Nam People's Army at Dien Bien Phu, as on all other fronts.

THE WAR OF LIBERATION OF OUR PEOPLE WAS ONE LONG AND GREAT DIEN BIEN PHU BATTLE

The victory of Dien Bien Phu and, generally speaking, the Winter 1953-Spring 1954 victories were the greatest victories won by our army and people in their long war of liberation against aggressive imperialism.

At Dien Bien Phu, our army annihilated the enemy's strongest fortified camp in Indo-China, and wiped out 16,000 of his crack troops. During this Winter-Spring campaign, for all the fronts operating in co-ordination with Dien Bien Phu, the total losses of the enemy amounted to 110,000 men.

The "Navarre plan" was smashed to pieces. The French and American imperialists failed in their attempt to prolong and extend the war in Indo-China. The Dien Bien Phu victory had very great influence. Thanks to it, and to the success scored on the other co-ordinated fronts, we liberated the capital Hanoi and also the North of Viet Nam. Thanks to it, we achieved brilliant success at the Geneva Conference, and peace was restored in Indo-China.

With the "Navarre plan", the French and American imperialists wanted to launch a decisive battle. In fact, the battle of Dien Bien Phu was decisive. Dien Bien Phu was a great victory for our army and people. Dien Bien Phu decided the humiliating defeat of the aggressive imperialists.

Dien Bien Phu was a battle in which our people and their army coped with the expeditionary corps of the French imperialists, backed by the U.S. warmongers. We won the war and the aggressive imperialists were the losers. Dien Bien Phu will forever symbolize the indomitable spirit of our people who opposed to the powerful army of an imperialist country, the unity and heroism of a weak nation and of a people's army still in its early days. This heroic spirit was the spirit of our people and army throughout the long resistance. Thus, we can assert that each of our struggles, however big or small, was imbued with "the spirit of Dien Bien Phu", that the war of liberation of our people was one long and great Dien Bien Phu battle.

We were victorious at Dien Bien Phu. Our national war ended with a great victory which showed the very clear-sighted and heroic leadership of our Party. This was a great victory of Marxism-Leninism in the liberation war of a small and heroic nation. Our people could say with pride: under the leadership of our Party headed by President Ho Chi Minh we established a great historic truth: *a colonized and weak people once it has risen up and is united in the struggle and determined to fight for its independence and peace, has the full power to defeat the strong aggressive army of an imperialist country.*

Thus, Dien Bien Phu was a victory not only for our people, but also for all weak peoples who are struggling to throw off the yoke of the colonialists and imperialists. That is the great significance of the Dien Bien Phu victory.

Therefore, its anniversary is a day for rejoicing for our whole people and also a day of great joy for the brother peoples, for the peoples who have just won back their independence, and for those who are fighting for their liberation.

Dien Bien Phu is written down forever in the annals of the struggle for national liberation of our people and of oppressed peoples all over the world. History will record it as one of the crucial events in the great movement of Asian, African and Latin American peoples who are rising up to liberate themselves and to be masters of their own destiny.

Solidarity in the struggle under the leadership of our Party led our people to the Dien Bien Phu victory. It will surely lead us to new and greater victories in the building of north Viet Nam on the way to socialism and in the struggle for national reunification by peaceful means.

APPENDIX

I — MILITARY SITUATION IN SUMMER 1953

The winter of 1940 marked a new change in the military situation in Viet Nam. After their great victory in the Border campaign, our forces undertook a series of important campaigns: the Midland campaign, the road No. 18 campaign and the Ha Nam - Nam Dinh-Ninh Binh campaign in 1951; the Hoa Binh campaign in winter 1951 and spring 1952; the North-West campaign in winter 1952.

In these victorious campaigns, we put hundreds of thousands of enemy troops out of action and liberated vast areas in the mountainous regions of north Viet Nam. The important provinces on the Viet Nam-China border — Cao Bang, Lang Son, Lao Cai — the province of Hoa Binh on the road joining the Viet Bac to the Fourth zone, the great part of the North-West region from the Red River to the Viet Nam-Laos border, were successively liberated. Our rear was greatly expanded. In the mountainous regions of the north, the enemy occupied only Hai Ninh province in the North-East, and the town of Lai Chau and the fortified camp of Na San in the North-West.

While our main force scored successive victories on the main front, guerilla warfare strongly developed in all the areas behind the enemy's lines in north Viet Nam. Especially during the Hoa Binh campaign, our main force penetrated deep into the enemy rear on both sides of the

Red River, combined its action with the local armed and semi-armed forces, enlarged the guerilla bases and zones and freed millions of our compatriots. The temporarily occupied zones of the enemy were limited to only one third of the land and villages near the communication lines and important cities.

On the other fronts, in the enemy rear at Binh-Tri-Thien *, in the south of central Viet Nam and in Nam Bo, guerilla warfare was going on and developing, causing heavy losses to the enemy.

In summer 1953, the Pathet Lao forces, combined with the Viet Nam People's volunteers, launched a sudden attack on the town of Sam Neua. The bulk of the garrison was annihilated; the town of Sam Neua and vast zones of Upper Laos were liberated, thus creating a new threat to the enemy.

Throughout north Viet Nam, we observed that from winter 1950 onwards, our forces constantly held the initiative in operations, driving the enemy more and more on to the defensive. To save this situation, the enemy made an urgent appeal to the American imperialists whose intervention in the aggressive war in Indo China had been constantly on the increase. During this period the French Government had several times changed the commanders of the French Expeditionary Corps. After the Border campaign, it sent to Indo-China the famous General de Lattre de Tassigny. As is known, Tassigny strove to concentrate his troops, fortify his defence lines, and launch an attack in the direction of

---

* Provinces in central Viet Nam : Quang Binh. Quang Tri. Thua Thien.

Prisoners of war on their way to the rear

A general view of the entrenched camp following the surrender of the garrison

Hoa Binh in order to recapture the initiative in the operations, but he was finally defeated. His successor, **General Salan**, was, in his turn, an impotent witness to severe defeats of the Expeditionary Corps on the North-West and Upper Laos fronts.

It was in this critical situation, that the American imperialists availed themvelves of the armistice in Korea to step up their intervention in Indo-China. And the "Navarre plan" expressed the new Franco-American scheme to prolong and extend the aggressive war in our country.

II — THE ENEMY'S NEW SCHEME THE "NAVARRE PLAN"

In mid-1953, with the consent of Washington, the French Government appointed General Navarre Commander in-Chief of the French Expeditionary Corps in Indo-China.

Navarre and the French and American generals estimated that the more and more critical situation of the French Expeditionary Corps was due to the extreme dispersal of French forces in thousands of posts and garrisons scattered on all fronts to cope with our guerilla warfare ; as a result, they lacked a strong mobile force to face the attacks of our main force. During that time, our forces were constantly growing, our mobile forces increased day by day, the scale of our campaigns became larger and larger.

Basing themselves upon this estimation, Navarre and the French and American generals mapped out a plan to save the day, hoping to reverse the situation and to win, in a short period of time, a decisive strategic success.

The "Navarre plan" envisaged the organisation of a very strong strategic mobile force, capable of breaking all our offensives and annihilating the main part of our forces later on. For this purpose, Navarre ordered the regroupment of his picked European and African units, which were to be withdrawn from a number of posts. At the same time, new units from France, West Germany, North Africa and Korea were rushed to the Indo-China front.

In the carrying out of this plan, the enemy met a great contradiction, a serious difficulty: if they kept their forces scattered in order to occupy territory, it would be impossible for them to organise a strong mobile force; but if they reduced their occupation forces to regroup them, our guerillas would take advantage of the new weakness of their position to increase their activity, their posts and garrisons would be threatened or annihilated, the local puppet authorities overthrown, and the occupied zones reduced. Navarre sought to get round the difficulty by developing the puppet forces on a large scale to replace European and African troops transferred towards the re-grouping points. In fact, this treacherous idea was nothing new, and had already been applied by Tassigny. Faced with the new dangerous situation, Navarre and the French and American generals decided to organise 54 new battalions of puppet troops immediately and to double this number in the following year. Later on, the enemy had to acknowledge that this expedient did not help, because the increase in the puppet forces really only represented a quantitative increase at the expense of the quality of the units.

With their great mobile forces, the Franco-American imperialists conceived a rather audacious plan, aimed at annihilating our main force and ending the war within 18 months.

On the one hand, they decided to concentrate their forces in the Red River delta in autumn and winter 1953 to open barbarous mopping-up operations to destroy our guerilla bases; on the other hand, they planned to launch attacks on our free zone in order to attract and exhaust our main forces. Simultaneously, they intended to create new battalions of puppet soldiers and re-group new units.

After winter, that is after the season of big operations in north Viet Nam, at the beginning of 1954, availing themselves of the fact that our army could at this time be resting, they would transfer to the south the greater part of their mobile forces. At this period, the climatic conditions in the south were favourable to their activity. Their intention was to open big operations to occupy all our free zones, particularly the Fifth and Ninth zones. To occupy all these regions would be for them tantamount to removing the gravest threats faced by them. Due to the impetus provided by these victories, they would recruit new puppet units, while continuing the regrouping of their mobile forces to prepare a decisive offensive on the front in the north.

If the plan were working well, in autumn and winter 1954, they would bring back to north Viet Nam their greatly increased forces, still under the influence of the enthusiasm created by their recent victories. In launching a major offensive against our bases, they would have occupied new territories, annihilated the bulk of our main forces to end the aggressive war and permanently transform the whole of Viet Nam into a colony and a Franco American military base.

According to his plan, in summer 1953, the enemy concentrated their forces. At the beginning of autumn, enemy mobile forces reached a total of 84 battalions in the whole of Indo-China.

To carry out the first phase of the "Navarre plan" the enemy concentrated in the Red River delta more than 50 per cent of their mobile forces, and declared that they were passing over to the offensive in order to regain the initiative in the operations. Scores of battalions launched savage mopping-up operations in the delta in order to consolidate the rear. Units of paratroops attacked Lang Son and it was announced that we had suffered heavy losses, although in fact our losses were insignificant. They launched a great attack on Nho Quan and on the region bordering Ninh Binh and Thanh Hoa provinces, and declared that the occupation of these provinces was imminent. But their troops had to withdraw with heavy losses.

In the North-West, the enemy withdraw from Na San to the delta. Formerly, Na San had been considered by them as "the second Verdun", "blocking the road to the Southward advance of communism", but when they had to evacuate it in order to escape destruction, they declared that Na San had lost all military interest. Before the evacuation, they saw that their myrmidons organised gangs of bandits in rather extensive areas to the north of this locality.

On November 20, 1953, the enemy dropped considerable paratroop forces into the valley of Dien Bien Phu. Their plan was to reinforce Dien Bien Phu, then go to Tuan Giao and Son La, re-occupy Na San and join it to Lai Chau. Thus Dien Bien Phu would become a strongpost threatening the flank of our North-West base. This new entrenched position would force us to scatter our troops between the delta and the mountains, and would protect

Upper Laos. It would constitute a spring-board for their next big offensive, one column pushing from the plain, the other from Dien Bien Phu to the delta. Thus, Dien Bien Phu became little by little a key position in the Navarre plan.

It was clear that in this autumn-winter period all enemy activities had one aim : to re-group forces, to strengthen the rear, to exhaust and scatter our forces, to prepare conditions for their coming great attacks. They thought that the first phase of their plan had been successful when our Autumn-Winter campaign began.

III — OUR PLAN IN WINTER 1953 - SPRING 1954, AND THE EVOLUTION OF THE MILITARY SITUATION ON THE VARIOUS FRONTS

After the cease-fire in Korea, we anticipated that the new Franco-American scheme war to expand their forces and extend the aggressive war in Indo-China. Early in summer 1953, their military situation deteriorated. Taking advantage of the serious difficulties met by the French Expeditionary Corps following its successive defeats from 1950 onwards, the American imperialists intervened more openly and more actively in the war in Indo-China.

On our side, the army and people were transported with the impetus of the victorious big campaign ; guerilla warfare was developing in all regions under enemy control. Our army had accumulated more fighting experience, and its tactical and technical level had been raised through the summing-up of the experience of the military campaign and the training courses. Moreover, a new factor appeared : this was the policy of systematic rent reduction, and the carrying out of land reform decided by the Party and Government. After the political course on the mobilisation of the peasant masses, our cadres and armymen saw more clearly that the objective of our struggle was: national independence, and land to the tillers. Hence, their combativeness increased greatly. More than ever our army was transported with enthusiasm, ready to go to the front to annihilate the enemy.

We were determined to break the " Navarre plan " and hold the new plot of the Franco-American imperialists in check. But how to do it? Faced with the new difficulties, it was neccessary to analyse the situation to determine a correct line of action which ensured success.

The concrete problem was: the enemy was concentrating forces in the Red River delta, and launching attacks on our free zones. Now, had we to concentrate our forces to face the enemy, or to mobilise them for attacks in other directions? The problem was difficult. In concentrating our forces to fight the enemy in the delta we could defend our free zone ; but here the enemy was still strong and we could easily be decimated. On the other hand, in attacking in other directions with our main forces, we could exploit the vulnerable points of the enemy to annihilate the bulk of their forces; but our free zone would thus be threatened.

After a careful study of the situation, the Party's Central Committee issued the following slogan to break the " Navarre plan " : " dynamism, initiative, mobility, and rapidity of decision in face of new situations." Keeping the initiative, we should concentrate our forces to attack strategic points which were relatively vulnerable. If we succeeded in keeping the initiative, we could achieve successes and compel the enemy to scatter their forces, and finally, their plan to threaten our free zone could not be realised. On the other hand, if we were driven on to the defensive, not only could we not annihilate many enemy forces, but our own force could easily suffer losses, and, finally, it would be difficult for us to break the enemy threat.

On all fronts, our Winter-Spring plan was the expression of this strategic conception. In October 1953, hundreds of thousands of persons were mobilised to quicken the preparations. In mid-November, our main forces went to the front. The Winter-Spring campaign began.

Liberation of Lai Chau

On December 10, 1953, we opened fire on the Lai Chau front. Formerly, we had annihilated or forced to surrender thousands of bandits in the regions of Muong La and Chau Thuan. On that very night, we wiped out the outpost of Paham, about 30 kilometres from Lai Chau. Aware of the presence of our main forces, the enemy was very afraid and ordered the garrison to withdraw from Lai Chau and to rally to Dien Bien Phu by the mountain tracks.

Our troops were ordered to march on to liberate Lai Chau, while one column attacked westward, cutting of the enemy's retreat to encircle and annihilate him.

On December 12, Lai Chau was liberated.

On December 13, we annihilated the enemy in retreat at Muong Pon. After ten days and ten nights of fighting, pursuit and encirclement in a mountainous region, we liberated the remaining part of the zone occupied by the enemy in Lai Chau province. The enemy lost 24 companies.

It was the first great success of our Winter-Spring campaign. It strengthened the faith of our army and people. Moreover, it obliged the enemy to send reinforcements to Dien Bien Phu. It was the first miscarriage of Navarre's regrouping plan. Our troops began to encircle the fortified entrenched camp of Dien Bien Phu.

Liberation of Thakhek and several regions in Middle Laos

Parallel with the preparations to attack Lai Chau orders were given to the Viet Nam People's volunteers to cooperate with the Pathet Lao troops to launch an offensive on the Middle Laos front, where the enemy was relatively vulnerable. At the beginning of December, the enemy became aware of our activity, and quickly rushed reinforcements to this sector. On December 22, the Vietnamese and Laotian units carried by storm the post of Banaphao, a strong entrenched position which controlled the frontier. Other units struck deep into the enemy's rear. After a series of victories, the Vietnamese and Laotian units made very quick progress towards Thakhek, at the same time pursuing the enemy in his flight along Road No. 9.

Bewildered, the enemy withdrew from Thakhek to Seno, a military base near Savannakhet, losing on the way three battalions of infantry and one artillery unit. On December 27, the Pathet Lao units and the Viet Nam People's volunteers entered Thakhek, and reached the bank of the Mekong. The liberated zones were extended to Road No. 9.

This was the second important victory in the Winter-Spring campaign. To face our activity in time, the enemy

THE DIEN BIEN PHU CAMPAIGN
(The main front and the co-ordinated battlefronts)

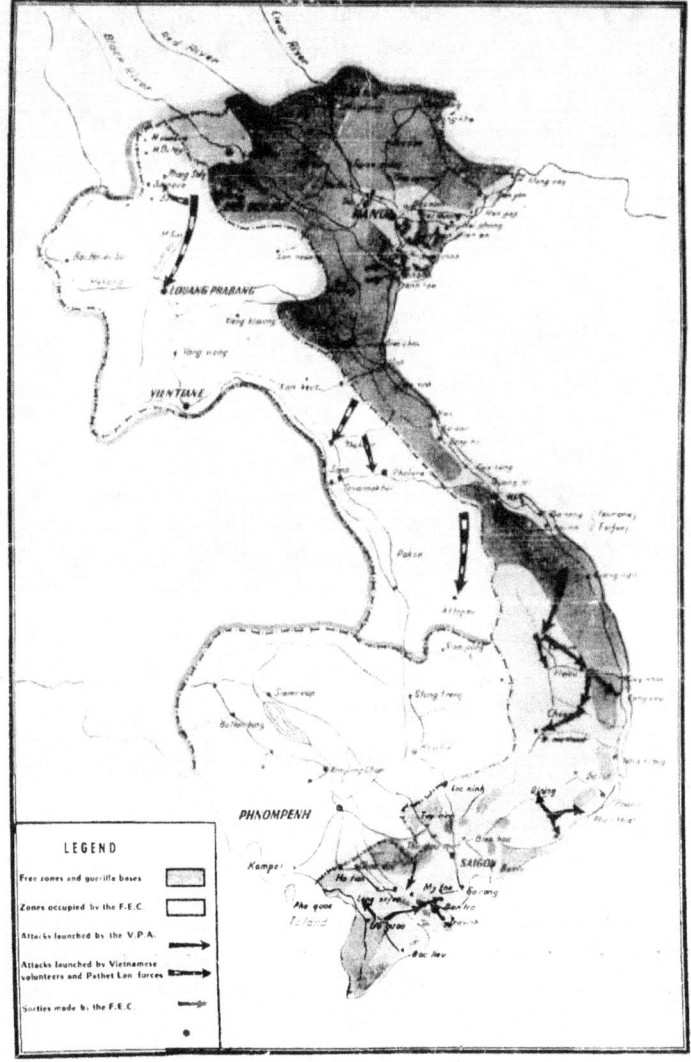

had to withdraw mobile forces from the Red River delta and from the South, to send them to Seno. To impede the Vietnamese and Laotian units in an advance into Lower Laos, they strengthened this base. Navarre was obliged to scatter his forces over several points.

Liberation of the Bolovens Highland and the town of Attopeu

Simultaneously with the attack on the Middle Laos front, one unit of the Laotian and Vietnamese forces crossed dangerous mountainous regions, and advanced into Lower Laos where it effected a junction with local armed forces.

On December 30 and 31, the Laotian and Vietnamese units defeated an enemy battalion in the region of Attopeu and liberated this town. Exploiting their victory, they advanced towards Saravane, and liberated the whole Bolovens Highland to south of Road No. 9. The enemy had to send reinforcements to Pakse.

Liberation of Kontum and the north of the Western Highlands of central Viet Nam

In spite of defeats at various points, the enemy remained subjective in making estimations. Due to the easy occupation of Dien Bien Phu, the enemy thought we were incapable of attacking it. According to them, the entrenched camp was too strong for our troops. Moreover they thought that the distance which separated it from our rear created insuperable obstacles for us in the supply of food. They thought we had passed to the attack at different points because we did not know how to deal with Dien Bien Phu; they thought that shortly, we should be obliged to evacuate the North-West because of supply difficulties; then they would find the means to destroy a part of our main forces and would continue execution of their plan: the occupation of Tuan Giao and Son La and the return to Na San.

It was this same subjective estimation which made them launch the Atlanta operation against the south of Phu Yen in the Fifth zone. This well-prepared attack was the first step in the occupation of our whole free zone in the south of central Viet Nam, as foreseen by the "Navarre plan".

Our strategic principle was: "dynamism, initiative." Our troops in the Fifth zone received the order to leave behind only a small part of their forces to cope with the enemy, while the bulk would continue their regroupment and pass to the offensive in the north of the Western Highlands. We opened the campaign on January 26. The following day, we took the Mandel sub-sector, the strongest sub-sector of the enemy. The post of Dakto was taken and we liberated the whole north of Kontum province. On February 17, we liberated the town of Kontum, wiped out the enemy in the whole north of the Western Highlands, and advanced as far as Road No. 19. Meanwhile, we attacked Pleiku. The enemy was at a loss, and had to stop the offensive in the coastal plains of the Fifth zone and withdraw many units from Middle Laos and the three Vietnamese provinces of Quang Binh, Quang Tri and Thua Thien to reinforce the Western Highlands.

This was another victory for our forces in the Winter-Spring campaign. It proved once more the correctness of the guiding principle of the Central Committee. The enemy was more and more obviously driven on to the defensive. They had to mobilise forces from the Red River delta to reinforce Middle Laos, and afterwards from Middle Laos to reinforce the Western Highlands. They had concentrated forces to make a lightning offensive against our Fifth zone but had to stop their action in order to protect themselves against our blows.

Our offensive on the Western Highlands was victoriously carried on till June 1954, and scored many more successes, particularly the resounding victory at An Khe where we cut to pieces the mobile regiment No. 100 which had just returned from Korea, thus liberating An Khe. Our troops captured in this battle a large number of vehicles and a great quantity of ammunition.

Liberation of Phong Saly and the Nam Hu river basin, the push forward towards Luang Prabang

Dien Bien Phu was encircled after the defeat of Lai Chau. The French High Command tried to effect junction between the Dien Bien Phu entrenched camp and Upper Laos by increasing their occupation forces along the Nam Hu river basin as far as Muong Khoa, intending to establish liaison with Dien Bien Phu.

To put them on the wrong track, to annihilate more of their forces, to weaken them more, and oblige them to continue to scatter their troops in order to create favourable conditions for our preparations at Dien Bien Phu, orders were given to our units to combine with the Pathet Lao forces to launch an offensive in the Nam Hu river basin.

On January 26, the Vietnamese and Laotian forces attacked Muong Khoa where they destroyed one European regiment; then, exploiting this success, they wiped out the enemy in the Nam Hu river basin, and came within striking distance of Luang Prabang, while one column pushed northward and liberated Phong Saly.

Before our strong offensive, the enemy had to withdraw mobile units from the Red River delta to send them to Upper Laos. Thus, Navarre was obliged to scatter his forces still further.

Our successes in the enemy rear in the Red River delta, the three provinces of Quang Binh, Quang Tri, Thua Thien, and Nam Bo

While the enemy was in difficulties on all fronts, our local armed forces, people's militia and guerillas effectively exploited the situation in the enemy rear and strongly combined activity with the front.

In the Red River delta, a series of enemy fortified camps was destroyed; and Road No. 5 was seriously threatened, being sometimes cut for weeks together. In two great attacks on Cat Bi (March 7, 1954) and Gia Lam airfields (March 8, 1954) our armymen destroyed 78 enemy planes.

At Binh-Tri-Thien, and in the southernmost part of central Viet Nam, our armymen's activity was intense; they expanded the guerilla bases, increased propaganda work directed to the enemy and won many successes.

In Nam Bo, through the whole winter-spring period, our armymen pushed forward their combined action, and obtained very great successes: more than one hundred enemy posts and watch-towers were either destroyed or evacuated, many localities were liberated, and the number of soldiers crossing to our side amounted to several thousands.

The development of hostilities until March 1954 showed that to a great extent the "Navarre plan" had collapsed. The enemy's plan to concentrate was essentially foiled. At this moment, enemy mobile forces were no longer concentrated in the Red River delta; they were scattered over several points: Luang Prabang and Muong Sai in Upper Laos, Seno in Middle Laos, the south of the Western Highlands in the Fifth zone, and large forces were pinned down at Dien Bien Phu. In the Red River delta, what was left of their mobile forces amounted to only 20 regiments, but a great part of these forces was no longer mobile and had to be scattered in order to protect the communication lines, particularly Road No. 5.

The situation of hostilities developed contrary to the enemy's will.

Navarre intended to concentrate his forces in the Red River delta with a view to recovering the initiative, but we obliged him to scatter his forces everywhere and passively take measures to protect himself.

He intended to annihilate a part of our main forces, but it was not our main forces but his that suffered heavy losses. He intended to attack our free zone, but instead his rear was severely attacked by us. Thus we threatened his whole system of disposition of forces.

However, the Franco-American generals did not want to recognise this disastrous truth. They still thought our activity in winter 1953-spring 1954 had reached its peak, that our withdrawal was beginning, that we lacked the strength to continue our activity, and that their favourable moment was approaching.

As a result, in order to get back the initiative, on March 12, the enemy resumed the Atlanta plan which had been interrupted, and opened an attack by landing at Quy Nhon.

Not for a moment did they believe that on the following day, March 13, 1954, we would launch a large-scale attack on the Dien Bien Phu entrenched camp. Thus, the historic Dien Bien Phu Campaign began.

IV — THE HISTORIC DIEN BIEN PHU CAMPAIGN

Dien Bien Phu is a large plain 18 kilometres long and six to eight kilometres wide in the mountainous zone of the North-West. It is the biggest and richest of the four plains in this hilly region close to the Viet Nam-Laos frontier. It is situated at the junction of important roads, running to the North-East towards Lai Chau, to the East and South-East towards Tuan Giao, Son La, Na San; to the West towards Luang Prabang and to the South towards Sam Neua. In the theatre of operations of Bac Bo and Upper Laos, Dien Bien Phu is a strategic position of first importance, capable of becoming an infantry and air base of extreme efficiency.

At the beginning there were at Dien Bien Phu only ten enemy battalions but they were gradually reinforced to cope with our offensive. When we launched the attack, the enemy forces totalled 17 battalions and 10 companies, comprising chiefly Europeans and Africans and units of highly-trained paratroops. Moreover the camp had three battalions of artillery, one battalion of sappers, one armoured company, a transport unit of 200 lorries and a permanent squadron of 12 aircraft. Altogether 16,200 men.

The forces were distributed in three sub-sectors which had to support one another and comprised 49 strong-points. Each had defensive autonomy, several were grouped in "complex defence centres" equipped with mobile forces and artillery, and surrounded by trenches and barbed wire, hundreds of metres thick. Each sub-sector comprised several strongly fortified defence centres.

But the most important was the Central sub-sector situated in the middle of the Muong Thanh village, the chief town of Dien Bien Phu. Two-thirds of the forces of the garrison were concentrated here. It had several connected defence centres protecting the command post, the artillery and commissariat bases, and at the same time the airfield. To the East, well-situated hills formed the most important defence system of the sub-sector, Dien Bien Phu was considered by the enemy to be an unassailable and impregnable fortress.

In fact, the central sub-sector did have rather strong forces, and the heights in the East could not be attacked easily. Besides, the artillery and armoured forces could break every attempt at intervention through the plain, a system of barbed wire and trenches would permit the enemy to decimate and repel any assault, and the mobile forces formed by the battalions of paratroops, whose action was combined with that of the defence centres, could counter-attack and break any offensive. The Northern sub-sector comprised the defence centres of Him Lam, Doc Lap and Ban Keo. The very strong positions of Him Lam and Doc Lap were required to check all attacks of our troops coming from Tuan Giao and Lai Chau.

As for the Southern sub-sector, also known as Hong Cum sub-sector, its purpose was to break any offensive coming from the South and to protect the communication way with Upper Laos.

Their artillery was divided between two bases: one at Muong Thanh, the other at Hong Cum, arranged in such a way as to support each other and to support all the surrounding strong-points.

Dien Bien Phu had two airfields: besides the main field at Muong Thanh, there was a reserve field at Hong Cum; they linked with Hanoi and Haiphong in an airlift which ensured 70 to 80 transports of supply daily.

The reconnaissance planes and fighters of the permanent squadron constantly flew over the entire region. The planes from the Gia Lam and Cat Bi airbases had the task of strafing and bombing our army. Navarre asserted that with such powerful forces and so strong a defence system, Dien Bien Phu was "an impregnable fortress..." The American general O'Daniel who paid a visit to the base shared this opinion. From this subjective point of view the enemy came to the conclusion that our troops had little chance in an attack on Dien Bien Phu. They even considered that an attack on our part would be a good opportunity for them to inflict a defeat on us.

On our side, after the liberation of Lai Chau, the attack upon Dien Bien Phu was on the agenda. We considered that the base, well entrenched as it was, also had

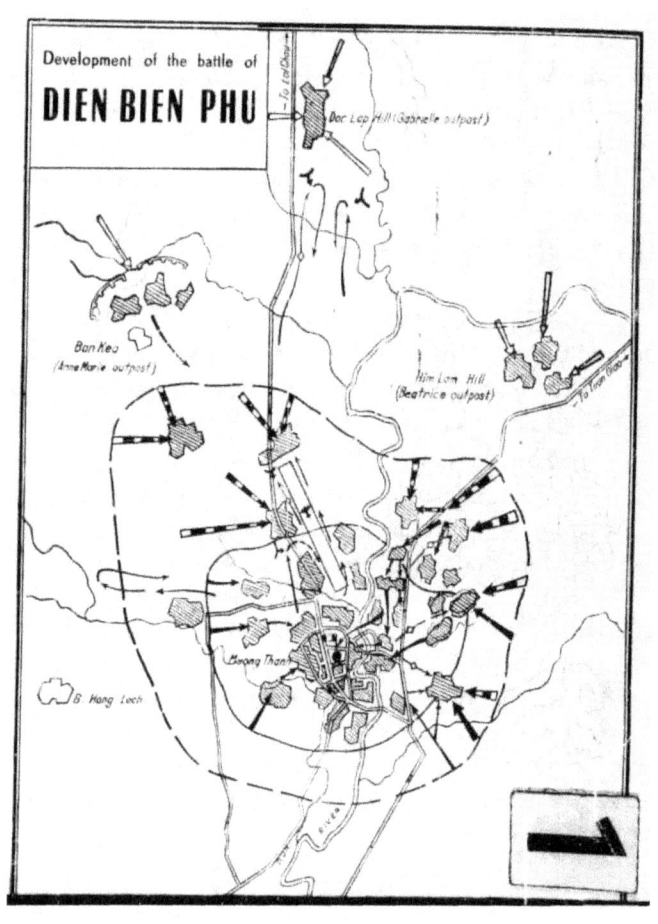

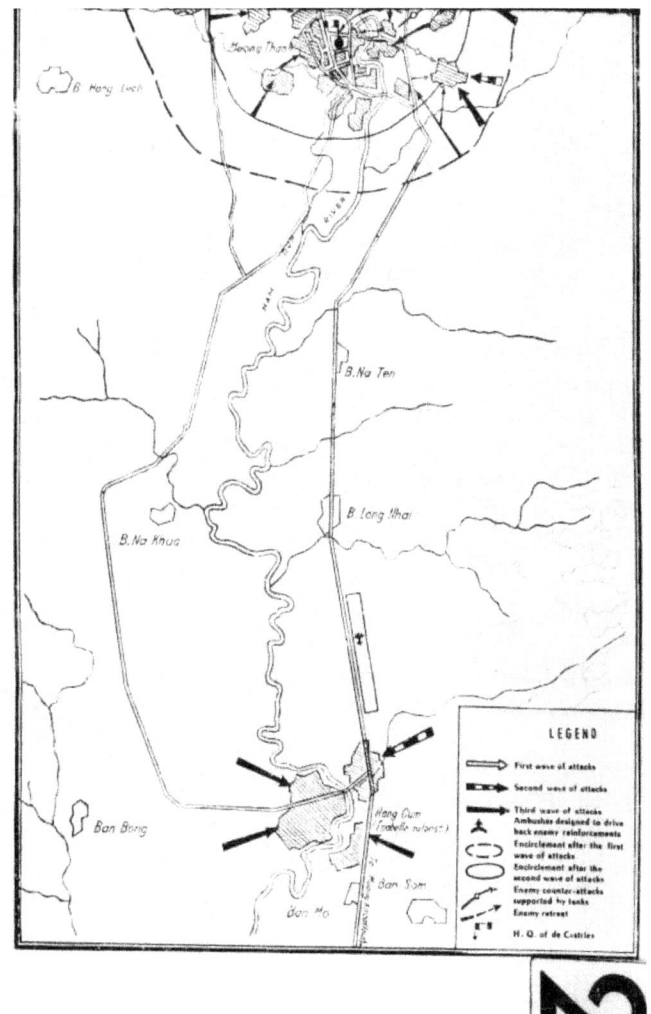

vulnerable points. In attacking it, we faced difficulties in strategy, tactics and supply; but these difficulties could be overcome. After having analysed the situation, and weighed the pros and cons, we decided to attack Dien Bien Phu according to the method: to take no risks. Our tactic would be to attack each enemy defence centre, each part of the entrenched camp, thus creating conditions for the launching of a general offensive to annihilate the whole camp.

Three months had passed from the occupation of Dien Bien Phu enemy paratroops to the launching of our campaign. During that time, the enemy did their utmost to consolidate their defence system, gather reinforcements, dig new trenches, and strengthen their entrenchments.

On our side, the army and people actively prepared the offensive, carrying out the orders of the Party's Central Committee and the Government; the army and people mustered all their strength to guarantee the success of the Winter-Spring campaign, to which Dien Bien Phu was the key. Our troops succeeded in liberating the surrounding regions, isolating Dien Bien Phu, obliging the enemy to scatter forces and thus reduce their possibilities of sending reinforcements to the battlefield. We made motor roads, cleared the tracks to haul up artillery pieces, built casemates for the artillery, prepared the ground for the offensive and encirclement; in short, transformed the relief of the battlefield terrain with a view to solving the tactical problems. We overcame very great difficulties. We called upon our local compatriots to supply food, to set up supply lines hundreds of kilometres long from Thanh Hoa or Phu Tho to the North-West, crossing very dangerous areas and very high hills. We used every means to carry food and ammunition to the front. Our troops and voluntary workers ceaselessly went to the front and actively participated in the preparations under the bombs and bullets of enemy aircraft.

In the first week of March, the preparations were completed: the artillery had solid casemates, the operational bases were established, food and ammunition were available in sufficient quantity. After having been educated in the aim and significance of the campaign, all officers and soldiers were filled with a very high determination to annihilate the enemy, as they were persuaded that only the destruction of the Dien Bien Phu entrenched camp would bring the "Navarre plan" to complete failure.

On March 13, 1954, our troops received the order to launch an offensive against Dien Bien Phu.

The campaign proceeded in three phases: in the first phase we destroyed the Northern sub-sector; in the second, the longest and bitterest one, we took the heights in the East of the Central sub-sector and tightened our encirclement; in the third, we launched the general offensive and annihilated the enemy.

First phase: destruction of Northern sub-sector

This phase began on March 13th and ended on March 17th. On the night of March 13th, we annihilated the very strong defence centre of Him Lam which overlooked the road from Tuan Giao to Dien Bien Phu. The battle was very sharp, the enemy artillery concentrated its fire, and poured scores of thousands of shells on our assaulting waves. Our troops carried the position in the night. This first victory had very deep repercussions on the development of the whole campaign.

In the night of the 14th, we concentrated our forces to attack the defence centre of Doc Lap, the second strong defence sector of the Northern sub sector which overlooked the road from Lai Chau to Dien Bien Phu. The battle went on till dawn. The enemy used every means to repel our forces, fired scores of thousands of shells and sent their mobile forces protected by tanks from Muong Thanh to support their position. Our troops fought heroically, took the strong-point and repelled the enemy reinforcements.

The third and last defence centre of the Northern sub-sector, the Ban Keo post, became isolated and was threatened by us. This was a less strong position, manned by a garrison chiefly made up of puppet soldiers. On March 17th, the whole garrison left its positions and surrendered. After the loss of the Northern sub-sector, the Central sub-sector, now exposed on its eastern, northern flanks, was threatened.

In the fighting in the first phase, the correctness of our tactical decisions, the good organisation of our defence and anti-aircraft activity reduced the efficiency of the enemy artillery and air force. Besides, our artillery fire, which was very accurate, inflicted heavy losses on the enemy. The main airfield was threatened. Our anti-aircraft batteries went into action for the first time and brought down enemy planes. But above all, it was by their heroic spirit, their high spirit of sacrifice and their will to win, that our troops distinguished themselves during these battles.

The great and resounding victory which ended the first phase of operations stirred our army and people and gave each and every one faith in final victory.

As for the enemy, despite their heavy losses, they still had confidence in the power of resistance of the Central sub-sector, in the strength of their artillery and air force. They even expected that we would suffer heavy losses and would be obliged to give up the offensive; and especially, that if the campaign was protected our supply lines would be cut and that the great logistic difficulties thus created would force us to withdraw.

Second phase: Occupation of the hills in the East and encirclement of the Central sub-sector

The second phase was the most important of the campaign. We had to deal with the Central sub-sector, in the middle of the Muong Thanh plain, and new difficulties arose in the conduct of the operations. Our troops had to work actively to complete the operations; they had to dig a vast network of trenches, from the neighbouring hills to the plain, to encircle the Central sub-sector and cut it off from the Southern sector. This advance of our lines which encircled the enemy positions was made at the cost of fierce fighting. By every means the enemy tried to upset our preparations by the fire of their air force and artillery. However, our troops drew closer to their positions with irresistible power in the course of uninterrupted fighting.

During the night of March 30th, the second phase began. We launched a large scale attack of long duration to annihilate the heights in the East and a certain number of strong points in the West in order to tighten our encirclement, and to hamper and cut off the supplies to the garrison. On this night of March 30th, we concentrated important forces to attack simultaneously the five fortified heights in the East. On this same night, we succeeded in capturing hills E-1, D-1 and C-1, but could not take hill

212

A-1, the most important of all. The defence line constituted by these heights was the key to the defensive system of the Central sub-sector: its loss would lead to the fall of Dien Bien Phu. Consequently, the fight here was at its fiercest. Particularly on hill A-1, the last height which protected the commant post, the battle lasted until April 4th. Every inch of ground was hotly disputed. Finally, we occupied half of the positsion while the enemy, entrenched in casemates and trenches, continued to resist in the other half. While this fighting was going on, the garrison received paratroop reinforcements.

On April 9th, the enemy launched a counter-attack to re-occupy hill C-1. The battle went on for four days and nights, and the position was occupied half by the enemy and half by us.

While the situation in the East was static, in the North and in the West our encirclement grew tighter and tighter. The lines of both sides drew nearer and nearer, in some points they were only 10 to 15 metres away from each other. From the occupied positions to the battlefields northward and westward, the fire of our artillery and mortars pounded the enemy without let-up. Day and night the fighting went on. We exhausted the enemy by harassing them, firing constantly at their lines, and at the same time tried to take their strong-points one by one with a tactic of combined nibbling advance and full scale attack.

In mid-April, after the destruction of several enemy positions in the North and West, our lines reached the airfield, then cut it from West to East. Our encirclement grew still tighter, the fighting was still more fierce. The enemy launched several violent counter-attacks supported by tanks and aircraft aimed at taking ground from us and obliging us to loosen our encirclement. On April 24th, the most violent counter-attack was launched with the aim of driving us off the airfield: after inflicting heavy losses on the enemy, we remained the master, and the airfield stayed in our control.

The territory occupied by the enemy shrank in size day by day, and they were driven into a two kilometres square. It was threatened by our heavy fire. The enemy's supply problem became more and more critical. The airfield had been out of action for a long time, all supplies were being droppped by parachute. But as the enemy zone was so narrow, and their pilots feared our anti-aircraft fire and dared not fly low, only a part of the parachutes carrying food and ammunition fell into the enemy position, and the bulk of them fell on our ground; thus we poured shells parachuted by the enemy on the entrenched camp.

Throughout the second phase, the situation was extremely tense. The American interventionists sent more bombers and transport planes to support the Dien Bien Phu base. The enemy bombers were very active; they ceaselessly bombed our positions, dropped napalm bombs to burn down the vegetation on the heights surrounding Dien Bien Phu, and bombed points that they took for our artillery bases. Day and night they shelled our supply lines, dropped blockbusters on the roads, showered the roads with delayed action and "butterfly" bombs, in an endeavour to cut our supply lines. These desperate efforts did not achieve the desired results. They could not check the flow of hundreds of thousands of voluntary workers, pack-horses and transport cars carrying food and ammunition to the front. They could not stop us from carrying out our plan of encirclement, the condition of their destruction.

The French and American generals clearly saw the danger of the destruction of the Dien Bien Phu entrenched camp.

At this moment, the High Command of the French Expeditionary Corps thought of gathering together the remaining forces for an attack on our rear and in the direction of the Viet Bac, to cut our supply lines and oblige us to withdraw for lack of food and ammunition. But it could not carry out this plan. Moreover, it feared that a still more severe defeat could be the result of so foolhardy an action. At another time it intended to regroup the Dien Bien Phu garrison in several columns which would try to break through our encirclement and open at all costs a way towards Upper Laos. Finally, it had to give up this plan and continue to defend its positions.

Third phase : Annihilation of the enemy

On May 1st, began the third phase. From May 1st to May 6th, following several successive attacks, we occupied hill C-1, hill A-1 which was the key of the last defensive system of the Central sub sector, and several other strongpoints from the foot of the hills in the East to the Nam Gion river, and, finally some positions in the West.

The enemy was driven into a square kilometre, entirely exposed to our fire. There was no fortified height to protect them. The problem of supply became very grave. Their situation was critical : the last hour of the entrenched camp had come.

In the afternoon of May 7th, from the East and West, we launched a massive combined attack upon the headquarters at Muong Thanh. At several posts, the enemy hoisted the white flag and surrendered. At 5.30 p.m. we seized the headquarters: General de Castries and his staff were captured.

The remaining forces at Dien Bien Phu surrendered. The prisoners of war were well treated by our troops.

The " Determined to fight and to win " banner of our army fluttered high in the valley of Dien Bien Phu. On this very night, we attacked the South sub-sector. The whole garrison of more than 2,000 men was captured.

The historic Dien Bien Phu campaign ended in our complete victory. Our troops had fought with an unprecedented heroism for 55 days and 55 nights.

During this time, our troops were very active in all theatres of operation in co-ordination with the main front.

In the enemy rear in the Red River delta, they destroyed, one after another, a large number of positions and seriously threatened road No. 5.

In the Fifth zone, they attacked road No. 19, annihilated the mobile regiment No. 100, liberated An Khe, penetrated deep into the region of Cheo Reo, and threatened Pleiku and Banmethuot.

Our troops were also very active in the region of Hue and in Nam Bo.

In Middle Laos, the Vietnamese and Laotian units increased their activity on road No. 9 and advanced southward.

Our troops won victories on all fronts.

* *

Such are the broad outlines of the military situation in Winter 1953 and Spring 1954.

On all fronts, we put out of action 112,000 enemy troops and brought down or destroyed on the ground 177 planes.

At Dien Bien Phu, we put out of action 16,200 enemy troops, including the whole staff of the entrenched camp, one general, 16 colonels, 1,749 officers and warrant-officers, brought down or destroyed 62 planes of all types, seized all the enemy's armaments, ammumition and equipment, and more than 30,000 parachutes.

These great victories of the Viet Nam People's Army and people as a whole at Dien Bien Phu and on the other fronts had smashed to pieces the "Navarre plan", and impeded the attempts of the Franco-American imperialists to prolong and extend the war. These great victories liberated the North of Viet Nam, contributed to the success of the Geneva Conference and the restoration of peace in Indo-China on the basis of recpect of the sovereignty, independence, national unity and territorial integrity of Viet Nam and of the two friendly countries, Cambodia and Laos.

These are glorious pages of our history, of our People's Army and our people. They illustrate the striking success of our Party in leading the movement for national liberation against the French imperialists and the American interventionists.

SITUATION OF THE BELLIGERENT FORCES FOLLOWING THE DIEN BIEN PHU BATTLE

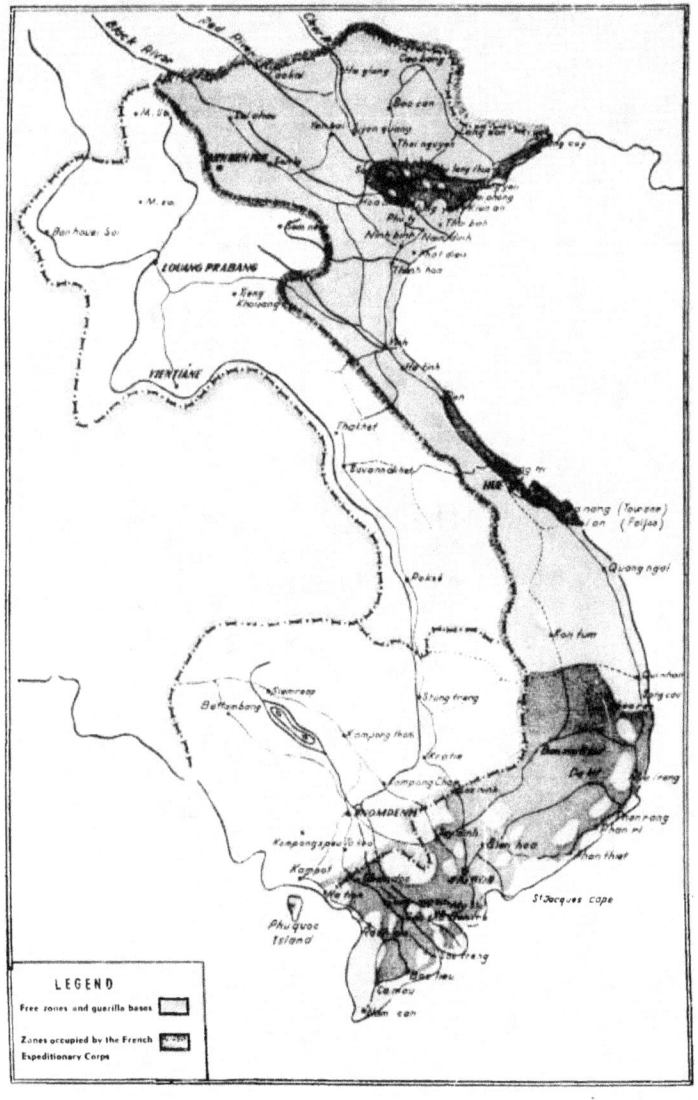

ERRATA

Page	Line	Instead of	Please read
18	25	war which	war, which
22	10	begining	beginning
33	18	war	was
34	12	national	nation
78	30	take part	takes part
80	8	battefields	battlefields
80	15	ont	out
80	22	serveral	several
85	19	difficuties	difficulties
110	14	wast	vast
135	9	revolutionmary	revolutionary
172	4	engineer	engineers and
172	22	possiblity	possibility
181	22	of duty	of its duty
199	22	cutting of	cutting off
203	10	riously	riously
207	21	sub sector,	sub-sector.
213	8	positsion	position
214	11	droppped	dropped
217	10	ammumition	ammunition
217	19	recpect	respect

PART 2

ON GUERRILLA WARFARE

MAO TSE TUNG

1937

CONTENTS

Introduction

I	The Nature of Revolutionary Guerrilla War	3
II	Profile of a Revolutionist	12
III	Strategy, Tactics, and Logistics in Revolutionary War	20
IV	Some Conclusions	27

Yu Chi Chan (Guerrilla Warfare)

	Translator's Note	37
	A Further Note	39
1	What Is Guerrilla Warfare?	41
2	The Relation of Guerrilla Hostilities to Regular Operations	51
3	Guerrilla Warfare in History	58
4	Can Victory Be Attained by Guerrilla Operations?	66
5	Organization for Guerrilla Warfare	71
	How Guerrilla Units Are Originally Formed	71
	The Method of Organizing Guerrilla Regimes	77
	Equipment of Guerrillas	82
	Elements of the Guerrilla Army	85
6	The Political Problems of Guerrilla Warfare	88
7	The Strategy of Guerrilla Resistance Against Japan	94
	Appendix	*116*

INTRODUCTION

I

THE NATURE OF REVOLUTIONARY GUERRILLA WAR

> ... the guerrilla campaigns being waged in China today are a page in history that has no precedent. Their influence will be confined not solely to China in her present anti-Japanese struggle, but will be world-wide.
> —MAO TSE-TUNG, *Yu Chi Chan*, 1937

AT ONE END OF THE SPECTRUM, ranks of electronic boxes buried deep in the earth hungrily consume data and spew out endless tapes. Scientists and engineers confer in air-conditioned offices; missiles are checked by intense men who move about them silently, almost reverently. In forty minutes, countdown begins.

At the other end of this spectrum, a tired man wearing a greasy felt hat, a tattered shirt, and soiled shorts is seated, his back against a tree. Barrel pressed between his knees, butt resting on the moist earth between sandaled feet, is a Browning automatic rifle. Hooked to his belt, two dirty canvas sacks—one holding three home-made bombs, the other four magazines loaded with .30-caliber ammunition. Draped around his neck, a sausage-like cloth tube with

Mao Tse-tung on Guerrilla Warfare

three days' supply of rice. The man stands, raises a water bottle to his lips, rinses his mouth, spits out the water. He looks about him carefully, corks the bottle, slaps the stock of the Browning three times, pauses, slaps it again twice, and disappears silently into the shadows. In forty minutes, his group of fifteen men will occupy a previously prepared ambush.

It is probable that guerrilla war, nationalist and revolutionary in nature, will flare up in one or more of half a dozen countries during the next few years. These outbreaks may not initially be inspired, organized, or led by local Communists; indeed, it is probable that they will not be. But they will receive the moral support and vocal encouragement of international Communism, and where circumstances permit, expert advice and material assistance as well.

As early as November, 1949, we had this assurance from China's Number Two Communist, Liu Shao-ch'i, when, speaking before the Australasian Trade Unions Conference in Peking, he prophesied that there would be other Asian revolutions that would follow the Chinese pattern. We paid no attention to this warning.

In December, 1960, delegates of eighty-one Communist and Workers' Parties resolved that the tempo of "wars of liberation" should be stepped up. A month later (January 6, 1961), the Soviet Premier, an unimpeachable authority on "national liberation wars," propounded an interesting series of questions to which he provided equally interesting answers:

Introduction

> Is there a likelihood of such wars recurring? Yes, there is. Are uprisings of this kind likely to recur? Yes, they are. But wars of this kind are popular uprisings. Is there the likelihood of conditions in other countries reaching the point where the cup of the popular patience overflows and they take to arms? Yes, there is such a likelihood. What is the attitude of the Marxists to such uprisings? A most favorable attitude. . . . These uprisings are directed against the corrupt reactionary regimes, against the colonialists. The Communists support just wars of this kind wholeheartedly and without reservations.*

Implicit is the further assurance that any popular movement infiltrated and captured by the Communists will develop an anti-Western character definitely tinged, in our own hemisphere at least, with a distinctive anti-American coloration.

This should not surprise us if we remember that several hundred millions less fortunate than we have arrived, perhaps reluctantly, at the conclusion that the Western peoples are dedicated to the perpetuation of the political, social, and economic *status quo*. In the not too distant past, many of these millions looked hopefully to America, Britain, or France for help in the realization of their justifiable aspirations. But today many of them feel that these aims can be achieved only by a desperate revolutionary struggle that we will probably oppose. This is not a hypothesis; it is fact.

A potential revolutionary situation exists in any country where the government consistently fails in its obligation to ensure at least a minimally decent standard of life for the

---

* *World Marxist Review*, January, 1961.

great majority of its citizens. If there also exists even the nucleus of a revolutionary party able to supply doctrine and organization, only one ingredient is needed: the instrument for violent revolutionary action.

In many countries, there are but two classes, the rich and the miserably poor. In these countries, the relatively small middle class—merchants, bankers, doctors, lawyers, engineers—lacks forceful leadership, is fragmented by unceasing factional quarrels, and is politically ineffective. Its program, which usually posits a socialized society and some form of liberal parliamentary democracy, is anathema to the exclusive and tightly knit possessing minority. It is also rejected by the frustrated intellectual youth, who move irrevocably toward violent revolution. To the illiterate and destitute, it represents a package of promises that experience tells them will never be fulfilled.

People who live at subsistence level want first things to be put first. They are not particularly interested in freedom of religion, freedom of the press, free enterprise as we understand it, or the secret ballot. Their needs are more basic: land, tools, fertilizers, something better than rags for their children, houses to replace their shacks, freedom from police oppression, medical attention, primary schools. Those who have known only poverty have begun to wonder why they should continue to wait passively for improvements. They see—and not always through Red-tinted glasses—examples of peoples who have changed the structure of their societies, and they ask, "What have we to lose?" When a great many people begin to ask themselves this question, a revolutionary guerrilla situation is incipient.

Introduction

A revolutionary war is never confined within the bounds of military action. Because its purpose is to destroy an existing society and its institutions and to replace them with a completely new state structure, any revolutionary war is a unity of which the constituent parts, in varying importance, are military, political, economic, social, and psychological. For this reason, it is endowed with a dynamic quality and a dimension in depth that orthodox wars, whatever their scale, lack. This is particularly true of revolutionary guerrilla war, which is not susceptible to the type of superficial military treatment frequently advocated by antediluvian doctrinaires.

It is often said that guerrilla warfare is primitive. This generalization is dangerously misleading and true only in the technological sense. If one considers the picture as a whole, a paradox is immediately apparent, and the primitive form is understood to be in fact more sophisticated than nuclear war or atomic war or war as it was waged by conventional armies, navies, and air forces. Guerrilla war is not dependent for success on the efficient operation of complex mechanical devices, highly organized logistical systems, or the accuracy of electronic computers. It can be conducted in any terrain, in any climate, in any weather; in swamps, in mountains, in farmed fields. Its basic element is man, and man is more complex than any of his machines. He is endowed with intelligence, emotions, and will. Guerrilla warfare is therefore suffused with, and reflects, man's admirable qualities as well as his less pleasant ones. While it is not always humane, it is human, which is more than can be said for the strategy of extinction.

Mao Tse-tung on Guerrilla Warfare

In the United States, we go to considerable trouble to keep soldiers out of politics, and even more to keep politics out of soldiers. Guerrillas do exactly the opposite. They go to great lengths to make sure that their men are politically educated and thoroughly aware of the issues at stake. A trained and disciplined guerrilla is much more than a patriotic peasant, workman, or student armed with an antiquated fowling-piece and a home-made bomb. His indoctrination begins even before he is taught to shoot accurately, and it is unceasing. The end product is an intensely loyal and politically alert fighting man.

Guerrilla leaders spend a great deal more time in organization, instruction, agitation, and propaganda work than they do fighting, for their most important job is to win over the people. "We must patiently explain," says Mao Tse-tung. "Explain," "persuade," "discuss," "convince"—these words recur with monotonous regularity in many of the early Chinese essays on guerrilla war. Mao has aptly compared guerrillas to fish, and the people to the water in which they swim. If the political temperature is right, the fish, however few in number, will thrive and proliferate. It is therefore the principal concern of all guerrilla leaders to get the water to the right temperature and to keep it there.

More than ten years ago, I concluded an analysis of guerrilla warfare with the suggestion that the problem urgently demanded further "serious study of all historical experience." Although a wealth of material existed then, and much more has since been developed, no such study

Introduction

has yet been undertaken in this country, so far as I am aware. In Indochina and Cuba, Ho Chi Minh and Ernesto (Che) Guevara were more assiduous. One rather interesting result of their successful activities has been the common identification of guerrilla warfare with Communism. But guerrilla warfare was not invented by the Communists; for centuries, there have been guerrilla fighters.

One of the most accomplished of them all was our own Revolutionary hero Francis Marion, "the Swamp Fox." Those present at his birth would probably not have foretold a martial future for him; the baby was "not larger than a New England lobster and might easily enough have been put into a quart pot." Marion grew up in South Carolina and had little formal schooling. He worked as a farmer. In 1759, at the age of twenty-seven, he joined a regiment raised to fight the Cherokees, who were then ravaging the borders of the Carolinas. He served for two years and in the course of these hostilities stored away in his mind much that was later to be put to good use against the British.

When the Revolution broke out, Marion immediately accepted a commission in the Second South Carolina Regiment. By 1780, he had seen enough of the war to realize that the Continentals were overlooking a very profitable field—that of partisan warfare. Accordingly, he sought and obtained permission to organize a company that at first consisted of twenty ill-equipped men and boys (Castro's "base" was twelve men). The appearance of this group, with a heterogeneous assortment of arms and ragged and poorly fitting clothes, provoked considerable jesting among the regulars of General Gates, but Marion's men were not long in proving that the appearance of a combat soldier is not necessarily a reliable criterion of his fighting abilities.

Marion's guerrilla activities in South Carolina soon told heavily on the British, especially Cornwallis, whose plans were continually disrupted by them. Marion's tactics were those of all successful guerrillas. Operating with the greatest speed from inaccessible bases, which he changed frequently, he struck his blows in rapid succession at isolated garrisons, convoys, and trains. His information was always timely and accurate, for the people supported him.

The British, unable to cope with Marion, branded him a criminal, and complained bitterly that he fought neither "like a gentleman" nor like "a Christian," a charge orthodox soldiers are wont to apply in all lands and in all wars to such ubiquitous, intangible, and deadly antagonists as Francis Marion.*

However, the first example of guerrilla operations on a grand scale was in Spain between 1808 and 1813. The Spaniards who fled from Napoleon's invading army to the

---

* Bryant, in the "Song of Marion's Men," wrote some lines that showed that he had a better understanding of guerrilla tactics and psychology than many who have followed more martial pursuits:

> Woe to the English soldiery,
> That little dreads us near!
> On them shall come at midnight
> A strange and sudden fear;
> When, waking to their tents on fire,
> They grasp their arms in vain,
> And they who stand to face us
> Are beat to earth again;
> And they who fly in terror deem
> A mighty host behind,
> And hear the tramp of thousands
> Upon the hollow wind.

Introduction

mountains were patriots loyal to the ruler whose crown had been taken from him by the Emperor of the French. They were not revolutionists. Most did not desire a change in the form of their government. Their single objective was to help Wellington force the French armies to leave Spain.

A few years later, thousands of Russian Cossacks and peasants harried Napoleon's Grande Armée as Kutuzov pushed it, stumbling, starving and freezing, down the ice-covered road to Smolensk. This dying army felt again and again the cudgel of the people's war, which, as Tolstoi later wrote, "was raised in all its menacing and majestic power; and troubling itself about no question of anyone's tastes or rules, about no fine distinctions, with stupid simplicity, with perfect consistency, it rose and fell and belabored the French until the whole invading army had been driven out."

A little more than a century and a quarter later, Hitler's armies fell back along the Smolensk road. They too would feel the fury of an aroused people. But in neither case were those who wielded the cudgel revolutionists. They were patriotic Russians.

Only when Lenin came on the scene did guerrilla warfare receive the potent political injection that was to alter its character radically. But it remained for Mao Tse-tung to produce the first systematic study of the subject, almost twenty-five years ago. His study, now endowed with the authority that deservedly accrues to the works of the man who led the most radical revolution in history, will continue to have a decisive effect in societies ready for change.

II

PROFILE OF A REVOLUTIONIST

> Political power comes out of the barrel of a gun.
> —Mao Tse-tung, 1938

Mao tse-tung, the man who was to don the mantle of Lenin, was born in Hunan Province, in central China, in 1893. His father, an industrious farmer, had managed to acquire several acres, and with this land, the status of a "middle" peasant. He was a strict disciplinarian, and Mao's youth was not a happy one. The boy was in constant conflict with his father but found an ally in his mother, whose "indirect tactics" (as he once described her methods of coping with her husband) appealed to him. But the father gave his rebellious son educational opportunities that only a tiny minority of Chinese were then able to enjoy. Mao's primary and secondary schooling was thorough. His literary taste was catholic; while a pupil at the provincial normal school he read omnivorously. His indiscriminate diet included Chinese philosophy, poetry, history, and romances as well as translations of many Western historians, novelists, and biographers. However, history and political sciences par-

Introduction

ticularly appealed to him; in them, he sought, but without success, the key to the future of China.

His studies had led him to reject both democratic liberalism and parliamentary socialism as unsuited to his country. Time, he realized, was running out for China. History would not accord her the privilege of gradual political, social, and economic change, of a relatively painless and orderly evolution. To survive in the power jungle, China had to change, to change radically, to change fast. But how?

Shortly after graduating from normal school, in 1917, Mao accepted a position as assistant in the Peking University library. Here he associated himself with the Marxist study groups set up by Li Ta-chao and Ch'en Tu-hsiu; here he discovered Lenin, read his essays, pored over Trotsky's explosive speeches, and began to study Marx and Engels. By 1920, Mao was a convinced Communist and a man who had discovered his mission: to create a new China according to the doctrine of Marx and Lenin. When the CCP was organized in Shanghai, in 1921, Mao joined.

The China Mao decided to change was not a nation in the accepted sense of the word. Culturally, China was, of course, homogeneous; politically and economically, China was chaos. The peasants, 400 million of them, lived from day to day at subsistence level. Tens of millions of peasant families owned no land at all. Other millions cultivated tiny holdings from which they scraped out just enough food to sustain life.

The peasant was fair game for everyone. He was pillaged by tax collectors, robbed by landlords and usurers, at the mercy of rapacious soldiery and bandits, afflicted by blights, droughts, floods, and epidemics. His single stark problem was simply to survive. The tough ones did. The others slowly starved, died of disease, and in the fierce winters of North China and Manchuria, froze to death.

It is difficult for an American today to conceive tens of thousands of small communities in which no public services existed, in which there were no doctors, no schools, no running water, no electricity, no paved streets, and no sewage disposal. The inhabitants of these communities were with few exceptions illiterate; they lived in constant fear of army press gangs and of provincial officials who called them out summer and winter alike to work on military roads and dikes. The Chinese peasant, in his own expressive idiom, "ate bitterness" from the time he could walk until he was laid to rest in the burial plot beneath the cypress trees. This was feudal China. Dormant within this society were the ingredients that were soon to blow it to pieces.

An external factor had for almost a century contributed to the chaos of China: the unrelenting pressure and greed of foreign powers. French, British, Germans, and Russians vied with one another in exacting from a succession of corrupt and feeble governments commercial, juridical, and financial concessions that had, in fact, turned China into an international colony. (The American record in these respects was a reasonably good one.) Mao once described the China he knew in his youth as "semicolonial and feudal." He was right.

Introduction

Shortly after Chiang Kai-shek took command of the National Revolutionary Army, in 1926, Mao went to Hunan to stir up the peasants. The campaign he waged for land reform in his native province can be described as almost a one-man show. The fundamental requisite in China was then, as it had long been, to solve the land question. Reduced to elementary terms, the problem was how to get rid of the gentry landowners who fastened themselves to the peasants like leeches and whose exactions kept the people constantly impoverished. In the circumstances, there was only one way to accomplish this necessary reform: expropriation and redistribution of the land. Naturally, the Nationalists, eager to retain the support of the gentry (historically the stabilizing element in Chinese society), considered such a radical solution social dynamite. But in Mao's view, there could be no meaningful revolution unless and until the power of this class had been completely eliminated.

While Mao was making himself extremely unpopular with the landed gentry in Hunan, the revolutionary armies of the Kuomintang were marching north from Canton to Wuhan, on the Yangtze, where a Nationalist Government was established in December, 1926. These armies incorporated a number of Communist elements. But by the time the vanguard divisions of Chiang's army reached the outskirts of Shanghai, in March, 1927, the honeymoon was almost over. In April, Chiang's secret police captured and executed the radical labor leaders in Shanghai and began to purge the army of its Communist elements. In the

Mao Tse-tung on Guerrilla Warfare

meantime the left-wing government in Wuhan had broken up. The Communists walked out; the Soviet advisers packed their bags and started for home.

During this period, the Communists were having their own troubles, and these were serious. The movement was literally on the verge of extinction. Those who managed to escape Chiang's secret police had fled to the south and assembled at Ching Kang Shan, a rugged area in the Fukien-Kiangsi borderlands. One of the first to reach this haven was the agrarian agitator from Hunan. As various groups drifted in to the mountain stronghold, Mao and Chu Teh (who had arrived in April, 1928) began to mold an army. Several local bandit chieftains were induced to join the Communists, whose operations gradually became more extensive. Principally these activities were of a propaganda nature. District soviets were established; landlords were dispossessed; wealthy merchants were "asked" to make patriotic contributions. Gradually, the territory under Red control expanded, and from a temporarily secure base area, operations commenced against provincial troops who were supposed to suppress the Reds.

In the early summer of 1930, an ominous directive was received at Ching Kang Shan from the Central Committee of the Party, then dominated by Li Li-san. This directive required the Communist armies to take the offensive against cities held by the Nationalists. The campaigns that followed were not entirely successful and culminated in a serious Communist defeat at Changsha in September. On the thirteenth of that month, the single most vital decision

Introduction

in the history of the Chinese Communist Party was taken; the ultimate responsibility for it rested equally on the shoulders of Mao and Chu Teh. These two agreed that the only hope for the movement was to abandon immediately the line laid down by Moscow in favor of one of Mao's own devising. Basically the conflict that split the Chinese Communist Party wide open and alienated the traditionalists in Moscow revolved about this question: Was the Chinese revolution to be based on the industrial proletariat—as Marxist dogma prescribed—or was it to be based on the peasant? Mao, who knew and trusted the peasants, and had correctly gauged their revolutionary potential, was convinced that the Chinese urban proletariat were too few in number and too apathetic to make a revolution. This decision, which drastically reoriented the policy of the Chinese Communist Party, was thereafter to be carried out with vigorous consistency. History has proved that Mao was right, Moscow wrong. And it is for this reason that the doctrine of Kremlin infallibility is so frequently challenged by Peking.

In October, 1930, the Generalissimo, in the misguided belief that he could crush the Communists with no difficulty, announced with great fanfare a "Bandit Suppression Campaign." This was launched in December. How weak the Nationalists really were was now to become apparent. The campaign was a complete flop. Government troops ran away or surrendered to the Communists by platoons, by companies, by battalions. Three more Suppression Campaigns, all failures, followed this fiasco. Finally, in 1933, the Generalissimo reluctantly decided to adopt the plans of his German advisers and to commit well-equipped, well-trained, and loyal "Central" divisions to a coordinated and methodical compression of the Communist-controlled area. As the Nationalists inched southward, supported by artillery and aviation, they evacuated peasants from every village and town and constructed hundreds of mutually supporting wired-in blockhouses. The Communists, isolated from the support of the peasants they had laboriously converted, found themselves for the first time almost completely deprived of food and information. Chiang's troops were slowly strangling the Communists. For the first time, Communist morale sagged. It was in this context that the bold decision to shift the base to Shensi Province was taken, and the now celebrated march of almost 6,000 miles was begun.

This was indeed one of the fateful migrations of history: its purpose, to preserve the military power of the Communist Party. How many pitched battles and skirmishes the Reds fought during this epic trek cannot now be established. It is known, however, that for days on end their columns were under air attack. They crossed innumerable mountains and rivers and endured both tropical and subarctic climates. As they marched toward the borders of Tibet and swung north, they sprinkled the route with cadres and caches of arms and ammunition.

The Reds faced many critical situations, but they were tough and determined. Every natural obstacle, and there were many, was overcome. Chiang's provincial troops, ineffective as usual, were unable to bar the way, and the

Introduction

exhausted remnants of the Reds eventually found shelter in the loess caves of Pao An.

Later, after the base was shifted to Yenan, Mao had time to reflect on his experiences and to derive from them the theory and doctrine of revolutionary guerrilla war which he embodied in *Yu Chi Chan.*

III

STRATEGY, TACTICS, AND LOGISTICS IN REVOLUTIONARY WAR

> The first law of war is to preserve ourselves and destroy the enemy.
> —MAO TSE-TUNG, 1937

MAO HAS NEVER CLAIMED that guerrilla action alone is decisive in a struggle for political control of the state, but only that it is a possible, natural, and necessary development in an agrarian-based revolutionary war.

Mao conceived this type of war as passing through a series of merging phases, the first of which is devoted to organization, consolidation, and preservation of regional base areas situated in isolated and difficult terrain. Here volunteers are trained and indoctrinated, and from here, agitators and propagandists set forth, individually or in groups of two or three, to "persuade" and "convince" the inhabitants of the surrounding countryside and to enlist their support. In effect, there is thus woven about each base a protective belt of sympathizers willing to supply

Introduction

food, recruits, and information. The pattern of the process is conspiratorial, clandestine, methodical, and progressive. Military operations will be sporadic.

In the next phase, direct action assumes an ever-increasing importance. Acts of sabotage and terrorism multiply; collaborationists and "reactionary elements" are liquidated. Attacks are made on vulnerable military and police outposts; weak columns are ambushed. The primary purpose of these operations is to procure arms, ammunition, and other essential material, particularly medical supplies and radios. As the growing guerrilla force becomes better equipped and its capabilities improve, political agents proceed with indoctrination of the inhabitants of peripheral districts soon to be absorbed into the expanding "liberated" area.

One of the primary objectives during the first phases is to persuade as many people as possible to commit themselves to the movement, so that it gradually acquires the quality of "mass." Local "home guards" or militia are formed. The militia is not primarily designed to be a mobile fighting force; it is a "back-up" for the better-trained and better-equipped guerrillas. The home guards form an indoctrinated and partially trained reserve. They function as vigilantes. They collect information, force merchants to make "voluntary" contributions, kidnap particularly obnoxious local landlords, and liquidate informers and collaborators. Their function is to protect the revolution.

Following Phase I (organization, consolidation, and preservation) and Phase II (progressive expansion) comes Phase III: decision, or destruction of the enemy. It is dur-

Mao Tse-tung on Guerrilla Warfare

ing this period that a significant percentage of the active guerrilla force completes its transformation into an orthodox establishment capable of engaging the enemy in conventional battle. This phase may be protracted by "negotiations." Such negotiations are not originated by revolutionists for the purpose of arriving at amicable arrangements with the opposition. Revolutions rarely compromise; compromises are made only to further the strategic design. Negotiation, then, is undertaken for the dual purpose of gaining time to buttress a position (military, political, social, economic) and to wear down, frustrate, and harass the opponent. Few, if any, essential concessions are to be expected from the revolutionary side, whose aim is only to create conditions that will preserve the unity of the strategic line and guarantee the development of a "victorious situation."

Intelligence is the decisive factor in planning guerrilla operations. Where is the enemy? In what strength? What does he propose to do? What is the state of his equipment, his supply, his morale? Are his leaders intelligent, bold, and imaginative or stupid and impetuous? Are his troops tough, efficient, and well disciplined, or poorly trained and soft? Guerrillas expect the members of their intelligence service to provide the answers to these and dozens more detailed questions.

Guerrilla intelligence nets are tightly organized and pervasive. In a guerrilla area, every person without exception must be considered an agent—old men and women, boys driving ox carts, girls tending goats, farm laborers, storekeepers, schoolteachers, priests, boatmen, scavengers.

Introduction

The local cadres "put the heat" on everyone, without regard to age or sex, to produce all conceivable information. And produce it they do.

As a corollary, guerrillas deny all information of themselves to their enemy, who is enveloped in an impenetrable fog. Total inability to get information was a constant complaint of the Nationalists during the first four Suppression Campaigns, as it was later of the Japanese in China and of the French in both Indochina and Algeria. This is a characteristic feature of all guerrilla wars. The enemy stands as on a lighted stage; from the darkness around him, thousands of unseen eyes intently study his every move, his every gesture. When he strikes out, he hits the air; his antagonists are insubstantial, as intangible as fleeting shadows in the moonlight.

Because of superior information, guerrillas always engage under conditions of their own choosing; because of superior knowledge of terrain, they are able to use it to their advantage and the enemy's discomfiture. Guerrillas fight only when the chances of victory are weighted heavily in their favor; if the tide of battle unexpectedly flows against them, they withdraw. They rely on imaginative leadership, distraction, surprise, and mobility to create a victorious situation before battle is joined. The enemy is deceived and again deceived. Attacks are sudden, sharp, vicious, and of short duration. Many are harassing in nature; others designed to dislocate the enemy's plans and to agitate and confuse his commanders. The mind of the enemy and the will of his leaders is a target of far more importance than the bodies of his troops. Mao once remarked, not entirely facetiously, that guerrillas must be expert at running away since they do it so often. They avoid static dispositions; their effort is always to keep the situation as fluid as possible, to strike where and when the enemy least expects them. Only in this way can they retain the initiative and so be assured of freedom of action. Usually designed to lure the enemy into a baited trap, to confuse his leadership, or to distract his attention from an area in which a more decisive blow is imminent, "running away" is thus, paradoxically, offensive.

Guerrilla operations conducted over a wide region are necessarily decentralized. Each regional commander must be familiar with local conditions and take advantage of local opportunities. The same applies to commands in subordinate districts. This decentralization is to some extent forced upon guerrillas because they ordinarily lack a well-developed system of technical communications. But at the same time, decentralization for normal operations has many advantages, particularly if local leaders are ingenious and bold.

The enemy's rear is the guerrillas' front; they themselves have no rear. Their logistical problems are solved in a direct and elementary fashion: The enemy is the principal source of weapons, equipment, and ammunition.

Mao once said:

> We have a claim on the output of the arsenals of London as well as of Hanyang, and what is more, it is to be delivered to us by the enemy's own transport corps. This is the sober truth, not a joke.

Introduction

If it is a joke, it is a macabre one as far as American taxpayers are concerned. Defectors to the Communists from Chiang Kai-shek's American-equipped divisions were numbered in the tens of thousands. When they surrendered, they turned in mountains of American-made individual arms, jeeps, tanks, guns, bazookas, mortars, radios, and automatic weapons.

It is interesting to examine Mao's strategical and tactical theories in the light of his principle of "unity of opposites." This seems to be an adaptation to military action of the ancient Chinese philosophical concept of *Yin-Yang*. Briefly, the *Yin* and the *Yang* are elemental and pervasive. Of opposite polarities, they represent female and male, dark and light, cold and heat, recession and aggression. Their reciprocal interaction is endless. In terms of the dialectic, they may be likened to the thesis and antithesis from which the synthesis is derived.

An important postulate of the *Yin-Yang* theory is that concealed within strength there is weakness, and within weakness, strength. It is a weakness of guerrillas that they operate in small groups that can be wiped out in a matter of minutes. But because they do operate in small groups, they can move rapidly and secretly into the vulnerable rear of the enemy.

In conventional tactics, dispersion of forces invites destruction; in guerrilla war, this very tactic is desirable both to confuse the enemy and to preserve the illusion that the guerrillas are ubiquitous.

It is often a disadvantage not to have heavy infantry

weapons available, but the very fact of having to transport them has until recently tied conventional columns to roads and well-used tracks. The guerrilla travels light and travels fast. He turns the hazards of terrain to his advantage and makes an ally of tropical rains, heavy snow, intense heat, and freezing cold. Long night marches are difficult and dangerous, but the darkness shields his approach to an unsuspecting enemy.

In every apparent disadvantage, some advantage is to be found. The converse is equally true: In each apparent advantage lie the seeds of disadvantage. The *Yin* is not wholly *Yin*, nor the *Yang* wholly *Yang*. It is only the wise general, said the ancient Chinese military philosopher Sun Tzu, who is able to recognize this fact and to turn it to good account.

Guerrilla tactical doctrine may be summarized in four Chinese characters pronounced "*Sheng Tung, Chi Hsi*," which mean "Uproar [in the] East; Strike [in the] West." Here we find expressed the all-important principles of distraction on the one hand and concentration on the other; to fix the enemy's attention and to strike where and when he least anticipates the blow.

Guerrillas are masters of the arts of simulation and dissimulation; they create pretenses and simultaneously disguise or conceal their true semblance. Their tactical concepts, dynamic and flexible, are not cut to any particular pattern. But Mao's first law of war, to preserve oneself and destroy the enemy, is always governing.

IV

SOME CONCLUSIONS

> Historical experience is written in blood and iron.
> —Mao Tse-tung, 1937

THE FUNDAMENTAL DIFFERENCE between patriotic partisan resistance and revolutionary guerrilla movements is that the first usually lacks the ideological content that always distinguishes the second.

A resistance is characterized by the quality of spontaneity; it begins and then is organized. A revolutionary guerrilla movement is organized and then begins.

A resistance is rarely liquidated and terminates when the invader is ejected; a revolutionary movement terminates only when it has succeeded in displacing the incumbent government or is liquidated.

Historical experience suggests that there is very little hope of destroying a revolutionary guerrilla movement *after it has survived the first phase and has acquired the sympathetic support of a significant segment of the population*. The size of this "significant segment" will vary; a decisive figure might range from 15 to 25 per cent.

In addition to an appealing program and popular support, such factors as terrain; communications; the quality

Mao Tse-tung on Guerrilla Warfare

of the opposing leadership; the presence or absence of material help, technical aid, advisers, or "volunteers" from outside sources; the availability of a sanctuary; the relative military efficiency and the political flexibility of the incumbent government are naturally relevant to the ability of a movement to survive and expand.

In specific aspects, revolutionary guerrilla situations will of course differ, but if the Castro movement, for example, had been objectively analyzed in the light of the factors suggested during the latter period of its first phase, a rough "expectation of survival and growth" might have looked something like Figure I.

Had an impartial analyst applied such criteria to Vietnam six to eight months before the final debacle, he might have produced a chart somewhat like Figure II.

Here Determinants A, B, H, and I definitely favored the guerrillas, who also (unlike Castro) had an available sanctuary. Two others, C and F, might have been considered in balance. Although the Vietminh had demonstrated superior tactical ability in guerrilla situations, an experienced observer might have been justified in considering "military efficiency" equal; the French were learning.

While other determinants may no doubt be adduced, those used are, I believe, valid so far as they go, and the box scores indicative. These show that Castro's chances of success might have been estimated as approximately three to two, Ho Chi Minh's as approximately four to three.

These analyses may be criticized as having been formulated after the event; it is, however, my belief that the outcome in Cuba and Indochina could have been pre-

FIGURE I. THE REVOLUTIONARY GUERRILLA SITUATION IN CUBA

Determinants*	Castro	Incumbent (Batista)	Remarks
A. Appeal of program	Progressive, plus (8)	Static, minus (3)	Batista government oppressive and reactionary
B. Popular support	Growing, active (7)	Diminishing, passive (3)	
C. Quality of leadership	Excellent, dedicated (8)	Mediocre to poor (4)	
D. Quality of troops	Good, improving to excellent (8)	Good, decreasing to fair (5)	
E. Military efficiency	Growing (6)	Mediocre to poor (4)	In guerrilla situations
F. Internal unity	Positive, strong (8)	Weak (3)	
G. Equipment	Poor, improving to good as taken (4)	Largely U.S., excellent (8)	Radios, transport, medical supplies, etc., available from incumbent
H. Base area terrain	Operationally favorable (10)	Unfavorable (3)	
I. Base area communications	Operationally favorable (10)	Unfavorable (3)	
J. Sanctuary	None (0)	Remainder of island (10)	Available for rest, retraining, equipment
AGGREGATE	69	46	

*Determinants are arbitrarily weighted on a scale of 0-10.

Figure II. The Revolutionary Guerrilla Situation in Vietnam

Determinants*	Ho Chi Minh	Incumbent (French)	Remarks
A. Appeal of program	Dynamic (7)	No program (0)	
B. Popular support	Growing (7)	Diminishing, slight (3)	
C. Quality of leadership	Good (7)	Good (7)	
D. Quality of troops	Good, improving (6)	Very good (7)	
E. Military efficiency	Very good (8)	Good (6)	In guerrilla situations
F. Internal unity	Excellent (8)	Excellent (8)	
G. Equipment	Fair but improving (7)	Generally well equipped (9)	Received from China and taken from French
H. Operational terrain	Favorable (10)	Unfavorable (5)	
I. Operational area communications	Favorable (10)	Unfavorable (5)	
J. Sanctuary	Available in China (8)	Remainder of Indochina (10)	
AGGREGATE	78	60	

* Determinants are arbitrarily weighted on a scale of 0–10.

Introduction

dicted some time before the respective movements had emerged from the stage of organization and consolidation—Phase I.

At the present time, much attention is being devoted to the development of "gadgetry." A good example of this restricted approach to the problem was reported in *Newsweek*:*

> ***PENTAGON***—A new and fiendishly ingenious antiguerrilla weapon is being tested by the Navy. It's a delayed-action liquid explosive, squirted from a flame-thrower-like gun, that seeps into foxholes and bunkers. Seconds later, fed by oxygen from the air, it blows up with terrific force.

Apparently we are to assume that guerrillas will conveniently ensconce themselves in readily identifiable "foxholes and bunkers" awaiting the arrival of half a dozen admirals armed with "flame-thrower-like guns" to march up, squirt, and retire to the nearest officers' club. To anyone even remotely acquainted with the philosophy and doctrine of revolutionary guerrilla war, this sort of thing is not hilariously funny. There are no mechanical panaceas.

I do not mean to suggest that proper weapons and equipment will not play an important part in antiguerrilla operations, for of course they will. Constant efforts should be made to improve communication, food, medical, and surgical "packs." Weapons and ammunition must be drastically reduced in weight; there seems to be no technical reason why a sturdy, light, accurate automatic rifle weigh-

---

* July 3, 1961, "The Periscope."

ing a maximum of four to five pounds cannot be developed. And the search for new and effective weapons must continue. But we must realize that "flame-thrower-like guns" and bullets are only a very small part of the answer to a challenging and complex problem.

The position of active third parties in a revolutionary guerrilla war and the timing, nature, and scope of the assistance given to one side or the other has become of great importance. Basically, this is a political matter; responsibility for a decision to intervene would naturally devolve upon the head of state. Any assistance given should, however, stop short of participation in combat. The role of a third party should be restricted to advice, materials, and technical training.

The timing of aid is often critical. If extended to the incumbent government, aid must be given while it is still possible to isolate and eradicate the movement; if to the revolutionary side, aid must be made available during the same critical period, that is, when the movement is vulnerable and its existence quite literally a matter of life and death.

From a purely military point of view, antiguerrilla operations may be summed up in three words: location, isolation, and eradication. In the brief definitions of each term, it will be well to bear in mind that these activities are not rigidly compartmented.

Location of base area or areas requires careful terrain studies, photographic and physical reconnaissance, and possibly infiltration of the movement. *Isolation* involves sepa-

Introduction

ration of guerrillas from their sources of information and food. It may require movement and resettlement of entire communities. *Eradication* presupposes reliable information and demands extreme operational flexibility and a high degree of mobility. Parachutists and helicopter-borne commando-type troops are essential.

The tactics of guerrillas must be used against the guerrillas themselves. They must be constantly harried and constantly attacked. Every effort must be made to induce defections and take prisoners. The best source of information of the enemy is men who know the enemy situation.

Imaginative, intelligent, and bold leadership is absolutely essential. *Commanders and leaders at every echelon* must be selected with these specific qualities in mind. Officers and NCO's who are more than competent under normal conditions will frequently be hopelessly ineffective when confronted with the dynamic and totally different situations characteristic of guerrilla warfare.

Finally, there is the question of whether it is possible to create effective counterguerrilla forces. Can two shoals of fish, each intent on destruction of the other, flourish in the same medium? Mao is definite on this point; he is convinced they cannot, that "counterrevolutionary guerrilla war" is impossible. If the guerrilla experiences of the White Russians (which he cites) or of Mikhailovitch are valid criteria, he is correct. But, on the other hand, the history of the movement in Greece during the German occupation indicates that under certain circumstances, his thesis will not stand too close an examination. This sug-

Mao Tse-tung on Guerrilla Warfare

gests the need for a careful analysis of relevant political factors in each individual situation.

Mao Tse-tung contends that the phenomena we have considered are subject to their own peculiar laws, and are predictable. If he is correct (and I believe he is), it is possible to prevent such phenomena from appearing, or, if they do, to control and eradicate them. And if historical experience teaches us anything about revolutionary guerrilla war, it is that military measures alone will not suffice.

YU CHI CHAN
(Guerrilla Warfare)

TRANSLATOR'S NOTE,

IN JULY, 1941, the undeclared war between China and Japan will enter its fifth year. One of the most significant features of the struggle has been the organization of the Chinese people for unlimited guerrilla warfare. The development of this warfare has followed the pattern laid out by Mao Tse-tung and his collaborators in the pamphlet *Yu Chi Chan* (*Guerrilla Warfare*), which was published in 1937 and has been widely distributed in "Free China" at 10 cents a copy.

Mao Tse-tung, a member of the Chinese Communist Party and formerly political commissar of the Fourth Red Army, is no novice in the art of war. Actual battle experience with both regular and guerrilla troops has qualified him as an expert.

The influence of the ancient military philosopher Sun Tzu on Mao's military thought will be apparent to those who have read *The Book of War*. Sun Tzu wrote that speed, surprise, and deception were the primary essentials of the attack and his succinct advice, *"Sheng Tung, Chi Hsi"* ("Uproar [in the] East, Strike [in the] West"), is no less valid today than it was when he wrote it 2,400 years ago. The tactics of Sun Tzu are in large measure the tactics of China's guerrillas today.

Mao Tse-tung on Guerrilla Warfare

Mao says that unlimited guerrilla warfare, with vast time and space factors, established a new military process. This seems a true statement since there are no other historical examples of guerrilla hostilities as thoroughly organized from the military, political, and economic point of view as those in China. We in the Marine Corps have as yet encountered nothing but relatively primitive and strictly limited guerrilla war. Thus, what Mao has written of this new type of guerrilla war may be of interest to us.

I have tried to present the author's ideas accurately, but as the Chinese language is not a particularly suitable medium for the expression of technical thought, the translation of some of the modern idioms not yet to be found in available dictionaries is probably arguable. I cannot vouch for the accuracy of retranslated quotations. I have taken the liberty to delete from the translation matter that was purely repetitious.

SAMUEL B. GRIFFITH
Captain, USMC

Quantico, Virginia
1940

A FURTHER NOTE

THE PRECEDING NOTE was written twenty-one years ago, but I see no need to amplify it.

Yu Chi Chan (1937) is frequently confused with one of Mao's later (1938) essays entitled *K'ang Jih Yu Chi Chan Cheng Ti Chan Lueh Wen T'i* (*Strategic Problems in the Anti-Japanese Guerrilla War*), which was issued in an English version in 1952 by the People's Publishing House, Peking. There are some similarities in these two works.

I had hoped to locate a copy of *Yu Chi Chan* in the Chinese to check my translation but have been unable to do so. Some improvement is always possible in any rendering from the Chinese. I have not been able to identify with standard English titles all the works cited by Mao.

Mao wrote *Yu Chi Chan* during China's struggle against Japan; consequently there are, naturally, numerous references to the strategy to be used against the Japanese. These in no way invalidate Mao's fundamental thesis. For instance, when Mao writes, "The moment that this war of resistance dissociates itself from the masses of the people is the precise moment that it dissociates itself from hope of ultimate victory over the Japanese," he might have added, "and from hope of ultimate victory over the forces of Chiang Kai-shek." However, he did not do so, because at that time both sides were attempting to preserve the illusion of a "united front." "Our basic policy," he said, "is the creation of a national united anti-Japanese front." This was, of course, *not* the basic policy of the Chinese Communist Party then, or at any other time. Its basic policy was to seize state power; the type of revolutionary guerrilla war described by Mao was the basic weapon in the protracted and ultimately successful process of doing so.

SAMUEL B. GRIFFITH
Brigadier General, USMC (Ret.)

Mount Vernon, Maine
July, 1961

1

WHAT IS GUERRILLA WARFARE?

IN A WAR OF REVOLUTIONARY CHARACTER, guerrilla operations are a necessary part. This is particularly true in a war waged for the emancipation of a people who inhabit a vast nation. China is such a nation, a nation whose techniques are undeveloped and whose communications are poor. She finds herself confronted with a strong and victorious Japanese imperialism. Under these circumstances, the development of the type of guerrilla warfare characterized by the quality of mass is both necessary and natural. This warfare must be developed to an unprecedented degree and it must coordinate with the operations of our regular armies. If we fail to do this, we will find it difficult to defeat the enemy.

These guerrilla operations must not be considered as an independent form of warfare. They are but one step in the total war, one aspect of the revolutionary struggle. They are the inevitable result of the clash between oppressor and oppressed when the latter reach the limits of their endurance. In our case, these hostilities began at a time when the people were unable to endure any more from the Japanese imperialists. Lenin, in *People and Revolution*, said: "A people's insurrection and a people's revolution are not only natural but inevitable." We consider guerrilla operations as but one aspect of our total or mass war because they, lacking the quality of independence, are of themselves incapable of providing a solution to the struggle.

Guerrilla warfare has qualities and objectives peculiar to itself. It is a weapon that a nation inferior in arms and military equipment may employ against a more powerful aggressor nation. When the invader pierces deep into the heart of the weaker country and occupies her territory in a cruel and oppressive manner, there is no doubt that conditions of terrain, climate, and society in general offer obstacles to his progress and may be used to advantage by those who oppose him. In guerrilla warfare, we turn these advantages to the purpose of resisting and defeating the enemy.

During the progress of hostilities, guerrillas gradually develop into orthodox forces that operate in conjunction with other units of the regular army. Thus the regularly organized troops, those guerrillas who have attained that status, and those who have not reached that level of development combine to form the military power of a national revolutionary war. There can be no doubt that the ultimate result of this will be victory.

Both in its development and in its method of application, guerrilla warfare has certain distinctive characteristics. We first discuss the relationship of guerrilla warfare to national policy. Because ours is the resistance of a semicolonial country against an imperialism, our hostilities must have a clearly defined political goal and firmly established political responsibilities. Our basic policy is the creation of a national

Yu Chi Chan (Guerrilla Warfare)

united anti-Japanese front. This policy we pursue in order to gain our political goal, which is the complete emancipation of the Chinese people. There are certain fundamental steps necessary in the realization of this policy, to wit:

1. Arousing and organizing the people.
2. Achieving internal unification politically.
3. Establishing bases.
4. Equipping forces.
5. Recovering national strength.
6. Destroying enemy's national strength.
7. Regaining lost territories.

There is no reason to consider guerrilla warfare separately from national policy. On the contrary, it must be organized and conducted in complete accord with national anti-Japanese policy. It is only those who misinterpret guerrilla action who say, as does Jen Ch'i Shan, "The question of guerrilla hostilities is purely a military matter and not a political one." Those who maintain this simple point of view have lost sight of the political goal and the political effects of guerrilla action. Such a simple point of view will cause the people to lose confidence and will result in our defeat.

What is the relationship of guerrilla warfare to the people? Without a political goal, guerrilla warfare must fail, as it must if its political objectives do not coincide with the aspirations of the people and their sympathy, cooperation, and assistance cannot be gained. The essence of guerrilla warfare is thus revolutionary in character. On the other hand, in a war of counterrevolutionary nature, there is no place for guerrilla hostilities. Because guerrilla warfare basically derives from the masses and is supported by them, it can neither exist nor flourish if it separates itself from their sympathies and cooperation. There are those who do not comprehend guerrilla action, and who therefore do not understand the distinguishing qualities of a people's guerrilla war, who say: "Only regular troops can carry on guerrilla operations." There are others who, because they do not believe in the ultimate success of guerrilla action, mistakenly say: "Guerrilla warfare is an insignificant and highly specialized type of operation in which there is no place for the masses of the people" (Jen Ch'i Shan). Then there are those who ridicule the masses and undermine resistance by wildly asserting that the people have no understanding of the war of resistance (Yeh Ch'ing, for one). The moment that this war of resistance dissociates itself from the masses of the people is the precise moment that it dissociates itself from hope of ultimate victory over the Japanese.

What is the organization for guerrilla warfare? Though all guerrilla bands that spring from the masses of the people suffer from lack of organization at the time of their formation, they all have in common a basic quality that makes organization possible. All guerrilla units must have political and military leadership. This is true regardless of the source or size of such units. Such units may originate locally, in the masses of the people; they may be formed from an admixture of regular troops with groups of the people, or they may consist of regular army units intact.

Yu Chi Chan (Guerrilla Warfare)

And mere quantity does not affect this matter. Such units may consist of a squad of a few men, a battalion of several hundred men, or a regiment of several thousand men.

All these must have leaders who are unyielding in their policies--resolute, loyal, sincere, and robust. These men must be well educated in revolutionary technique, self-confident, able to establish severe discipline, and able to cope with counterpropaganda. In short, these leaders must be models for the people. As the war progresses, such leaders will gradually overcome the lack of discipline, which at first prevails; they will establish discipline in their forces, strengthening them and increasing their combat efficiency. Thus eventual victory will be attained.

Unorganized guerrilla warfare cannot contribute to victory and those who attack the movement as a combination of banditry and anarchism do not understand the nature of guerrilla action. They say: "This movement is a haven for disappointed militarists, vagabonds and bandits" (Jen Ch'i Shan), hoping thus to bring the movement into disrepute. We do not deny that there are corrupt guerrillas, nor that there are people who under the guise of guerrillas indulge in unlawful activities. Neither do we deny that the movement has at the present time symptoms of a lack of organization, symptoms that might indeed be serious were we to judge guerrilla warfare solely by the corrupt and temporary phenomena we have mentioned. We should study the corrupt phenomena and attempt to eradicate them in order to encourage guerrilla warfare, and to increase its military efficiency. "This is hard work, there is no help for it, and the problem cannot be solved immediately. The whole people must try to reform themselves during the course of the war. We must educate them and reform them in the light of past experience. Evil does not exist in guerrilla warfare but only in the unorganized and undisciplined activities that are anarchism," said Lenin, in *On Guerrilla Warfare.**

What is basic guerrilla strategy? Guerrilla strategy must be based primarily on alertness, mobility, and attack. It must be adjusted to the enemy situation, the terrain, the existing lines of communication, the relative strengths, the weather, and the situation of the people.

In guerrilla warfare, select the tactic of seeming to come from the east and attacking from the west; avoid the solid, attack the hollow; attack; withdraw; deliver a lightning blow, seek a lightning decision. When guerrillas engage a stronger enemy, they withdraw when he advances; harass him when he stops; strike him when he is weary; pursue him when he withdraws. In guerrilla strategy, the enemy's rear, flanks, and other vulnerable spots are his vital points, and there he must be harassed, attacked, dispersed, exhausted and annihilated. Only in this way can guerrillas carry out their mission of independent guerrilla action and coordination with the effort of the regular armies. But, in spite of the most complete preparation, there can be no victory if mistakes are made in the matter of command. Guerrilla warfare based on the principles we have mentioned and carried on over a vast extent of territory in which

---

* Presumably, Mao refers here to the essay that has been translated into English under the title "Partisan Warfare." See *Orbis*, II (Summer, 1958), No. 2, 194–208.—S.B.G.

Yu Chi Chan (Guerrilla Warfare)

communications are inconvenient will contribute tremendously towards ultimate defeat of the Japanese and consequent emancipation of the Chinese people.

A careful distinction must be made between two types of guerrilla warfare. The fact that revolutionary guerrilla warfare is based on the masses of the people does not in itself mean that the organization of guerrilla units is impossible in a war of counterrevolutionary character. As examples of the former type we may cite Red guerrilla hostilities during the Russian Revolution; those of the Reds in China; of the Abyssinians against the Italians for the past three years; those of the last seven years in Manchuria, and the vast anti-Japanese guerrilla war that is carried on in China today. All these struggles have been carried on in the interests of the whole people or the greater part of them; all had a broad basis in the national manpower, and all have been in accord with the laws of historical development. They have existed and will continue to exist, flourish, and develop as long as they are not contrary to national policy.

The second type of guerrilla warfare directly contradicts the law of historical development. Of this type, we may cite the examples furnished by the White Russian guerrilla units organized by Denikin and Kolchak; those organized by the Japanese; those organized by the Italians in Abyssinia; those supported by the puppet governments in Manchuria and Mongolia, and those that will be organized here by Chinese traitors. All such have oppressed the masses and have been contrary to the true interests of the people. They must be firmly opposed. They are easy to destroy because they lack a broad foundation in the people.

Mao Tse-tung on Guerrilla Warfare

If we fail to differentiate between the two types of guerrilla hostilities mentioned, it is likely that we will exaggerate their effect when applied by an invader. We might arrive at the conclusion that "the invader can organize guerrilla units from among the people." Such a conclusion might well diminish our confidence in guerrilla warfare. As far as this matter is concerned, we have but to remember the historical experience of revolutionary struggles.

Further, we must distinguish general revolutionary wars from those of a purely "class" type. In the former case, the whole people of a nation, without regard to class or party, carry on a guerrilla struggle that is an instrument of the national policy. Its basis is, therefore, much broader than is the basis of a struggle of class type. Of a general guerrilla war, it has been said: "When a nation is invaded, the people become sympathetic to one another and all aid in organizing guerrilla units. In civil war, no matter to what extent guerrillas are developed, they do not produce the same results as when they are formed to resist an invasion by foreigners" (*Civil War in Russia*).* The one strong feature of guerrilla warfare in a civil struggle is its quality of internal purity. One class may be easily united and perhaps fight with great effect, whereas in a national revolutionary war, guerrilla units are faced with the problem of internal unification of different class groups. This necessitates the use of propaganda. Both types of guerrilla

---

* Presumably, Mao refers here to *Lessons of Civil War*, by S. I. Gusev; first published in 1918 by the Staff Armed Forces, Ukraine; revised in 1921 and published by GIZ, Moscow; reprinted in 1958 by the Military Publishing House, Moscow.—S.B.G.

Yu Chi Chan (Guerrilla Warfare)

war are, however, similar in that they both employ the same military methods.

National guerrilla warfare, though historically of the same consistency, has employed varying implements as times, peoples, and conditions differ. The guerrilla aspects of the Opium War, those of the fighting in Manchuria since the Mukden incident, and those employed in China today are all slightly different. The guerrilla warfare conducted by the Moroccans against the French and the Spanish was not exactly similar to that which we conduct today in China. These differences express the characteristics of different peoples in different periods. Although there is a general similarity in the quality of all these struggles, there are dissimilarities in form. This fact we must recognize. Clausewitz wrote, in *On War:* "Wars in every period have independent forms and independent conditions, and, therefore, every period must have its independent theory of war." Lenin, in *On Guerrilla Warfare,* said: "As regards the form of fighting, it is unconditionally requisite that history be investigated in order to discover the conditions of environment, the state of economic progress, and the political ideas that obtained, the national characteristics, customs, and degree of civilization." Again: "It is necessary to be completely unsympathetic to abstract formulas and rules and to study with sympathy the conditions of the actual fighting, for these will change in accordance with the political and economic situations and the realization of the people's aspirations. These progressive changes in conditions create new methods."

If, in today's struggle, we fail to apply the historical

truths of revolutionary guerrilla war, we will fall into the error of believing with T'ou Hsi Sheng that under the impact of Japan's mechanized army, "the guerrilla unit has lost its historical function." Jen Ch'i Shan writes: "In olden days, guerrilla warfare was part of regular strategy but there is almost no chance that it can be applied today." These opinions are harmful. If we do not make an estimate of the characteristics peculiar to our anti-Japanese guerrilla war, but insist on applying to it mechanical formulas derived from past history, we are making the mistake of placing our hostilities in the same category as all other national guerrilla struggles. If we hold this view, we will simply be beating our heads against a stone wall and we will be unable to profit from guerrilla hostilities.

To summarize: What is the guerrilla war of resistance against Japan? It is one aspect of the entire war, which, although alone incapable of producing the decision, attacks the enemy in every quarter, diminishes the extent of area under his control, increases our national strength, and assists our regular armies. It is one of the strategic instruments used to inflict defeat on our enemy. It is the one pure expression of anti-Japanese policy, that is to say, it is military strength organized by the active people and inseparable from them. It is a powerful special weapon with which we resist the Japanese and without which we cannot defeat them.

2

THE RELATION OF GUERRILLA HOSTILITIES TO REGULAR OPERATIONS

THE GENERAL FEATURES of orthodox hostilities, that is, the war of position and the war of movement, differ fundamentally from guerrilla warfare. There are other readily apparent differences such as those in organization, armament, equipment, supply, tactics, command; in conception of the terms "front" and "rear"; in the matter of military responsibilities.

When considered from the point of view of total numbers, guerrilla units are many; as individual combat units, they may vary in size from the smallest, of several score or several hundred men, to the battalion or the regiment, of several thousand. This is not the case in regularly organized units. A primary feature of guerrilla operations is their dependence upon the people themselves to organize battalions and other units. As a result of this, organization depends largely upon local circumstances. In the case of guerrilla groups, the standard of equipment is of a low order, and they must depend for their sustenance primarily upon what the locality affords.

Mao Tse-tung on Guerrilla Warfare

The strategy of guerrilla warfare is manifestly unlike that employed in orthodox operations, as the basic tactic of the former is constant activity and movement. There is in guerrilla warfare no such thing as a decisive battle; there is nothing comparable to the fixed, passive defense that characterizes orthodox war. In guerrilla warfare, the transformation of a moving situation into a positional defensive situation never arises. The general features of reconnaissance, partial deployment, general deployment, and development of the attack that are usual in mobile warfare are not common in guerrilla war.

There are differences also in the matter of leadership and command. In guerrilla warfare, small units acting independently play the principal role, and there must be no excessive interference with their activities. In orthodox warfare, particularly in a moving situation, a certain degree of initiative is accorded subordinates, but in principle, command is centralized. This is done because all units and all supporting arms in all districts must coordinate to the highest degree. In the case of guerrilla warfare, this is not only undesirable but impossible. Only adjacent guerrilla units can coordinate their activities to any degree. Strategically, their activities can be roughly correlated with those of the regular forces, and tactically, they must cooperate with adjacent units of the regular army. But there are no strictures on the extent of guerrilla activity nor is it primarily characterized by the quality of cooperation of many units.

When we discuss the terms "front" and "rear," it must be remembered, that while guerrillas do have bases, their

Yu Chi Chan (Guerrilla Warfare)

primary field of activity is in the enemy's rear areas. They themselves have no rear. Because an orthodox army has rear installations (except in some special cases as during the 10,000-mile* march of the Red Army or as in the case of certain units operating in Shansi Province), it cannot operate as guerrillas can.

As to the matter of military responsibilities, those of the guerrillas are to exterminate small forces of the enemy; to harass and weaken large forces; to attack enemy lines of communication; to establish bases capable of supporting independent operations in the enemy's rear; to force the enemy to disperse his strength; and to coordinate all these activities with those of the regular armies on distant battle fronts.

From the foregoing summary of differences that exist between guerrilla warfare and orthodox warfare, it can be seen that it is improper to compare the two. Further distinction must be made in order to clarify this matter. While the Eighth Route Army is a regular army, its North China campaign is essentially guerrilla in nature, for it operates in the enemy's rear. On occasion, however, Eighth Route Army commanders have concentrated powerful forces to strike an enemy in motion, and the characteristics of orthodox mobile warfare were evident in the battle at P'ing Hsing Kuan and in other engagements.

On the other hand, after the fall of Feng Ling Tu, the operations of Central Shansi, and Suiyuan, troops were more guerrilla than orthodox in nature. In this connection,

* It has been estimated that the Reds actually marched about 6,000 miles. See Introduction, Chapter II.--S.B.G.

the precise character of Generalissimo Chiang's instructions to the effect that independent brigades would carry out guerrilla operations should be recalled. In spite of such temporary activities, these orthodox units retained their identity and after the fall of Feng Ling Tu, they not only were able to fight along orthodox lines but often found it necessary to do so. This is an example of the fact that orthodox armies may, due to changes in the situation, temporarily function as guerrillas.

Likewise, guerrilla units formed from the people may gradually develop into regular units and, when operating as such, employ the tactics of orthodox mobile war. While these units function as guerrillas, they may be compared to innumerable gnats, which, by biting a giant both in front and in rear, ultimately exhaust him. They make themselves as unendurable as a group of cruel and hateful devils, and as they grow and attain gigantic proportions, they will find that their victim is not only exhausted but practically perishing. It is for this very reason that our guerrilla activities are a source of constant mental worry to Imperial Japan.

While it is improper to confuse orthodox with guerrilla operations, it is equally improper to consider that there is a chasm between the two. While differences do exist, similarities appear under certain conditions, and this fact must be appreciated if we wish to establish clearly the relationship between the two. If we consider both types of warfare as a single subject, or if we confuse guerrilla warfare with the mobile operations of orthodox war, we fall into this error: We exaggerate the function of guerrillas and minimize

Yu Chi Chan (Guerrilla Warfare)

that of the regular armies. If we agree with Chang Tso Hua, who says, "Guerrilla warfare is the primary war strategy of a people seeking to emancipate itself," or with Kao Kang, who believes that "Guerrilla strategy is the only strategy possible for an oppressed people," we are exaggerating the importance of guerrilla hostilities. What these zealous friends I have just quoted do not realize is this: If we do not fit guerrilla operations into their proper niche, we cannot promote them realistically. Then, not only would those who oppose us take advantage of our varying opinions to turn them to their own uses to undermine us, but guerrillas would be led to assume responsibilities they could not successfully discharge and that should properly be carried out by orthodox forces. In the meantime, the important guerrilla function of coordinating activities with the regular forces would be neglected.

Furthermore, if the theory that guerrilla warfare is our only strategy were actually applied, the regular forces would be weakened, we would be divided in purpose, and guerrilla hostilities would decline. If we say, "Let us transform the regular forces into guerrillas," and do not place our first reliance on a victory to be gained by the regular armies over the enemy, we may certainly expect to see as a result the failure of the anti-Japanese war of resistance. The concept that guerrilla warfare is an end in itself and that guerrilla activities can be divorced from those of the regular forces is incorrect. If we assume that guerrilla warfare does not progress from beginning to end beyond its elementary forms, we have failed to recognize the fact that guerrilla hostilities can, under specific conditions, develop and assume orthodox characteristics. An opinion that admits the existence of guerrilla war, but isolates it, is one that does not properly estimate the potentialities of such war.

Equally dangerous is the concept that condemns guerrilla war on the ground that war has no other aspects than the purely orthodox. This opinion is often expressed by those who have seen the corrupt phenomena of some guerrilla regimes, observed their lack of discipline, and have seen them used as a screen behind which certain persons have indulged in bribery and other corrupt practices. These people will not admit the fundamental necessity for guerrilla bands that spring from the armed people. They say, "Only the regular forces are capable of conducting guerrilla operations." This theory is a mistaken one and would lead to the abolition of the people's guerrilla war.

A proper conception of the relationship that exists between guerrilla effort and that of the regular forces is essential. We believe it can be stated this way: "Guerrilla operations during the anti-Japanese war may for a certain time and temporarily become its paramount feature, particularly insofar as the enemy's rear is concerned. However, if we view the war as a whole, there can be no doubt that our regular forces are of primary importance, because it is they who are alone capable of producing the decision. Guerrilla warfare assists them in producing this favorable decision. Orthodox forces may under certain conditions operate as guerrillas, and the latter may, under certain conditions, develop to the status of the former. However, both guerrilla forces and regular forces have their own respective development and their proper combinations."

Yu Chi Chan (Guerrilla Warfare)

To clarify the relationship between the mobile aspect of orthodox war and guerrilla war, we may say that general agreement exists that the principal element of our strategy must be mobility. With the war of movement, we may at times combine the war of position. Both of these are assisted by general guerrilla hostilities. It is true that on the battlefield mobile war often becomes positional; it is true that this situation may be reversed; it is equally true that each form may combine with the other. The possibility of such combination will become more evident after the prevailing standards of equipment have been raised. For example, in a general strategical counterattack to recapture key cities and lines of communication, it would be normal to use both mobile and positional methods. However, the point must again be made that our fundamental strategical form must be the war of movement. If we deny this, we cannot arrive at the victorious solution of the war. In sum, while we must promote guerrilla warfare as a necessary strategical auxiliary to orthodox operations, we must neither assign it the primary position in our war strategy nor substitute it for mobile and positional warfare as conducted by orthodox forces.

3

GUERRILLA WARFARE IN HISTORY

Guerrilla warfare is neither a product of China nor peculiar to the present day. From the earliest historical days, it has been a feature of wars fought by every class of men against invaders and oppressors. Under suitable conditions, it has great possibilities. The many guerrilla wars in history have their points of difference, their peculiar characteristics, their varying processes and conclusions, and we must respect and profit by the experience of those whose blood was shed in them. What a pity it is that the priceless experience gained during the several hundred wars waged by the peasants of China cannot be marshaled today to guide us. Our only experience in guerrilla hostilities has been that gained from the several conflicts that have been carried on against us by foreign imperialisms. But that experience should help the fighting Chinese recognize the necessity for guerrilla warfare and should confirm them in confidence of ultimate victory.

In September, 1812, the Frenchman Napoleon, in the course of swallowing all of Europe, invaded Russia at the head of a great army totaling several hundred thousand infantry, cavalry, and artillery. At that time, Russia was

Yu Chi Chan (Guerrilla Warfare)

weak and her ill-prepared army was not concentrated. The most important phase of her strategy was the use made of Cossack cavalry and detachments of peasants to carry on guerrilla operations. After giving up Moscow, the Russians formed nine guerrilla divisions of about five hundred men each. These, and vast groups of organized peasants, carried on partisan warfare and continually harassed the French Army. When the French Army was withdrawing, cold and starving, Russian guerrillas blocked the way and, in combination with regular troops, carried out counter-attacks on the French rear, pursuing and defeating them. The army of the heroic Napoleon was almost entirely annihilated, and the guerrillas captured many officers, men, cannon, and rifles. Though the victory was the result of various factors, and depended largely on the activities of the regular army, the function of the partisan groups was extremely important. "The corrupt and poorly organized country that was Russia defeated and destroyed an army led by the most famous soldier of Europe and won the war in spite of the fact that her ability to organize guerrilla regimes was not fully developed. At times, guerrilla groups were hindered in their operations and the supply of equipment and arms was insufficient. If we use the Russian saying, it was a case of a battle between 'the fist and the ax' " (Ivanov).

From 1918 to 1920, the Russian Soviets, because of the opposition and intervention of foreign imperialisms and the internal disturbances of White Russian groups, were forced to organize themselves in occupied territories and fight a real war. In Siberia and Alashan, in the rear of the army of the traitor Denikin and in the rear of the Poles, there were many Red Russian guerrillas. These not only disrupted and destroyed the communications in the enemy's rear but also frequently prevented his advance. On one occasion, the guerrillas completely destroyed a retreating White Army that had previously been defeated by regular Red forces. Kolchak, Denikin, the Japanese, and the Poles, owing to the necessity of staving off the attacks of guerrillas, were forced to withdraw regular troops from the front. "Thus not only was the enemy's manpower impoverished but he found himself unable to cope with the ever-moving guerrilla" (*The Nature of Guerrilla Action*).

The development of guerrillas at that time had only reached the stage where there were detached groups of several thousands in strength, old, middle aged, and young. The old men organized themselves into propaganda groups known as "silver-haired units"; there was a suitable guerrilla activity for the middle aged; the young men formed combat units, and there were even groups for the children. Among the leaders were determined Communists who carried on general political work among the people. These, although they opposed the doctrine of extreme guerrilla warfare, were quick to oppose those who condemned it. Experience tells us that "Orthodox armies are the fundamental and principal power; guerrilla units are secondary to them and assist in the accomplishment of the mission assigned the regular forces" (*Lessons of the Civil War in Russia*).* Many of the guerrilla regimes in Russia gradually developed until in battle they were able to dis-

* See p. 48 n.—S.B.G.

Yu Chi Chan (Guerrilla Warfare)

charge functions of organized regulars. The army of the famous General Galen was entirely derived from guerrillas.

During seven months in 1935 and 1936, the Abyssinians lost their war against Italy. The cause of defeat—aside from the most important political reasons that there were dissentient political groups, no strong government party, and unstable policy—was the failure to adopt a positive policy of mobile warfare. There was never a combination of the war of movement with large-scale guerrilla operations. Ultimately, the Abyssinians adopted a purely passive defense, with the result that they were unable to defeat the Italians. In addition to this, the fact that Abyssinia is a relatively small and sparsely populated country was contributory. Even in spite of the fact that the Abyssinian Army and its equipment were not modern, she was able to withstand a mechanized Italian force of 400,000 for seven months. During that period, there were several occasions when a war of movement was combined with large-scale guerrilla operations to strike the Italians heavy blows. Moreover, several cities were retaken and casualties totaling 140,000 were inflicted. Had this policy been steadfastly continued, it would have been difficult to have named the ultimate winner. At the present time, guerrilla activities continue in Abyssinia, and if the internal political questions can be solved, an extension of such activities is probable.

In 1841 and 1842, when brave people from San Yuan Li fought the English; again from 1850 to 1864, during the Taiping War, and for a third time in 1899, in the Boxer Uprising, guerrilla tactics were employed to a remarkable degree. Particularly was this so during the Taiping War, when guerrilla operations were most extensive and the Ch'ing troops were often completely exhausted and forced to flee for their lives.

In these wars, there were no guiding principles of guerrilla action. Perhaps these guerrilla hostilities were not carried out in conjunction with regular operations, or perhaps there was a lack of coordination. But the fact that victory was not gained was not because of any lack in guerrilla activity but rather because of the interference of politics in military affairs. Experience shows that if precedence is not given to the question of conquering the enemy in both political and military affairs, and if regular hostilities are not conducted with tenacity, guerrilla operations alone cannot produce final victory.

From 1927 to 1936, the Chinese Red Army fought almost continually and employed guerrilla tactics constantly. At the very beginning, a positive policy was adopted. Many bases were established, and from guerrilla bands, the Reds were able to develop into regular armies. As these armies fought, new guerrilla regimes were developed over a wide area. These regimes coordinated their efforts with those of the regular forces. This policy accounted for the many victories gained by guerrilla troops relatively few in number, who were armed with weapons inferior to those of their opponents. The leaders of that period properly combined guerrilla operations with a war of movement both strategically and tactically. They depended primarily upon alertness. They stressed the correct basis for both political affairs and military operations. They developed

Yu Chi Chan (Guerrilla Warfare)

their guerrilla bands into trained units. They then determined upon a ten-year period of resistance during which time they overcame innumerable difficulties and have only lately reached their goal of direct participation in the anti-Japanese war. There is no doubt that the internal unification of China is now a permanent and definite fact and that the experience gained during our internal struggles has proved to be both necessary and advantageous to us in the struggle against Japanese imperialism. There are many valuable lessons we can learn from the experience of those years. Principal among them is the fact that guerrilla success largely depends upon powerful political leaders who work unceasingly to bring about internal unification. Such leaders must work with the people; they must have a correct conception of the policy to be adopted as regards both the people and the enemy.

After September 18, 1931, strong anti-Japanese guerrilla campaigns were opened in each of the three northeast provinces. Guerrilla activity persists there in spite of the cruelties and deceits practiced by the Japanese at the expense of the people, and in spite of the fact that her armies have occupied the land and oppressed the people for the last seven years. The struggle can be divided into two periods. During the first, which extended from September 18, 1931, to January, 1933, anti-Japanese guerrilla activity exploded constantly in all three provinces. Ma Chan Shan and Ssu Ping Wei established an anti-Japanese regime in Heilungkiang. In Chi Lin, the National Salvation Army and the Self-Defense Army were led by Wang Te Lin and Li Tu respectively. In Feng T'ien, Chu Lu and others

Mao Tse-tung on Guerrilla Warfare

commanded guerrilla units. The influence of these forces was great. They harassed the Japanese unceasingly, but because there was an indefinite political goal, improper leadership, failure to coordinate military command and operations and to work with the people, and, finally, failure to delegate proper political functions to the army, the whole organization was feeble, and its strength was not unified. As a direct result of these conditions, the campaigns failed and the troops were finally defeated by our enemy.

During the second period, which has extended from January, 1933, to the present time, the situation has greatly improved. This has come about because great numbers of people who have been oppressed by the enemy have decided to resist him, because of the participation of the Chinese Communists in the anti-Japanese war, and because of the fine work of the volunteer units. The guerrillas have finally educated the people to the meaning of guerrilla warfare, and in the northeast, it has again become an important and powerful influence. Already seven or eight guerrilla regiments and a number of independent platoons have been formed, and their activities make it necessary for the Japanese to send troops after them month after month. These units hamper the Japanese and undermine their control in the northeast, while, at the same time, they inspire a Nationalist revolution in Korea. Such activities are not merely of transient and local importance but directly contribute to our ultimate victory.

However, there are still some weak points. For instance: National defense policy has not been sufficiently developed;

Yu Chi Chan (Guerrilla Warfare)

participation of the people is not general; internal political organization is still in its primary stages, and the force used to attack the Japanese and the puppet governments is not yet sufficient. But if present policy is continued tenaciously, all these weaknesses will be overcome. Experience proves that guerrilla war will develop to even greater proportions and that, in spite of the cruelty of the Japanese and the many methods they have devised to cheat the people, they cannot extinguish guerrilla activities in the three northeastern provinces.

The guerrilla experiences of China and of other countries that have been outlined prove that in a war of revolutionary nature such hostilities are possible, natural and necessary. They prove that if the present anti-Japanese war for the emancipation of the masses of the Chinese people is to gain ultimate victory, such hostilities must expand tremendously.

Historical experience is written in iron and blood. We must point out that the guerrilla campaigns being waged in China today are a page in history that has no precedent. Their influence will not be confined solely to China in her present anti-Japanese war but will be world-wide.

4

CAN VICTORY BE ATTAINED BY GUERRILLA OPERATIONS?

Guerrilla hostilities are but one phase of the war of resistance against Japan and the answer to the question of whether or not they can produce ultimate victory can be given only after investigation and comparison of all elements of our own strength with those of the enemy. The particulars of such a comparison are several. First, the strong Japanese bandit nation is an absolute monarchy. During the course of her invasion of China, she had made comparative progress in the techniques of industrial production and in the development of excellence and skill in her army, navy, and air force. But in spite of this industrial progress, she remains an absolute monarchy of inferior physical endowments. Her manpower, her raw materials, and her financial resources are all inadequate and insufficient to maintain her in protracted warfare or to meet the situation presented by a war prosecuted over a vast area. Added to this is the antiwar feeling now manifested by the Japanese people, a feeling that is shared by the junior officers and, more extensively, by the soldiers of the invading army. Furthermore, China is not Japan's

Yu Chi Chan (Guerrilla Warfare)

only enemy. Japan is unable to employ her entire strength in the attack on China; she cannot, at most, spare more than a million men for this purpose, as she must hold any in excess of that number for use against other possible opponents. Because of these important primary considerations, the invading Japanese bandits can hope neither to be victorious in a protracted struggle nor to conquer a vast area. Their strategy must be one of lightning war and speedy decision. If we can hold out for three or more years, it will be most difficult for Japan to bear up under the strain.

In the war, the Japanese brigands must depend upon lines of communication linking the principal cities as routes for the transport of war materials. The most important considerations for her are that her rear be stable and peaceful and that her lines of communication be intact. It is not to her advantage to wage war over a vast area with disrupted lines of communication. She cannot disperse her strength and fight in a number of places, and her greatest fears are thus eruptions in her rear and disruption of her lines of communication. If she can maintain communications, she will be able at will to concentrate powerful forces speedily at strategic points to engage our organized units in decisive battle. Another important Japanese objective is to profit from the industries, finances, and manpower in captured areas and with them to augment her own insufficient strength. Certainly, it is not to her advantage to forgo these benefits, nor to be forced to dissipate her energies in a type of warfare in which the gains will not compensate for the losses. It is for these reasons that guerrilla warfare conducted in each bit of conquered territory over a wide area will be a heavy blow struck at the Japanese bandits. Experience in the five northern provinces as well as in Kiangsu, Chekiang, and Anhwei has absolutely established the truth of this assertion.

China is a country half colonial and half feudal; it is a country that is politically, militarily, and economically backward. This is an inescapable conclusion. It is a vast country with great resources and tremendous population, a country in which the terrain is complicated and the facilities for communication are poor. All these factors favor a protracted war; they all favor the application of mobile warfare and guerrilla operations. The establishment of innumerable anti-Japanese bases behind the enemy's lines will force him to fight unceasingly in many places at once, both to his front and his rear. He thus endlessly expends his resources.

We must unite the strength of the army with that of the people; we must strike the weak spots in the enemy's flanks, in his front, in his rear. We must make war everywhere and cause dispersal of his forces and dissipation of his strength. Thus the time will come when a gradual change will become evident in the relative position of ourselves and our enemy, and when that day comes, it will be the beginning of our ultimate victory over the Japanese.

Although China's population is great, it is unorganized. This is a weakness which must be taken into account.

The Japanese bandits have invaded our country not

Yu Chi Chan (Guerrilla Warfare)

merely to conquer territory but to carry out the violent, rapacious, and murderous policy of their government, which is the extinction of the Chinese race. For this compelling reason, we must unite the nation without regard to parties or classes and follow our policy of resistance to the end. China today is not the China of old. It is not like Abyssinia. China today is at the point of her greatest historical progress. The standards of literacy among the masses have been raised; the *rapprochement* of Communists and Nationalists has laid the foundation for an anti-Japanese war front that is constantly being strengthened and expanded; government, army, and people are all working with great energy; the raw-material resources and the economic strength of the nation are waiting to be used; the unorganized people is becoming an organized nation.

These energies must be directed toward the goal of protracted war so that should the Japanese occupy much of our territory or even most of it, we shall still gain final victory. Not only must those behind our lines organize for resistance but also those who live in Japanese-occupied territory in every part of the country. The traitors who accept the Japanese as fathers are few in number, and those who have taken oath that they would prefer death to abject slavery are many. If we resist with this spirit, what enemy can we not conquer and who can say that ultimate victory will not be ours?

The Japanese are waging a barbaric war along uncivilized lines. For that reason, Japanese of all classes oppose the policies of their government, as do vast international groups. On the other hand, because China's cause is righteous, our countrymen of all classes and parties are united to oppose the invader; we have sympathy in many foreign countries, including even Japan itself. This is perhaps the most important reason why Japan will lose and China will win.

The progress of the war for the emancipation of the Chinese people will be in accord with these facts. The guerrilla war of resistance will be in accord with these facts, and that guerrilla operations correlated with those of our regular forces will produce victory is the conviction of the many patriots who devote their entire strength to guerrilla hostilities.

5

ORGANIZATION FOR GUERRILLA WARFARE

Four points must be considered under this subject. These are:

1. How are guerrilla bands formed?
2. How are guerrilla bands organized?
3. What are the methods of arming guerrilla bands?
4. What elements constitute a guerrilla band?

These are all questions pertaining to the organization of armed guerrilla units; they are questions which those who have had no experience in guerrilla hostilities do not understand and on which they can arrive at no sound decisions; indeed, they would not know in what manner to begin.

How Guerrilla Units Are Originally Formed

The unit may originate in any one of the following ways:

a) From the masses of the people.
b) From regular army units temporarily detailed for the purpose.
c) From regular army units permanently detailed.

d) From the combination of a regular army unit and a unit recruited from the people.

e) From the local militia.

f) From deserters from the ranks of the enemy.

g) From former bandits and bandit groups.

In the present hostilities, no doubt, all these sources will be employed.

In the first case above, the guerrilla unit is formed from the people. This is the fundamental type. Upon the arrival of the enemy army to oppress and slaughter the people, their leaders call upon them to resist. They assemble the most valorous elements, arm them with old rifles or bird guns, and thus a guerrilla unit begins. Orders have already been issued throughout the nation that call upon the people to form guerrilla units both for local defense and for other combat. If the local governments approve and aid such movements, they cannot fail to prosper. In some places, where the local government is not determined or where its officers have all fled, the leaders among the masses (relying on the sympathy of the people and their sincere desire to resist Japan and succor the country) call upon the people to resist, and they respond. Thus, many guerrilla units are organized. In circumstances of this kind, the duties of leadership usually fall upon the shoulders of young students, teachers, professors, other educators, local soldiery, professional men, artisans, and those without a fixed profession, who are willing to exert themselves to the last drop of their blood. Recently, in Shansi, Hopeh, Chahar, Suiyuan, Shantung, Chekiang, Anhwei, Kiangsu,

Yu Chi Chan (Guerrilla Warfare)

and other provinces, extensive guerrilla hostilities have broken out. All these are organized and led by patriots. The amount of such activity is the best proof of the foregoing statement. The more such bands there are, the better will the situation be. Each district, each county, should be able to organize a great number of guerrilla squads, which, when assembled, form a guerrilla company.

There are those who say: "I am a farmer," or, "I am a student"; "I can discuss literature but not military arts." This is incorrect. There is no profound difference between the farmer and the soldier. You must have courage. You simply leave your farms and become soldiers. That you are farmers is of no difference, and if you have education, that is so much the better. When you take your arms in hand, you become soldiers; when you are organized, you become military units.

Guerrilla hostilities are the university of war, and after you have fought several times valiantly and aggressively, you may become a leader of troops, and there will be many well-known regular soldiers who will not be your peers. Without question, the fountainhead of guerrilla warfare is in the masses of the people, who organize guerrilla units directly from themselves.

The second type of guerrilla unit is that which is organized from small units of the regular forces temporarily detached for the purpose. For example, since hostilities commenced, many groups have been temporarily detached from armies, divisions, and brigades and have been assigned guerrilla duties. A regiment of the regular army may, if circumstances warrant, be dispersed into groups for the purpose of carrying on guerrilla operations. As an example of this, there is the Eighth Route Army, in North China. Excluding the periods when it carries on mobile operations as an army, it is divided into its elements and these carry on guerrilla hostilities. This type of guerrilla unit is essential for two reasons. First, in mobile-warfare situations, the coordination of guerrilla activities with regular operations is necessary. Second, until guerrilla hostilities can be developed on a grand scale, there is no one to carry out guerrilla missions but regulars. Historical experience shows us that regular army units are not able to undergo the hardships of guerrilla campaigning over long periods. The leaders of regular units engaged in guerrilla operations must be extremely adaptable. They must study the methods of guerrilla war. They must understand that initiative, discipline, and the employment of stratagems are all of the utmost importance. As the guerrilla status of regular units is but temporary, their leaders must lend all possible support to the organization of guerrilla units from among the people. These units must be so disciplined that they hold together after the departure of the regulars.

The third type of unit consists of a detachment of regulars who are permanently assigned guerrilla duties.. This type of small detachment does not have to be prepared to rejoin the regular forces. Its post is somewhere in the rear of the enemy, and there it becomes the backbone of guerrilla organization. As an example of this type of organization, we may take the Wu Tai Shan district in the heart of the Hopeh-Chahar-Shansi area. Along the borders of these provinces, units from the Eighth Route Army have

Yu Chi Chan (Guerrilla Warfare)

established a framework for guerrilla operations. Around these small cores, many detachments have been organized and the area of guerrilla activity greatly expanded. In areas in which there is a possibility of cutting the enemy's lines of supply, this system should be used. Severing enemy supply routes destroys his life line; this is one feature that cannot be neglected. If, at the time the regular forces withdraw from a certain area, some units are left behind, these should conduct guerrilla operations in the enemy's rear. As an example of this, we have the guerrilla bands now continuing their independent operations in the Shanghai-Woosung area in spite of the withdrawal of regular forces.

The fourth type of organization is the result of a merger between small regular detachments and local guerrilla units. The regular forces may dispatch a squad, a platoon, or a company, which is placed at the disposal of the local guerrilla commander. If a small group experienced in military and political affairs is sent, it becomes the core of the local guerrilla unit. These several methods are all excellent, and if properly applied, the intensity of guerrilla warfare can be extended. In the Wu Tai Shan area, each of these methods has been used.

The fifth type mentioned above is formed from the local militia, from police and home guards. In every North China province, there are now many of these groups, and they should be formed in every locality. The government has issued a mandate to the effect that the people are not to depart from war areas. The officer in command of the county, the commander of the peace-preservation unit, the chief of police are all required to obey this mandate. They cannot retreat with their forces but must remain at their stations and resist.

The sixth type of unit is that organized from troops that come over from the enemy—the Chinese "traitor troops" employed by the Japanese. It is always possible to produce disaffection in their ranks, and we must increase our propaganda efforts and foment mutinies among such troops. Immediately after mutinying, they must be received into our ranks and organized. The concord of the leaders and the assent of the men must be gained, and the units rebuilt politically and reorganized militarily. Once this has been accomplished, they become successful guerrilla units. In regard to this type of unit, it may be said that political work among them is of the utmost importance.

The seventh type of guerrilla organization is that formed from bands of bandits and brigands. This, although difficult, must be carried out with utmost vigor lest the enemy use such bands to his own advantage. Many bandit groups pose as anti-Japanese guerrillas, and it is only necessary to correct their political beliefs to convert them.

In spite of inescapable differences in the fundamental types of guerrilla bands, it is possible to unite them to form a vast sea of guerrillas. The ancients said, "Tai Shan is a great mountain because it does not scorn the merest handful of dirt; the rivers and seas are deep because they absorb the waters of small streams." Attention paid to the enlistment and organization of guerrillas of every type and from every source will increase the potentialities of guerrilla action in the anti-Japanese war. This is something that patriots will not neglect.

Yu Chi Chan (Guerrilla Warfare)

THE METHOD OF ORGANIZING GUERRILLA REGIMES

Many of those who decide to participate in guerrilla activities do not know the methods of organization. For such people, as well as for students who have no knowledge of military affairs, the matter of organization is a problem that requires solution. Even among those who have military knowledge, there are some who know nothing of guerrilla regimes because they are lacking in that particular type of experience. The subject of the organization of such regimes is not confined to the organization of specific units but includes all guerrilla activities within the area where the regime functions.

As an example of such organization, we may take a geographical area in the enemy's rear. This area may comprise many counties. It must be subdivided and individual companies or battalions formed to accord with the subdivisions. To this "military area," a military commander and political commissioners are appointed. Under these, the necessary officers, both military and political, are appointed. In the military headquarters, there will be the staff, the aides, the supply officers, and the medical personnel. These are controlled by the chief of staff, who acts in accordance with orders from the commander. In the political headquarters, there are bureaus of propaganda organization, people's mass movements, and miscellaneous affairs. Control of these is vested in the political chairmen.

The military areas are subdivided into smaller districts in accordance with local geography, the enemy situation locally, and the state of guerrilla development. Each of these smaller divisions within the area is a district, each of which may consist of from two to six counties. To each district, a military commander and several political commissioners are appointed. Under their direction, military and political headquarters are organized. Tasks are assigned in accordance with the number of guerrilla troops available. Although the names of the officers in the "district" correspond to those in the larger "area," the number of functionaries assigned in the former case should be reduced to the least possible. In order to unify control, to handle guerrilla troops that come from different sources, and to harmonize military operations and local political affairs, a committee of from seven to nine members should be organized in each area and district. This committee, the members of which are selected by the troops and the local political officers, should function as a forum for the discussion of both military and political matters.

All the people in an area should arm themselves and be organized into two groups. One of these groups is a combat group, the other a self-defense unit with but limited military quality. Regular combatant guerrillas are organized into one of three general types of unit. The first of these is the small unit, the platoon or company. In each county, three to six units may be organized. The second type is the battalion of from two to four companies. One such unit should be organized in each county. While the unit fundamentally belongs to the county in which it was organized, it may operate in other counties. While in areas other than its own, it must operate in conjunction with local units in order to take advantage of their manpower, their

Yu Chi Chan (Guerrilla Warfare)

knowledge of local terrain and local customs, and their information of the enemy.

The third type is the guerrilla regiment, which consists of from two to four of the above-mentioned battalion units. If sufficient manpower is available, a guerrilla brigade of from two to four regiments may be formed.

Each of the units has its own peculiarities of organization. A squad, the smallest unit, has a strength of from nine to eleven men, including the leader and the assistant leader. Its arms may be from two to five Western-style rifles, with the remaining men armed with rifles of local manufacture, bird guns, spears, or big swords. Two to four such squads form a platoon. This, too, has a leader and an assistant leader, and when acting independently, it is assigned a political officer to carry on political propaganda work. The platoon may have about ten rifles, with the remainder of its weapons being bird guns, lances, and big swords. Two to four of such units form a company, which, like the platoon, has a leader, an assistant leader, and a political officer. All these units are under the direct supervision of the military commanders of the areas in which they operate.

The battalion unit must be more thoroughly organized and better equipped than the smaller units. Its discipline and its personnel should be superior. If a battalion is formed from company units, it should not deprive subordinate units entirely of their manpower and their arms. If, in a small area, there is a peace-preservation corps, a branch of the militia, or police, regular guerrilla units should not be dispersed over it.

Mao Tse-tung on Guerrilla Warfare

The guerrilla unit next in size to the battalion is the regiment. This must be under more severe discipline than the battalion. In an independent guerrilla regiment, there may be ten men per squad, three squads per platoon, three platoons per company, three companies per battalion, and three battalions to the regiment. Two of such regiments form a brigade. Each of these units has a commander, a vice-commander, and a political officer.

In North China, guerrilla cavalry units should be established. These may be regiments of from two to four companies, or battalions.

All these units from the lowest to the highest are combatant guerrilla units and receive their supplies from the central government. Details of their organization are shown in the tables.*

All the people of both sexes from the ages of sixteen to forty-five must be organized into anti-Japanese self-defense units, the basis of which is voluntary service. As a first step, they must procure arms, then they must be given both military and political training. Their responsibilities are: local sentry duties, securing information of the enemy, arresting traitors, and preventing the dissemination of enemy propaganda. When the enemy launches a guerrilla-suppression drive, these units, armed with what weapons there are, are assigned to certain areas to deceive, hinder, and harass him. Thus, the self-defense units assist the combatant guerrillas. They have other functions. They furnish stretcher-bearers to transport the wounded, carriers to take food to the troops, and comfort missions to provide

* See Appendix.—S.B.G.

Yu Chi Chan (Guerrilla Warfare)

the troops with tea and rice. If a locality can organize such a self-defense unit as we have described, the traitors cannot hide nor can bandits and robbers disturb the peace of the people. Thus the people will continue to assist the guerrillas and supply manpower to our regular armies. "The organization of self-defense units is a transitional step in the development of universal conscription. Such units are reservoirs of manpower for the orthodox forces."

There have been such organizations for some time in Shansi, Shensi, Honan, and Suiyuan. The youth organizations in different provinces were formed for the purpose of educating the young. They have been of some help. However, they were not voluntary, and the confidence of the people was thus not gained. These organizations were not widespread, and their effect was almost negligible. This system was, therefore, supplanted by the new-type organizations, which are organized on the principles of voluntary cooperation and nonseparation of the members from their native localities. When the members of these organizations are in their native towns, they support themselves. Only in case of military necessity are they ordered to remote places, and when this is done, the government must support them. Each member of these groups must have a weapon even if the weapon is only a knife, a pistol, a lance, or a spear.

In all places where the enemy operates, these self-defense units should organize within themselves a small guerrilla group of perhaps from three to ten men armed with pistols or revolvers. This group is not required to leave its native locality.

The organization of these self-defense units is mentioned in this book because such units are useful for the purposes of inculcating the people with military and political knowledge, keeping order in the rear, and replenishing the ranks of the regulars. These groups should be organized not only in the active war zones but in every province in China. "The people must be inspired to cooperate voluntarily. We must not force them, for if we do, it will be ineffectual." This is extremely important. The organization of a self-defense army similar to that we have mentioned is shown in Table 5.*

In order to control anti-Japanese military organization as a whole, it is necessary to establish a system of military areas and districts along the lines we have indicated. The organization of such areas and districts is shown in Table 6.

EQUIPMENT OF GUERRILLAS

In regard to the problem of guerrilla equipment, it must be understood that guerrillas are lightly armed attack groups, which require simple equipment. The standard of equipment is based upon the nature of duties assigned; the equipment of low-class guerrilla units is not as good as that of higher-class units. For example, those who are assigned the task of destroying railroads are better-equipped than those who do not have that task. The equipment of guerrillas cannot be based on what the guerrillas want, or even what they need, but must be based on what is available for their use. Equipment cannot be furnished

---

* Unfortunately, this table, as well as Table 6, was omitted from the edition of *Yu Chi Chan* available to me.—S.B.G.

Yu Chi Chan (Guerrilla Warfare)

immediately but must be acquired gradually. These are points to be kept in mind.

The question of equipment includes the collection, supply, distribution, and replacement of weapons, ammunition, blankets, communication materials, transport, and facilities for propaganda work. The supply of weapons and ammunition is most difficult, particularly at the time the unit is established, but this problem can always be solved eventually. Guerrilla bands that originate in the people are furnished with revolvers, pistols, bird guns, spears, big swords, and land mines and mortars of local manufacture. Other elementary weapons are added and as many new-type rifles as are available are distributed. After a period of resistance, it is possible to increase the supply of equipment by capturing it from the enemy. In this respect, the transport companies are the easiest to equip, for in any successful attack, we will capture the enemy's transport.

An armory should be established in each guerrilla district for the manufacture and repair of rifles and for the production of cartridges, hand grenades, and bayonets. Guerrillas must not depend too much on an armory. The enemy is the principal source of their supply.

For destruction of railway trackage, bridges, and stations in enemy-controlled territory, it is necessary to gather together demolition materials. Troops must be trained in the preparation and use of demolitions, and a demolition unit must be organized in each regiment.

As for minimum clothing requirements, these are that each man shall have at least two summer-weight uniforms, one suit of winter clothing, two hats, a pair of wrap puttees, and a blanket. Each man must have a haversack or a bag for food. In the north, each man must have an overcoat. In acquiring this clothing, we cannot depend on captures made from the enemy, for it is forbidden for captors to take clothing from their prisoners. In order to maintain high morale in guerrilla forces, all the clothing and equipment mentioned should be furnished by the representatives of the government stationed in each guerrilla district. These men may confiscate clothing from traitors or ask contributions from those best able to afford them. In subordinate groups, uniforms are unnecessary.

Telephone and radio equipment is not necessary in lower groups, but all units from regiment up are equipped with both. This material can be obtained by contributions from the regular forces and by capture from the enemy.

In the guerrilla army in general, and at bases in particular, there must be a high standard of medical equipment. Besides the services of the doctors, medicines must be procured. Although guerrillas can depend on the enemy for some portion of their medical supplies, they must, in general, depend upon contributions. If Western medicines are not available, local medicines must be made to suffice.

The problem of transport is more vital in North China than in the south, for in the south all that are necessary are mules and horses. Small guerrilla units need no animals, but regiments and brigades will find them necessary. Commanders and staffs of units from companies up should be furnished a riding animal each. At times, two officers will have to share a horse. Officers whose duties are of minor nature do not have to be mounted.

Yu Chi Chan (Guerrilla Warfare)

Propaganda materials are very important. Every large guerrilla unit should have a printing press and a mimeograph stone. They must also have paper on which to print propaganda leaflets and notices. They must be supplied with chalk and large brushes. In guerrilla areas, there should be a printing press or a lead-type press.

For the purpose of printing training instructions, this material is of the greatest importance.

In addition to the equipment listed above, it is necessary to have field glasses, compasses, and military maps. An accomplished guerrilla group will acquire these things.

Because of the proved importance of guerrilla hostilities in the anti-Japanese war, the headquarters of the Nationalist Government and the commanding officers of the various war zones should do their best to supply the guerrillas with what they actually need and are unable to get for themselves. However, it must be repeated that guerrilla equipment will in the main depend on the efforts of the guerrillas themselves. If they depend on higher officers too much, the psychological effect will be to weaken the guerrilla spirit of resistance.

ELEMENTS OF THE GUERRILLA ARMY

The term "element" as used in the title to this section refers to the personnel, both officers and men, of the guerrilla army. Since each guerrilla group fights in a protracted war, its officers must be brave and positive men whose entire loyalty is dedicated to the cause of emancipation of the people. An officer should have the following qualities: great powers of endurance so that in spite of any hardship he sets an example to his men and is a model for them; he must be able to mix easily with the people; his spirit and that of the men must be one in strengthening the policy of resistance to the Japanese. If he wishes to gain victories, he must study tactics. A guerrilla group with officers of this caliber would be unbeatable. I do not mean that every guerrilla group can have, at its inception, officers of such qualities. The officers must be men naturally endowed with good qualities which can be developed during the course of campaigning. The most important natural quality is that of complete loyalty to the idea of people's emancipation. If this is present, the others will develop; if it is not present, nothing can be done. When officers are first selected from a group, it is this quality that should receive particular attention. The officers in a group should be inhabitants of the locality in which the group is organized, as this will facilitate relations between them and the local civilians. In addition, officers so chosen would be familiar with conditions. If in any locality there are not enough men of sufficiently high qualifications to become officers, an effort must be made to train and educate the people so these qualities may be developed and the potential officer material increased. There can be no disagreements between officers native to one place and those from other localities.

A guerrilla group ought to operate on the principle that only volunteers are acceptable for service. It is a mistake to impress people into service. As long as a person is willing to fight, his social condition or position is no consideration, but only men who are courageous and determined can bear the hardships of guerrilla campaigning in a protracted war.

Yu Chi Chan (Guerrilla Warfare)

A soldier who habitually breaks regulations must be dismissed from the army. Vagabonds and vicious people must not be accepted for service. The opium habit must be forbidden, and a soldier who cannot break himself of the habit should be dismissed. Victory in guerrilla war is conditioned upon keeping the membership pure and clean.

It is a fact that during the war the enemy may take advantage of certain people who are lacking in conscience and patriotism and induce them to join the guerrillas for the purpose of betraying them. Officers must, therefore, continually educate the soldiers and inculcate patriotism in them. This will prevent the success of traitors. The traitors who are in the ranks must be discovered and expelled, and punishment and expulsion meted out to those who have been influenced by them. In all such cases, the officers should summon the soldiers and relate the facts to them, thus arousing their hatred and detestation for traitors. This procedure will serve as well as a warning to the other soldiers. If an officer is discovered to be a traitor, some prudence must be used in the punishment adjudged. However, the work of eliminating traitors in the army begins with their elimination from among the people.

Chinese soldiers who have served under puppet governments and bandits who have been converted should be welcomed as individuals or as groups. They should be well treated and repatriated. But care should be used during their reorientation to distinguish those whose idea is to fight the Japanese from those who may be present for other reasons.

6

THE POLITICAL PROBLEMS OF GUERRILLA WARFARE

In chapter 1, I mentioned the fact that guerrilla troops should have a precise conception of the political goal of the struggle and the political organization to be used in attaining that goal. This means that both organization and discipline of guerrilla troops must be at a high level so that they can carry out the political activities that are the life of both the guerrilla armies and of revolutionary warfare.

First of all, political activities depend upon the indoctrination of both military and political leaders with the idea of anti-Japanism. Through them, the idea is transmitted to the troops. One must not feel that he is anti-Japanese merely because he is a member of a guerrilla unit. The anti-Japanese idea must be an ever-present conviction, and if it is forgotten, we may succumb to the temptations of the enemy or be overcome with discouragements. In a war of long duration, those whose conviction that the people must be emancipated is not deep rooted are likely to become shaken in their faith or actually revolt. Without the general education that enables everyone to understand

Yu Chi Chan (Guerrilla Warfare)

our goal of driving out Japanese imperialism and establishing a free and happy China, the soldiers fight without conviction and lose their determination.

The political goal must be clearly and precisely indicated to inhabitants of guerrilla zones and their national consciousness awakened. Hence, a concrete explanation of the political systems used is important not only to guerrilla troops but to all those who are concerned with the realization of our political goal. The Kuomintang has issued a pamphlet entitled *System of National Organization for War*, which should be widely distributed throughout guerrilla zones. If we lack national organization, we will lack the essential unity that should exist between the soldiers and the people.

A study and comprehension of the political objectives of this war and of the anti-Japanese front is particularly important for officers of guerrilla troops. There are some militarists who say: "We are not interested in politics but only in the profession of arms." It is vital that these simple-minded militarists be made to realize the relationship that exists between politics and military affairs. Military action is a method used to attain a political goal. While military affairs and political affairs are not identical, it is impossible to isolate one from the other.

It is to be hoped that the world is in the last era of strife. The vast majority of human beings have already prepared or are preparing to fight a war that will bring justice to the oppressed peoples of the world. No matter how long this war may last, there is no doubt that it will be followed by an unprecedented epoch of peace. The war that we are fighting today for the emancipation of the Chinese is a part of the war for the freedom of all human beings, and the independent, happy, and liberal China that we are fighting to establish will be a part of that new world order. A conception like this is difficult for the simple-minded militarist to grasp and it must therefore be carefully explained to him.

There are three additional matters that must be considered under the broad question of political activities. These are political activities, first, as applied to the troops; second, as applied to the people; and, third, as applied to the enemy. The fundamental problems are: first, spiritual unification of officers and men within the army; second, spiritual unification of the army and the people; and, last, destruction of the unity of the enemy. The concrete methods for achieving these unities are discussed in detail in pamphlet Number 4 of this series, entitled *Political Activities in Anti-Japanese Guerrilla Warfare*.

A revolutionary army must have discipline that is established on a limited democratic basis. In all armies, obedience of the subordinates to their superiors must be exacted. This is true in the case of guerrilla discipline, but the basis for guerrilla discipline must be the individual conscience. With guerrillas, a discipline of compulsion is ineffective. In any revolutionary army, there is unity of purpose as far as both officers and men are concerned, and, therefore, within such an army, discipline is self-imposed. Although discipline in guerrilla ranks is not as severe as in the ranks of orthodox forces, the necessity for discipline exists. This must be self-imposed, because only when it is, is the soldier

Yu Chi Chan (Guerrilla Warfare)

able to understand completely why he fights and why he must obey. This type of discipline becomes a tower of strength within the army, and it is the only type that can truly harmonize the relationship that exists between officers and soldiers.

In any system where discipline is externally imposed, the relationship that exists between officer and man is characterized by indifference of the one to the other. The idea that officers can physically beat or severely tongue-lash their men is a feudal one and is not in accord with the conception of a self-imposed discipline. Discipline of the feudal type will destroy internal unity and fighting strength. A discipline self-imposed is the primary characteristic of a democratic system in the army.

A secondary characteristic is found in the degree of liberties accorded officers and soldiers. In a revolutionary army, all individuals enjoy political liberty and the question, for example, of the emancipation of the people must not only be tolerated but discussed, and propaganda must be encouraged. Further, in such an army, the mode of living of the officers and the soldiers must not differ too much, and this is particularly true in the case of guerrilla troops. Officers should live under the same conditions as their men, for that is the only way in which they can gain from their men the admiration and confidence so vital in war. It is incorrect to hold to a theory of equality in all things, but there must be equality of existence in accepting the hardships and dangers of war. Thus we may attain to the unification of the officer and soldier groups, a unity both horizontal within the group itself, and vertical, that is, from lower to higher echelons. It is only when such unity is present that units can be said to be powerful combat factors.

There is also a unity of spirit that should exist between troops and local inhabitants. The Eighth Route Army put into practice a code known as "The Three Rules and the Eight Remarks," which we list here:

Rules:

1. All actions are subject to command.
2. Do not steal from the people.
3. Be neither selfish nor unjust.

Remarks:

1. Replace the door when you leave the house.*
2. Roll up the bedding on which you have slept.
3. Be courteous.
4. Be honest in your transactions.
5. Return what you borrow.
6. Replace what you break.
7. Do not bathe in the presence of women.
8. Do not without authority search the pocketbooks of those you arrest.

The Red Army adhered to this code for ten years and the Eighth Route Army and other units have since adopted it.

Many people think it impossible for guerrillas to exist for long in the enemy's rear. Such a belief reveals lack of com-

---

* In summer, doors were frequently lifted off and used as beds. —S.B.G.

Yu Chi Chan (Guerrilla Warfare)

prehension of the relationship that should exist between the people and the troops. The former may be likened to water and the latter to the fish who inhabit it. How may it be said that these two cannot exist together? It is only undisciplined troops who make the people their enemies and who, like the fish out of its native element, cannot live.

We further our mission of destroying the enemy by propagandizing his troops, by treating his captured soldiers with consideration, and by caring for those of his wounded who fall into our hands. If we fail in these respects, we strengthen the solidarity of our enemy.

7

THE STRATEGY OF GUERRILLA RESISTANCE AGAINST JAPAN

I T HAS BEEN DEFINITELY DECIDED that in the strategy of our war against Japan, guerrilla strategy must be auxiliary to fundamental orthodox methods. If this were a small country, guerrilla activities could be carried out close to the scene of operations of the regular army and directly complementary to them. In such a case, there would be no question of guerrilla strategy as such. Nor would the question arise if our country were as strong as Russia, for example, and able speedily to eject an invader. The question exists because China, a weak country of vast size, has today progressed to the point where it has become possible to adopt the policy of a protracted war characterized by guerrilla operations. Although these may at first glance seem to be abnormal or heterodox, such is not actually the case.

Because Japanese military power is inadequate, much of the territory her armies have overrun is without sufficient garrison troops. Under such circumstances the primary functions of guerrillas are three: first, to conduct a war on exterior lines, that is, in the rear of the enemy; second, to

Yu Chi Chan (Guerrilla Warfare)

establish bases; and, last, to extend the war areas. Thus, guerrilla participation in the war is not merely a matter of purely local guerrilla tactics but involves strategical considerations.

Such war, with its vast time and space factors, establishes a new military process, the focal point of which is China today. The Japanese are apparently attempting to recall a past that saw the Yüan extinguish the Sung and the Ch'ing conquer the Ming; that witnessed the extension of the British Empire to North America and India; that saw the Latins overrun Central and South America. As far as China today is concerned, such dreams of conquest are fantastic and without reality. Today's China is better equipped than was the China of yesterday, and a new type of guerrilla hostilities is a part of that equipment. If our enemy fails to take these facts into consideration and makes too optimistic an estimate of the situation, he courts disaster.

Though the strategy of guerrillas is inseparable from war strategy as a whole, the actual conduct of these hostilities differs from the conduct of orthodox operations. Each type of warfare has methods peculiar to itself, and methods suitable to regular warfare cannot be applied with success to the special situations that confront guerrillas.

Before we treat the practical aspects of guerrilla war, it might be well to recall the fundamental axiom of combat on which all military action is based. This can be stated: "Conservation of one's own strength; destruction of enemy strength." A military policy based on this axiom is consonant with a national policy directed towards the building of a free and prosperous Chinese state and the destruction of Japanese imperialism. It is in furtherance of this policy that government applies its military strength. Is the sacrifice demanded by war in conflict with the idea of self-preservation? Not at all. The sacrifices demanded are necessary both to destroy the enemy and to preserve ourselves; the sacrifice of a part of the people is necessary to preserve the whole. All the considerations of military action are derived from this axiom. Its application is as apparent in all tactical and strategical conceptions as it is in the simple case of the soldier who shoots at his enemy from a covered position.

All guerrilla units start from nothing and grow. What methods should we select to ensure the conservation and development of our own strength and the destruction of that of the enemy? The essential requirements are the six listed below:

1. Retention of the initiative; alertness; carefully planned tactical attacks in a war of strategical defense; tactical speed in a war strategically protracted; tactical operations on exterior lines in a war conducted strategically on interior lines.
2. Conduct of operations to complement those of the regular army.
3. The establishment of bases.
4. A clear understanding of the relationship that exists between the attack and the defense.
5. The development of mobile operations.
6. Correct command.

Yu Chi Chan (Guerrilla Warfare)

The enemy, though numerically weak, is strong in the quality of his troops and their equipment; we, on the other hand, are strong numerically but weak as to quality. These considerations have been taken into account in the development of the policy of tactical offense, tactical speed, and tactical operations on exterior lines in a war that, strategically speaking, is defensive in character, protracted in nature, and conducted along interior lines. Our strategy is based on these conceptions. They must be kept in mind in the conduct of all operations.

Although the element of surprise is not absent in orthodox warfare, there are fewer opportunities to apply it than there are during guerrilla hostilities. In the latter, speed is essential. The movements of guerrilla troops must be secret and of supernatural rapidity; the enemy must be taken unaware, and the action entered speedily. There can be no procrastination in the execution of plans; no assumption of a negative or passive defense; no great dispersion of forces in many local engagements. The basic method is the attack in a violent and deceptive form.

While there may be cases where the attack will extend over a period of several days (if that length of time is necessary to annihilate an enemy group), it is more profitable to launch and push an attack with maximum speed. The tactics of defense have no place in the realm of guerrilla warfare. If a delaying action is necessary, such places as defiles, river crossings, and villages offer the most suitable conditions, for it is in such places that the enemy's arrangements may be disrupted and he may be annihilated.

The enemy is much stronger than we are, and it is true

that we can hinder, distract, disperse, and destroy him only if we disperse our own forces. Although guerrilla warfare is the warfare of such dispersed units, it is sometimes desirable to concentrate in order to destroy an enemy. Thus, the principle of concentration of force against a relatively weaker enemy is applicable to guerrilla warfare.

We can prolong this struggle and make of it a protracted war only by gaining positive and lightning-like tactical decisions; by employing our manpower in proper concentrations and dispersions; and by operating on exterior lines in order to surround and destroy our enemy. If we cannot surround whole armies, we can at least partially destroy them; if we cannot kill the Japanese, we can capture them. The total effect of many local successes will be to change the relative strengths of the opposing forces. The destruction of Japan's military power, combined with the international sympathy for China's cause and the revolutionary tendencies evident in Japan, will be sufficient to destroy Japanese imperialism.

We will next discuss initiative, alertness, and the matter of careful planning. What is meant by initiative in warfare? In all battles and wars, a struggle to gain and retain the initiative goes on between the opposing sides, for it is the side that holds the initiative that has liberty of action. When an army loses the initiative, it loses its liberty; its role becomes passive; it faces the danger of defeat and destruction.

It is more difficult to obtain the initiative when defending on interior lines than it is while attacking on exterior

Yu Chi Chan (Guerrilla Warfare)

lines. This is what Japan is doing. There are, however, several weak points as far as Japan is concerned. One of these is lack of sufficient manpower for the task; another is her cruelty to the inhabitants of conquered areas; a third is the underestimation of Chinese strength, which has resulted in the differences between military cliques, which, in turn, have been productive of many mistakes in the direction of her military forces. For instance, she has been gradually compelled to increase her manpower in China while, at the same time, the many arguments over plans of operations and disposition of troops have resulted in the loss of good opportunities for improvement of her strategical position. This explains the fact that although the Japanese are frequently able to surround large bodies of Chinese troops, they have never yet been able to capture more than a few. The Japanese military machine is thus being weakened by insufficiency of manpower, inadequacy of resources, the barbarism of her troops, and the general stupidity that has characterized the conduct of operations. Her offensive continues unabated, but because of the weaknesses pointed out, her attack must be limited in extent. She can never conquer China. The day will come—indeed, already has in some areas—when she will be forced into a passive role. When hostilities commenced, China was passive, but as we enter the second phase of the war, we find ourselves pursuing a strategy of mobile warfare, with both guerrillas and regulars operating on exterior lines. Thus, with each passing day, we seize some degree of initiative from the Japanese.

Mao Tse-tung on Guerrilla Warfare

The matter of initiative is especially serious for guerrilla forces, who must face critical situations unknown to regular troops. The superiority of the enemy and the lack of unity and experience within our own ranks may be cited. Guerrillas can, however, gain the initiative if they keep in mind the weak points of the enemy. Because of the enemy's insufficient manpower, guerrillas can operate over vast territories; because he is a foreigner and a barbarian, guerrillas can gain the confidence of millions of their countrymen; because of the stupidity of enemy commanders, guerrillas can make full use of their own cleverness. Both guerrillas and regulars must exploit these enemy weaknesses while, at the same time, our own are remedied. Some of our weaknesses are apparent only and are, in actuality, sources of strength. For example, the very fact that most guerrilla groups are small makes it desirable and advantageous for them to appear and disappear in the enemy's rear. With such activities, the enemy is simply unable to cope. A similar liberty of action can rarely be obtained by orthodox forces.

When the enemy attacks the guerrillas with more than one column, it is difficult for the latter to retain the initiative. Any error, no matter how slight, in the estimation of the situation is likely to result in forcing the guerrillas into a passive role. They will then find themselves unable to beat off the attacks of the enemy.

It is apparent that we can gain and retain the initiative only by a correct estimation of the situation and a proper arrangement of all military and political factors. A too pessimistic estimate will operate to force us into a passive

Yu Chi Chan (Guerrilla Warfare)

position, with consequent loss of initiative; an overly optimistic estimate, with its rash ordering of factors, will produce the same result.

No military leader is endowed by heaven with an ability to seize the initiative. It is the intelligent leader who does so after a careful study and estimate of the situation and arrangement of the military and political factors involved. When a guerrilla unit, through either a poor estimate on the part of its leader or pressure from the enemy, is forced into a passive position, its first duty is to extricate itself. No method can be prescribed for this, as the method to be employed will, in every case, depend on the situation. One can, if necessary, run away. But there are times when the situation seems hopeless and, in reality, is not so at all. It is at such times that the good leader recognizes and seizes the moment when he can regain the lost initiative.

Let us revert to alertness. To conduct one's troops with alertness is an essential of guerrilla command. Leaders must realize that to operate alertly is the most important factor in gaining the initiative and vital in its effect on the relative situation that exists between our forces and those of the enemy. Guerrilla commanders adjust their operations to the enemy situation, to the terrain, and to prevailing local conditions. Leaders must be alert to sense changes in these factors and make necessary modifications in troop dispositions to accord with them. The leader must be like the fisherman, who, with his nets, is able both to cast them and to pull them out in awareness of the depth of the water, the strength of the current, or the presence of any obstructions that may foul them. As the fisherman controls

Mao Tse-tung on Guerrilla Warfare

his nets through the lead ropes, so the guerrilla leader maintains contact with and control over his units. As the fisherman must change his position, so must the guerrilla commander. Dispersion, concentration, constant change of position—it is in these ways that guerrillas employ their strength.

In general, guerrilla units disperse to operate:

1. When the enemy is in overextended defense, and sufficient force cannot be concentrated against him, guerrillas must disperse, harass him, and demoralize him.
2. When encircled by the enemy, guerrillas disperse to withdraw.
3. When the nature of the ground limits action, guerrillas disperse.
4. When the availability of supplies limits action, they disperse.
5. Guerrillas disperse in order to promote mass movements over a wide area.

Regardless of the circumstances that prevail at the time of dispersal, caution must be exercised in certain matters:

1. A relatively large group should be retained as a central force. The remainder of the troops should not be divided into groups of absolutely equal size. In this way, the leader is in a position to deal with any circumstances that may arise.
2. Each dispersed unit should have clear and definite responsibilities. Orders should specify a place to which to

Yu Chi Chan (Guerrilla Warfare)

proceed, the time of proceeding, and the place, time, and method of assembly.

Guerrillas concentrate when the enemy is advancing upon them, and there is opportunity to fall upon him and destroy him. Concentration may be desirable when the enemy is on the defensive and guerrillas wish to destroy isolated detachments in particular localities. By the term "concentrate," we do not mean the assembly of all manpower but rather of only that necessary for the task. The remaining guerrillas are assigned missions of hindering and delaying the enemy, of destroying isolated groups, or of conducting mass propaganda.

In addition to the dispersion and concentration of forces, the leader must understand what is termed "alert shifting." When the enemy feels the danger of guerrillas, he will generally send troops out to attack them. The guerrillas must consider the situation and decide at what time and at what place they wish to fight. If they find that they cannot fight, they must immediately shift. Then the enemy may be destroyed piecemeal. For example, after a guerrilla group has destroyed an enemy detachment at one place, it may be shifted to another area to attack and destroy a second detachment. Sometimes, it will not be profitable for a unit to become engaged in a certain area, and in that case, it must move immediately.

When the situation is serious, the guerrillas must move with the fluidity of water and the ease of the blowing wind. Their tactics must deceive, tempt, and confuse the enemy. They must lead the enemy to believe that they will attack him from the east and north, and they must then strike him from the west and the south. They must strike, then rapidly disperse. They must move at night.

Guerrilla initiative is expressed in dispersion, concentration, and the alert shifting of forces. If guerrillas are stupid and obstinate, they will be led to passive positions and severely damaged. Skill in conducting guerrilla operations, however, lies not in merely understanding the things we have discussed but rather in their actual application on the field of battle. The quick intelligence that constantly watches the ever-changing situation and is able to seize on the right moment for decisive action is found only in keen and thoughtful observers.

Careful planning is necessary if victory is to be won in guerrilla war, and those who fight without method do not understand the nature of guerrilla action. A plan is necessary regardless of the size of the unit involved; a prudent plan is as necessary in the case of the squad as in the case of the regiment. The situation must be carefully studied, then an assignment of duties made. Plans must include both political and military instruction; the matter of supply and equipment, and the matter of cooperation with local civilians. Without study of these factors, it is impossible either to seize the initiative or to operate alertly. It is true that guerrillas can make only limited plans, but even so, the factors we have mentioned must be considered.

The initiative can be secured and retained only following a positive victory that results from attack. The attack must be made on guerrilla initiative; that is, guerrillas must not permit themselves to be maneuvered into a position

Yu Chi Chan (Guerrilla Warfare)

where they are robbed of initiative and where the decision to attack is forced upon them. Any victory will result from careful planning and alert control. Even in defense, all our efforts must be directed toward a resumption of the attack, for it is only by attack that we can extinguish our enemies and preserve ourselves. A defense or a withdrawal is entirely useless as far as extinguishing our enemies is concerned and of only temporary value as far as the conservation of our forces is concerned. This principle is valid both for guerrillas and regular troops. The differences are of degree only; that is to say, in the manner of execution.

The relationship that exists between guerrillas and the orthodox forces is important and must be appreciated. Generally speaking, there are three types of cooperation between guerrillas and orthodox groups. These are:

1. Strategical cooperation.
2. Tactical cooperation.
3. Battle cooperation.

Guerrillas who harass the enemy's rear installations and hinder his transport are weakening him and encouraging the national spirit of resistance. They are cooperating strategically. For example, the guerrillas in Manchuria had no functions of strategical cooperation with orthodox forces until the war in China started. Since that time, their function of strategical cooperation is evident, for if they can kill one enemy, make the enemy expend one round of ammunition, or hinder one enemy group in its advance southward, our powers of resistance here are proportionately increased. Such guerrilla action has a positive action on the enemy nation and on its troops, while, at the same time, it encourages our own countrymen. Another example of strategical cooperation is furnished by the guerrillas who operate along the P'ing-Sui, P'ing-Han, Chin-P'u, T'ung-Pu, and Cheng-T'ai railways. This cooperation began when the invader attacked, continued during the period when he held garrisoned cities in the areas, and was intensified when our regular forces counterattacked, in an effort to restore the lost territories.

As an example of tactical cooperation, we may cite the operations at Hsing-K'ou, when guerrillas both north and south of Yeh Men destroyed the T'ung-P'u railway and the motor roads near P'ing Hsing Pass and Yang Fang K'ou. A number of small operating bases were established, and organized guerrilla action in Shansi complemented the activities of the regular forces both there and in the defense of Honan. Similarly, during the south Shantung campaign, guerrillas in the five northern provinces cooperated with the army's operation on the Hsuchow front.

Guerrilla commanders in rear areas and those in command of regiments assigned to operate with orthodox units must cooperate in accordance with the situation. It is their function to determine weak points in the enemy dispositions, to harass them, to disrupt their transport, and to undermine their morale. If guerrilla action were independent, the results to be obtained from tactical cooperation would be lost and those that result from strategical cooperation greatly diminished. In order to accomplish their

Yu Chi Chan (Guerrilla Warfare)

mission and improve the degree of cooperation, guerrilla units must be equipped with some means of rapid communication. For this purpose, two-way radio sets are recommended.

Guerrilla forces in the immediate battle area are responsible for close cooperation with regular forces. Their principal functions are to hinder enemy transport, to gather information, and to act as outposts and sentinels. Even without precise instructions from the commander of the regular forces, these missions, as well as any others that contribute to the general success, should be assumed.

The problem of establishment of bases is of particular importance. This is so because this war is a cruel and protracted struggle. The lost territories can be restored only by a strategical counterattack, and this we cannot carry out until the enemy is well into China. Consequently, some part of our country—or, indeed, most of it—may be captured by the enemy and become his rear area. It is our task to develop intensive guerrilla warfare over this vast area and convert the enemy's rear into an additional front. Thus the enemy will never be able to stop fighting. In order to subdue the occupied territory, the enemy will have to become increasingly severe and oppressive.

A guerrilla base may be defined as an area, strategically located, in which the guerrillas can carry out their duties of training, self-preservation and development. Ability to fight a war without a rear area is a fundamental characteristic of guerrilla action, but this does not mean that guerrillas can exist and function over a long period of time without the development of base areas. History shows us many examples of peasant revolts that were unsuccessful, and it is fanciful to believe that such movements, characterized by banditry and brigandage, could succeed in this era of improved communications and military equipment. Some guerrilla leaders seem to think that those qualities are present in today's movement, and before such leaders can comprehend the importance of base areas in the long-term war, their minds must be disabused of this idea.

The subject of bases may be better understood if we consider:

1. The various categories of bases.
2. Guerrilla areas and base areas.
3. The establishment of bases.
4. The development of bases.

Guerrilla bases may be classified according to their location as: first, mountain bases; second, plains bases; and, last, river, lake, and bay bases. The advantages of bases in mountainous areas are evident. Those which are now established are at Ch'ang P'o Chan, Wu Tai Shan, Taiheng Shan, Tai Shan, Yen Shan, and Mao Shan. These bases are strongly protected. Similar bases should be established in all enemy rear areas.

Plains country is generally not satisfactory for guerrilla operating bases, but this does not mean that guerrilla warfare cannot flourish in such country or that bases cannot be established there. The extent of guerrilla development in Hopeh and west Shantung proves the opposite to be the case. Whether we can count on the use of these bases

Yu Chi Chan (Guerrilla Warfare)

over long periods of time is questionable. We can, however, establish small bases of a seasonal or temporary nature. This we can do because our barbaric enemy simply does not have the manpower to occupy all the areas he has overrun and because the population of China is so numerous that a base can be established anywhere. Seasonal bases in plains country may be established in the winter when the rivers are frozen over, and in the summer when the crops are growing. Temporary bases may be established when the enemy is otherwise occupied. When the enemy advances, the guerrillas who have established bases in the plains area are the first to engage him. Upon their withdrawal into mountainous country, they should leave behind them guerrilla groups dispersed over the entire area. Guerrillas shift from base to base on the theory that they must be one place one day and another place the next.

There are many historical examples of the establishment of bases in river, bay, and lake country, and this is one aspect of our activity that has so far received little attention. Red guerrillas held out for many years in the Hungtze Lake region. We should establish bases in the Hungtze and Tai areas and along rivers and watercourses in territory controlled by the enemy so as to deny him access to, and free use of, the water routes.

There is a difference between the terms base area and guerrilla area. An area completely surrounded by territory occupied by the enemy is a "base area." Wu Tai Shan, Tai Shan, and Taiheng Shan are examples of base areas. On the other hand, the area east and north of Wu Tai Shan (the Shansi-Hopeh-Chahar border zone) is a guerrilla area. Such areas can be controlled by guerrillas only while they actually physically occupy them. Upon their departure, control reverts to a puppet pro-Japanese government. East Hopeh, for example, was at first a guerrilla area rather than a base area. A puppet government functioned there. Eventually, the people, organized and inspired by guerrillas from the Wu Tai mountains, assisted in the transformation of this guerrilla area into a real base area. Such a task is extremely difficult, for it is largely dependent upon the degree to which the people can be inspired. In certain garrisoned areas, such as the cities and zones contiguous to the railroads, the guerrillas are unable to drive the Japanese and puppets out. These areas remain guerrilla areas. At other times, base areas might become guerrilla areas due either to our own mistakes or to the activities of the enemy.

Obviously, in any given area in the war zone, any one of three situations may develop: The area may remain in Chinese hands; it may be lost to the Japanese and puppets; or it may be divided between the combatants. Guerrilla leaders should endeavor to see that either the first or the last of these situations is assured.

Another point essential in the establishment of bases is the cooperation that must exist between the armed guerrilla bands and the people. All our strength must be used to spread the doctrine of armed resistance to Japan, to arm the people, to organize self-defense units, and to train guerrilla bands. This doctrine must be spread among the people, who must be organized into anti-Japanese groups. Their political instincts must be sharpened and their mar-

and the encirclement broken by counterattack. As such enemy columns are without reserves, we should plan on using our main forces to attack one of them by surprise and devote our secondary effort to continual hindrance and harassment. At the same time, other forces should isolate enemy garrison troops and operate on their lines of supply and communication. When one column has been disposed of, we may turn our attention to one of the others. In a base area as large as Wu Tai Shan, for example, there are four or five military subdivisions. Guerrillas in these subdivisions must cooperate to form a primary force to counterattack the enemy, or the area from which he came, while a secondary force harasses and hinders him.

After defeating the enemy in any area, we must take advantage of the period he requires for reorganization to press home our attacks. We must not attack an objective we are not certain of winning. We must confine our operations to relatively small areas and destroy the enemy and traitors in those places.

When the inhabitants have been inspired, new volunteers accepted, trained, equipped, and organized, our operations may be extended to include cities and lines of communication not strongly held. We may hold these at least for temporary (if not for permanent) periods. All these are our duties in offensive strategy. Their object is to lengthen the period that the enemy must remain on the defensive. Then our military activities and our organization work among the masses of the people must be zealously expanded; and with equal zeal, the strength of the enemy attacked and diminished. It is of great importance that tial ardor increased. If the workers, the farmers, the lovers of liberty, the young men, the women, and the children are not organized, they will never realize their own anti-Japanese power. Only the united strength of the people can eliminate traitors, recover the measure of political power that has been lost, and conserve and improve what we still retain.

We have already touched on geographic factors in our discussion of bases, and we must also mention the economic aspects of the problem. What economic policy should be adopted? Any such policy must offer reasonable protection to commerce and business. We interpret "reasonable protection" to mean that people must contribute money in proportion to the money they have. Farmers will be required to furnish a certain share of their crops to guerrilla troops. Confiscation, except in the case of businesses run by traitors, is prohibited.

Our activities must be extended over the entire periphery of the base area if we wish to attack the enemy's bases and thus strengthen and develop our own. This will afford us opportunity to organize, equip, and train the people, thus furthering guerrilla policy as well as the national policy of protracted war. At times, we must emphasize the development and extension of base areas; at other times, the organization, training, or equipment of the people.

Each guerrilla base will have its own peculiar problems of attack and defense. In general, the enemy, in an endeavor to consolidate his gains, will attempt to extinguish guerrilla bases by dispatching numerous bodies of troops over a number of different routes. This must be anticipated

Yu Chi Chan (Guerrilla Warfare)

guerrilla units be rested and instructed. During such times as the enemy is on the defensive, the troops may get some rest and instruction may be carried out.

The development of mobile warfare is not only possible but essential. This is the case because our current war is a desperate and protracted struggle. If China were able to conquer the Japanese bandits speedily and to recover her lost territories, there would be no question of long-term war on a national scale. Hence, there would be no question of the relation of guerrilla warfare and the war of movement. Exactly the opposite is actually the case. In order to ensure the development of guerrilla hostilities into mobile warfare of an orthodox nature, both the quantity and quality of guerrilla troops must be improved. Primarily, more men must join the armies; then the quality of equipment and standards of training must be improved. Political training must be emphasized and our organization, the technique of handling our weapons, our tactics—all must be improved. Our internal discipline must be strengthened. The soldiers must be educated politically. There must be a gradual change from guerrilla formations to orthodox regimental organization. The necessary bureaus and staffs, both political and military, must be provided. At the same time, attention must be paid to the creation of suitable supply, medical, and hygiene units. The standards of equipment must be raised and types of weapons increased. Communication equipment must not be forgotten. Orthodox standards of discipline must be established.

Mao Tse-tung on Guerrilla Warfare

Because guerrilla formations act independently and because they are the most elementary of armed formations, command cannot be too highly centralized. If it were, guerrilla action would be too limited in scope. At the same time, guerrilla activities, to be most effective, must be coordinated, not only insofar as they themselves are concerned, but additionally with regular troops operating in the same areas. This coordination is a function of the war-zone commander and his staff.

In guerrilla base areas, the command must be centralized for strategical purposes and decentralized for tactical purposes. Centralized strategical command takes care of the general management of all guerrilla units, their coordination within war zones, and the general policy regarding guerrilla base areas. Beyond this, centralization of command will result in interference with subordinate units, as, naturally, the tactics to apply to concrete situations can be determined only as these various situations arise. This is true in orthodox warfare when communications between lower and higher echelons break down. In a word, proper guerrilla policy will provide for unified strategy and independent activity.

Each guerrilla area is divided into districts and these in turn are divided into subdistricts. Each subdivision has its appointed commander, and while general plans are made by higher commanders, the nature of actions is determined by inferior commanders. The former may suggest the nature of the action to be taken but cannot define it. Thus inferior groups have more or less complete local control.

APPENDIX

NOTES

1. Each squad consists of from 9 to 11 men. In case men or arms are not sufficient, the third platoon may be dispensed with or one squad organized as company headquarters.

2. The mobile propaganda unit consists of members of the company who are not relieved of primary duties except to carry out propaganda when they are not fighting.

3. If there is insufficient personnel, the medical section is not separately organized. If there are only two or three medical personnel, they may be attached to the administrative section.

4. If there is no barber, it is unimportant. If there is an insufficient number of cooks, any member of the company may be designated to prepare food.

5. Each combatant soldier should be armed with the rifle. If there are not enough rifles, each squad should have two or three. Shotguns, lances, and big swords can also be furnished. The distribution of rifles does not have to be equalized in platoons. As different missions are assigned to platoons, it may be necessary to give one platoon more rifles than the others.

6. The strength of a company should at the most be 180, divided into 12 squads of 11 men each. The minimum strength of a company should be 82 men, divided into 6 squads of 9 men each.

TABLE 1

ORGANIZATION OF AN INDEPENDENT GUERRILLA COMPANY

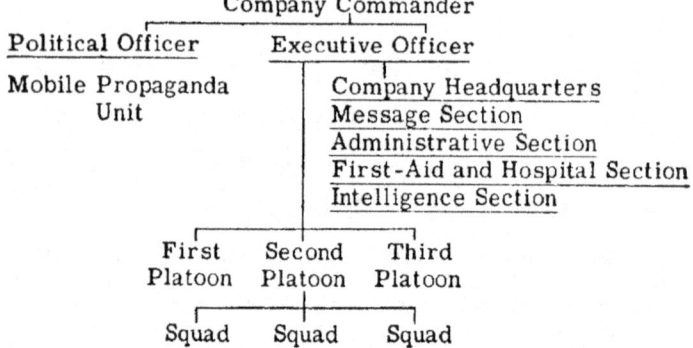

TABLE OF ORGANIZATION, GUERRILLA COMPANY

RANK	PERSONNEL	ARM
Company Leader	1	Pistol
Political Officer	1	Pistol
Executive Officer	1	Pistol
Company Headquarters		
Message Section Chief	1	
Signal	1	
Administrative Section Chief	1	Rifle
Public Relations	3	Rifle
Duty Personnel	2	
Barber	1	
Cooks	10	
Medical Section Chief	1	
Assistant	1	
First Aid and Nursing	4	
Intelligence Section Chief	1	Rifle
Intelligence	9	Rifle
Platoon Leaders	3	Rifle
Squad Leaders	9	Rifle
Nine Squads (8 each)	72	Rifle
TOTAL	122	3 Pistols 98 Rifles

TABLE 2

ORGANIZATION OF AN INDEPENDENT GUERRILLA BATTALION

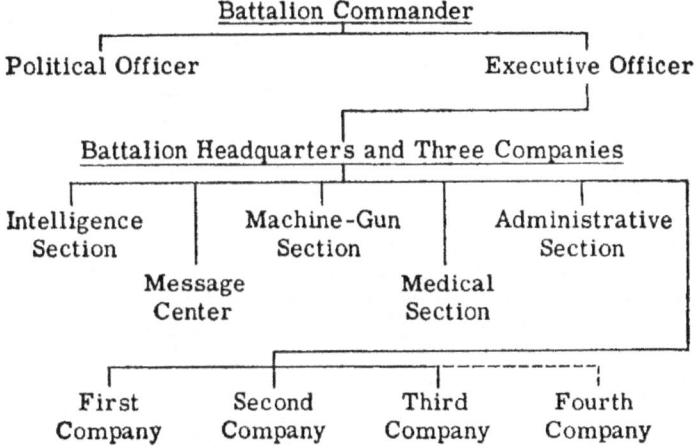

NOTES

1. Total headquarters of an independent guerrilla battalion may vary from a minimum of 46 to a maximum of 110.
2. When there are 4 companies to a battalion, regimental organization should be used.
3. Machine-gun squads may be heavy or light. A light machine-gun squad has from 5 to 7 men. A heavy machine-gun squad has from 7 to 9 men.
4. The intelligence section is organized in from 2 to 4 squads, at least one of which is made up of plain-clothes men. If horses are available, one squad should be mounted.
5. If no men are available for stretcher-bearers, omit them and use the cooks or ask aid from the people.
6. Each company must have at least 25 rifles. The remaining weapons may be bird guns, big swords, or locally made shotguns.

TABLE 3

ORGANIZATION OF AN INDEPENDENT GUERRILLA REGIMENT

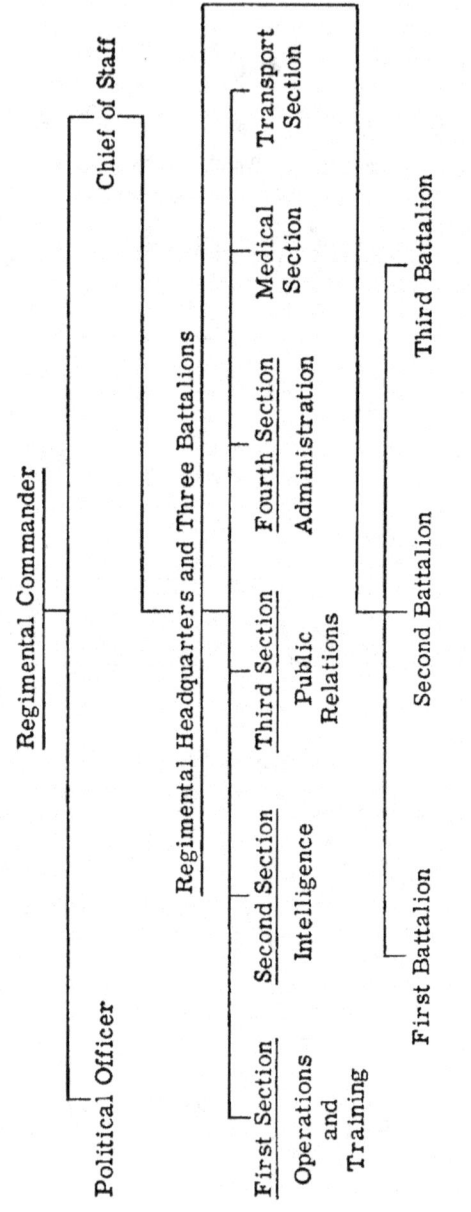

NOTES

1. See Tables 1 and 2 for company and battalion organization.
2. Battalions and companies have no transport sections.
3. The hand weapon may be either revolver or pistol. Of these, each battalion should have more than 100.

TABLE OF ORGANIZATION, GUERRILLA REGIMENT

RANK	PERSONNEL	ARM
Regimental Commander	1	Pistol
Political Officer	1	Pistol
Chief of Staff	1	Pistol
Operations Section		
Operations Officer	1	Pistol
Clerks	15	
Intelligence Section		
Intelligence Officer	1	Pistol
Personnel	36	Pistols
Public-Relations Section		
Public-Relations Officer	1	Pistol
Personnel	36	Carbines
Administrative Section		
Administrative Officer	1	Pistol
Clerks	15	Pistol
Runner	1	
Transport Section		
Chief of Section	1	Pistol
Finance	1	
Traffic Manager	1	Pistol
Supply	1	
Drivers	5	
Medical Section		
Chief of Section	1	
Doctors	2	
Nurses	15	
Total, Regimental Headquarters	137	60 Pistols 36 Carbines
Three Battalions (441 each)	1323	124 Pistols 900 Rifles
TOTAL	1460	184 Pistols 936 Rifles

TABLE OF ORGANIZATION, GUERRILLA BATTALION (INDEPENDENT)

RANK	PERSONNEL	ARM
Battalion Commander	1	Pistol
Political Officer	1	Pistol
Executive Officer	1	Pistol
Battalion Headquarters		
Signal Section	2	
Administrative Section		
Section Chief	1	Carbine
Runner	1	Carbine
Public Relations	10	Carbine
Duty Personnel	2	
Barbers	3	
Supply	1	
Cooks	10	
Medical Section		
Medical Officer	1	
Stretcher-Bearers	6	
Nursing	4	
Intelligence Section		
Section Chief	1	Pistol
Intelligence	30	Pistol
Machine-Gun Section	As Available	As Available
Total, Headquarters	75	34 Pistols 12 Carbines
Three Companies (122 each)	366	9 Pistols 288 Carbines
TOTAL	441	43 Pistols 300 Rifles

TABLE 4

ORGANIZATION OF INDEPENDENT GUERRILLA BRIGADE (OR DIVISION)

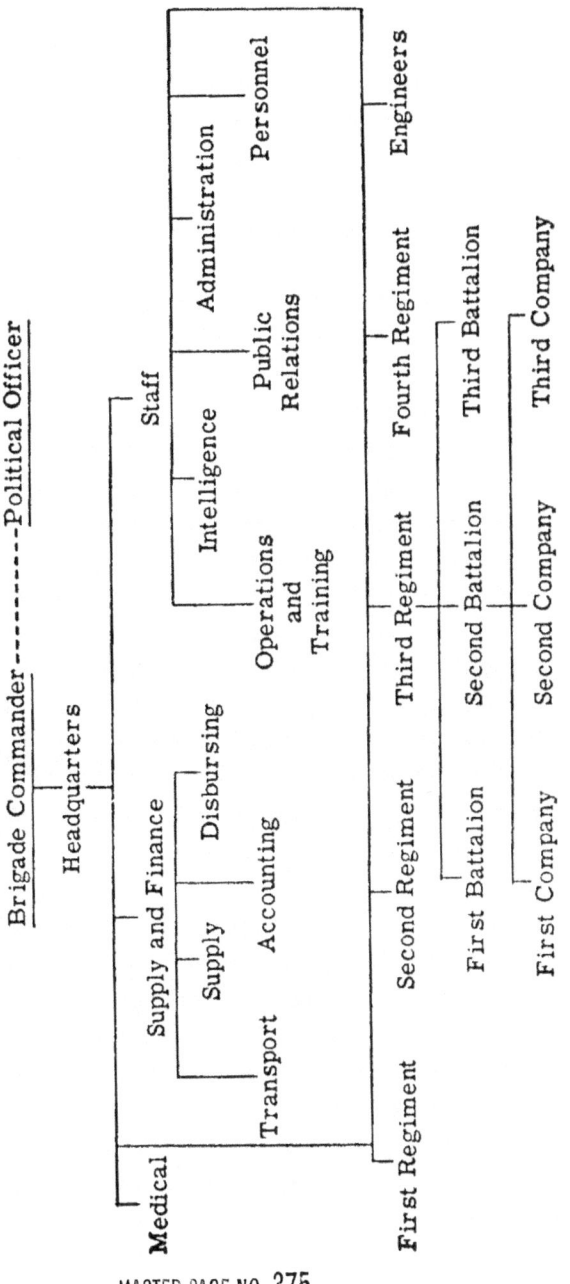

PART 3

GUERRILLA WARFARE

ERNESTO "CHE" GUEVARA

1961

Contents

Dedication to Camilo 2
CHAPTER I: GENERAL PRINCIPLES OF GUERRILLA WARFARE 4
    1 The Essence Of Guerrilla Warfare 4
    2 Guerrilla Strategy 7
    3 Guerrilla Tactics 9
    4 Warfare On Favorable Terrain 13
    5 Warfare On Unfavorable Terrain 15
    6 Urban Warfare 18
CHAPTER II: THE GUERRILLA BAND 20
    1 The Guerrilla Fighter: Social Reformer 20
    2 The Guerrilla Fighter As Combatant 21
    3 Organization Of A Guerrilla Band 28
    4 Combat 33
    5 Beginning, Development, And End Of A Guerrilla War 40
CHAPTER III: ORGANIZATION OF THE GUERRILLA FRONT 43
    1 supply 43
    2 Civil Organization 45
    3 The Role Of Women 48
    4 Health 50
    5 Sabotage 52
    6 War Industry 54
    7 Propaganda 55
    8 Intelligence 56
    9 Training And Indoctrination 57
    10   The Organizational Structure Of The Army Of A Revolutionary Movement 59
Appendices 62
    Appendix A - Underground Organization of the First Guerrilla Band 62
    Appendix 2 - Holding Power 64
    Appendix 3 - Epilogue: Analysis of the Situation in Cuba, Present and Future 66

Dedication to Camilo

This work is dedicated to Camilo Cienfuegos, who should have read and corrected it, but whose fate prevented him from carrying out the task.* These lines and those that follow may be considered a homage of the Rebel Army to its grand captain, to the greatest guerrilla chief that this revolution produced, to an outstanding revolutionary and a fraternal friend.

Camilo was the compañero of a hundred battles, the intimate confidant of Fidel in difficult moments of the war, the stoic fighter who always made sacrifice into an instrument for steeling his own character and forging the morale of his troops.

I believe he would have approved of this manual in which our guerrilla experiences are synthesized, because it is the product of life itself; but he added to the words outlined here the essential vitality of his temperament, his intelligence and his audacity, in such an exact measure as appears in few people in history.

But Camilo should not be seen as an isolated hero performing miraculous feats by virtue of the impulse of his individual genius, but rather as a true product of the people that formed him, as heroes, martyrs, and leaders are always selected through the exigencies of the struggle.

I don't know if Camilo knew Danton's maxim for revolutionary movements, "Audacity, audacity, and more audacity." Nevertheless, he practiced it in his actions, while adding other qualities necessary in a guerrilla fighter: an ability to assess a situation quickly and precisely and to anticipate problems to be solved in the future.

Although these lines may serve as the homage of the author and of a whole people to our hero, they are not a biography, and do not attempt to relate the many anecdotes about him. Camilo was the subject of a thousand anecdotes; inevitably, he created stories wherever he went. Camilo's easy manner, always appreciated by the people, was matched by a personality that naturally and almost unconsciously put a stamp on everything around him. Few men have succeeded in leaving such a distinctive personal mark on everything they do. As Fidel has said, his culture came not from books; he had the natural intelligence of the people, who had chosen him out of thousands for a privileged place because of the audacity of his strikes, his tenacity, his intelligence, and his unmatched devotion.

Camilo practiced loyalty like a religion; he was a devotee, both in his personal loyalty to Fidel, who embodied the will of the people as no one else, and in his loyalty to the people themselves. The people and Fidel march united, as did the devotions of the invincible guerrilla fighter.

Who killed him?

We should ask instead: Who destroyed his body? Because men like him live on in the people as long as the people will it to be so.

The enemy succeeded in killing him because they wanted him dead; they killed him because there are no safe airplanes; because pilots cannot acquire all the necessary experience; because, overburdened with work, Camilo wished to get back to Havana quickly. And his own character killed him, too. Camilo never calculated danger; he used it as a diversion, he mocked it, lured, toyed, and played with it. In the spirit of a guerrilla fighter a plan could not be postponed on account of a few clouds.

As it was, an entire people had come to know him, admire him, and love him. It might have happened earlier, and his story would be the simple tale of a guerrilla captain. "There will be many Camilos," said Fidel; to that I can add, there were other Camilos, Camilos whose lives ended before completing the magnificent circle that drew Camilo into history. Camilo and the other Camilos (those who did not make it and those who will come after) are the indicators of the power of the people. They are the highest expression of what a nation can produce in a time of war for the defense of its purest ideals, fought with faith in the achievement of its noblest goals.

Let's not typecast him or encase him in a mold — in other words, kill him. Let's leave him there, in general outline, without attributing to him a precise social and economic ideology that was never completely defined. Let's reiterate that there was no other soldier like Camilo in this war of liberation — a complete revolutionary, a man of the people, a product of this revolution that the Cuban nation made for itself. The slightest shadow of weariness or discouragement never entered his head. Camilo, the guerrilla fighter, who made everything "something of Camilo," who put his precise and indelible mark on the Cuban revolution, is a permanent and daily inspiration. He belongs to those others who did not make it and to those who are yet to come.

In his continual and immortal renewal, Camilo is the reflection of the people.

Ernesto Che Guevara

CHAPTER I: GENERAL PRINCIPLES OF GUERRILLA WARFARE

1 The Essence Of Guerrilla Warfare

The armed victory of the Cuban people over the Batista dictatorship has not only been the triumph of heroism reported by the world's newspapers; it has also forced a change in the old dogmas concerning the conduct of the popular masses of Latin America and clearly demonstrated the capacity of the people to free themselves through guerrilla warfare from an oppressive government.

We consider that the Cuban revolution contributed three fundamental lessons to the revolutionary movements in America.* They are:

1. Popular forces can win a war against the army.
2. It is not always necessary to wait until all the revolutionary conditions exist; the insurrectional foco can develop subjective conditions based on existing objective conditions
3. In underdeveloped America the countryside is the fundamental arena for armed struggle.

Of these three propositions, the first two challenge the defeatist attitude of revolutionaries or inactive pseudo-revolutionaries who take refuge in the argument that against a professional army nothing can be done, and who sit down to wait until all necessary objective and subjective conditions are attained somehow mechanically, without trying to accelerate them. Although now clear to everyone, these two undeniable truths were previously a matter for discussion in Cuba, and are probably still debated today in America.

Naturally, when speaking of the necessary conditions for a revolution, it should not be assumed they can be created solely through the stimulus of a guerrilla foco. It must always be understood that there are minimum conditions without which the establishment and consolidation of the foco is not practicable. Moreover, it is necessary to demonstrate clearly to people the futility of maintaining the fight for social gains within the framework of civil debate. When the oppressive forces maintain themselves in power against the laws they themselves established, peace must be considered already broken.

Under these conditions popular discontent expresses itself in more and more active forms, and resistance finally crystallizes, at a given moment, in an outbreak of the struggle. Where a government has come into power through some form of popular vote, fraudulent or not, and maintains at least an appearance of constitutional legality, the guerrilla movement will experience great difficulties, as the possibilities for civil struggle have not yet been exhausted.

The third proposition is of a fundamental strategic nature and must be noted by those who dogmatically argue that the mass struggle is based in cities, entirely ignoring the immense weight of the people from the countryside in the life of all the underdeveloped countries of America. This is not to underrate the struggles of the mass of organized workers; but simply to analyze the real possibilities for engaging in armed struggle where the guarantees that usually adorn our constitutions are suspended or ignored. In these conditions the workers' movement must function clandestinely without arms and face enormous dangers. The situation is less difficult in the open countryside, where the armed guerrillas can support the local people, and where there are places beyond the reach of the repressive forces.

Although later we will make a careful analysis, to begin this task we want to emphasize three conclusions that are features of the Cuban revolutionary experience and which are fundamental to our argument.

Guerrilla warfare, the basis of the struggle of a people to redeem themselves, has various characteristics, different aspects, even though the essential desire for liberation remains the same. It is obvious — and writers on this theme have said it many times — that war responds to certain laws4; and whoever disregards them will be defeated. These same laws must rule guerrilla warfare as a phase of war; but because of its special features, a series of corollary laws5 must also be recognized in order to carry it forward. Although geographical and social conditions in each country determine the mode and particular forms that guerrilla warfare will take, there are general laws that hold for all such struggles.

Our current task is to find the basic principles of this kind of war and the rules to be followed by peoples seeking liberation; to develop theory from facts; to generalize and give structure to our experience for the benefit of others.

Let us first consider the question: who are the combatants in a guerrilla war? On one side we have a group composed of the oppressor and his agents, the professional army, well armed and disciplined, in many cases receiving foreign aid as well as the help of the bureaucracy that is beholden to the oppressor. On the other side are the people of the nation or region. It is important to emphasize that guerrilla warfare is a war of the masses, a war of the people. The guerrilla band, as an armed nucleus, is the combative vanguard of the people. Its great force is drawn from the mass of the people themselves. The guerrilla band should not be considered inferior to the army against which it fights simply because it has inferior firepower. Guerrilla warfare is used by the side that is supported by a majority but which possesses a much smaller number of arms for use in defense against oppression.

The guerrilla fighter counts on the full support of the local people. This is an indispensable condition. And this is clearly seen by considering the case of bandit gangs that operate in a region; they have many characteristics of a guerrilla army, homogeneity, respect for the leader, bravery, knowledge of the terrain, and, often, even a good understanding of the tactics to be employed. The only thing lacking is the support of the people; and these gangs are inevitably captured and exterminated by the public force.

Analyzing the guerrilla band's mode of operation, its form of struggle, and understanding its mass base, we can answer the question: why does the guerrilla fighter fight? We must come to the inevitable conclusion that the guerrilla fighter is a social reformer, who takes up arms as the embodiment of the angry protest of the people against their oppressors; guerrillas fight in order to change the social system that keeps all their unarmed brothers and sisters in ignominy and misery. They launch themselves against the conditions of the ruling institutions at a particular moment and dedicate themselves with all the vigor that circumstances permit to smash the mold of those institutions.

When we analyze more deeply the tactic of guerrilla warfare, we see that guerrilla fighters must have good knowledge of the surrounding countryside; the paths of entry and escape, which will almost always have been constructed by the guerrillas themselves; the possibilities of rapid maneuver; good hiding places; naturally they must also count on the support of the people. All this indicates that the guerrilla fighter will carry out actions in rough, semi-populated areas. Since in such areas the struggle of the people for reforms is aimed primarily and almost exclusively at changing the form of land ownership, the guerrilla fighter is above all an agrarian revolutionary, who interprets the desires of the great peasant mass to be owners of land, owners of their means of production, of their animals, of everything they live for, which will also constitute their cemetery.

Current interpretations identify two different types of guerrilla warfare, one of these being a struggle that complements great regular armies, such as was the case of the Ukrainian fighters in the Soviet Union; but this type of warfare will not be considered in this analysis. We are interested in the other type, the case of an armed group engaged in struggle against a constituted power, colonial or otherwise, which establishes itself as a single base and which develops in rural areas. In all such cases, whatever the ideological aims inspiring the struggle, the economic goal is determined by the desire for land.

Mao's China begins as an uprising of groups of workers in the South that is defeated and almost annihilated. It succeeds in establishing itself and begins to advance only after the long march from Yenan, when it bases itself in rural areas and makes agrarian reform its fundamental goal. The struggle of Ho Chi Minh is based among the rice-growing peasants, who are oppressed by the French colonial yoke; with this force it is progressing toward the defeat of the colonialists.8 In both cases there was the framework of a patriotic war against the Japanese invader, but the economic basis of the fight for land has not disappeared. In the case of Algeria, the grand idea of Arab nationalism has its economic corollary in the fact that a million French settlers utilize nearly all the arable land. In some countries, such as Puerto Rico, where the special conditions of the island have not permitted a guerrilla movement, the nationalist spirit, deeply wounded by daily discrimination, is rooted in the aspiration of the peasants to recover the land that the Yankee invaders seized (even though many of these people are already proletarianized). This same central idea, though in different forms, inspired the small farmers, peasants, and slaves of the eastern estates of Cuba to close ranks and defend the right to possess land during the 30-year war of liberation.

Considering the possibilities for guerrilla warfare to become transformed into a conventional war as the operating potential of the guerrilla band increases, this special type of warfare should be considered as an embryo, a prelude, of the other. The possibilities for the growth of the guerrilla band and for changes in the mode of fighting until it becomes conventional warfare are as great as the possibilities for defeating the enemy in each of the different battles, combats, or skirmishes that occur. Therefore, the fundamental principle is that no battle, combat, or skirmish should be fought unless it can be won. There is a pejorative saying: "The guerrilla fighter is the Jesuit of warfare." This suggests qualities of treachery, of surprise, of secretiveness, that are obviously essential elements of guerrilla warfare. Naturally, it is a special kind of Jesuitism, promoted by circumstances, which necessitate acting at certain moments in ways different from the romantic and sporting conceptions with which we are taught to believe war is fought.

War is always a struggle in which each contender tries to annihilate the other. Besides using force, they will have recourse to all kinds of tricks and stratagems in order to achieve this goal. Military strategy and tactics are a representation of the objectives of the groups and of the means of achieving those objectives, taking advantage of all the enemy's weak points. In a war of positions, every platoon in a large army will display the same combative characteristics as those of the guerrilla band: treachery, secretiveness, and surprise. When these are not present, it is because vigilance on the other side prevents surprise. But since the guerrilla band is a division in itself, and since there are large areas of territory not controlled by the enemy, it is always possible to carry out guerrilla attacks in such a way as to guarantee surprise; and this is what the guerrilla fighter should do.

"Hit and run," some call this scornfully, and this is accurate. Hit and run, wait, lie in ambush, again hit and run, repeatedly, giving the enemy no rest. There would appear in all this a negative quality, an attitude of retreat, of avoiding frontal combat. This is, however, a consequence of the general strategy of guerrilla warfare, the ultimate aim of which is the same as in any war: to win, to annihilate the enemy.

Guerrilla warfare is therefore clearly a phase that does not afford in itself the opportunity to attain a complete victory, but rather is one of the initial phases of a war and will develop continuously until, through steady growth, the guerrilla army acquires the characteristics of a regular army. At that moment it will be ready to deal the enemy definitive blows and to achieve victory. The triumph will always be achieved by a regular army, even though its origins were in a guerrilla army.

So, just as the general of a division in a modern war does not have to die in front of his soldiers, the guerrilla fighter, who is the general of himself, need not die in every battle. The guerrilla is ready to give his or her life, but the positive feature of guerrilla warfare is that each guerrilla fighter is ready to die not just to defend an idea but to make that idea a reality. That is the essence of the guerrilla struggle. The miracle is that a small nucleus, the armed vanguard of a great popular movement that supports them, can proceed to realize that idea, to establish a new society, to break the old patterns of the past, to achieve, ultimately, the social justice for which they fight.

Viewed in this way, what was disparaged acquires true nobility — the nobility of the ends sought, and we are clearly not speaking of a distorted means to an end. This combative attitude, this attitude of never being discouraged, this resolution in confronting the great challenge presented by the final objective also epitomizes the nobility of the guerrilla fighter.

2 Guerrilla Strategy

In the terminology of war, strategy is understood as the analysis of the objectives to be achieved in the light of the total military situation, and the overall ways of accomplishing these objectives.

To have a correct strategic appreciation from the point of view of the guerrilla band, it is necessary to analyze fundamentally the enemy's likely mode of operation. If the final objective is always the complete destruction of the opposing force, in the case of a civil war the enemy is confronted with the standard task: the total destruction of each component of the guerrilla band. The guerrilla fighter, on the other hand, must analyze the resources that the enemy has for trying to achieve that outcome: in terms of men, mobility, popular support, armaments, and the leadership capacity which can be relied on. We must adjust our own strategy on the basis of these considerations, always bearing in mind the final objective of defeating the enemy army.

There are fundamental aspects to be studied: armaments, for example, and how they are used. The value of a tank, of an airplane in a fight of this nature must be assessed. The arms of the enemy, his ammunition, his habits must be considered, because the principal source of provision for the guerrilla force is precisely in enemy armaments. If there is a choice, we should choose the same type as that used by the enemy, since the greatest problem of the guerrilla band is the lack of ammunition, which the opponent must provide.

Once objectives have been fixed and analyzed, it is necessary to review the order of the steps leading to the achievement of the final objective. This should be planned in advance, even though it will be modified and adjusted as the struggle develops and unforeseen circumstances arise.

At the outset, the guerrilla fighter's essential task is to keep himself from being wiped out. Step by step, it will become easier for members of the guerrilla band or bands to adapt themselves to their lifestyle and to escape from the forces ranged against them, as this becomes a daily practice. When this is achieved, and it has been able to take up inaccessible positions that are very difficult for the enemy to reach, or it has assembled forces that deter the enemy from attacking, the guerrilla band should proceed to the gradual weakening of the enemy. At first, this can be done close to the points of active warfare against the guerrilla band, and later it can be taken deeper into enemy territory, attacking their communications, and then attacking or harassing their bases of operations and their central base, tormenting them wherever possible, to the full extent of the capabilities of the guerrilla forces.

The blows should be continuous. The enemy soldier in a zone of operations should not be allowed to sleep; his outposts should be attacked and destroyed systematically. At every moment the impression should be created that he is completely surrounded. In wooded areas and rough ground this effort should be maintained day and night; in open zones that are more easily penetrated by enemy patrols, only at night. For this, the absolute cooperation of the people and a perfect knowledge of the terrain are necessary. These two conditions affect every minute of the guerrilla fighter's life. Therefore, along with centers for the study of current and future zones of operations, intensive work must be undertaken among the local people to explain the motives of the revolution, its goals, and to spread the incontrovertible truth that the enemy's victory over the people is ultimately impossible. Whoever does not feel this indisputable truth cannot be a guerrilla fighter.

To begin with, this work among the people should be aimed at ensuring secrecy; that is, each peasant, each member of the community in which the action is taking place, will be asked not to mention what he or she sees and hears; later, help can be sought from those local residents whose loyalty to the revolution offers greater guarantees; later on, these persons can be used in contact missions, to transport goods or arms, or as guides in the zones familiar to them; after that, it is possible to establish organized mass action in workplaces, of which the final result will be the general strike.

The strike is a most important element in a civil war, but in order to achieve it a range of complementary conditions are necessary that do not always exist, and which very rarely emerge spontaneously. These fundamental conditions must be created, basically by explaining the purpose of the revolution and by demonstrating the power of the people and their capabilities.

It is also possible to have recourse to certain very homogeneous groups, which must have shown their efficacy previously in less dangerous tasks, in order to make use of sabotage, another of the terrible arms of the guerrilla band. Entire armies can be paralyzed, the industrial life of a zone suspended, leaving the inhabitants of a city without factories, without light, without water, without communications of any kind, without being able to risk travel by road except at certain times. If all this is achieved, the morale of the enemy declines, the morale of combatant units weakens, and the fruit ripens for picking at the precise moment.

All this presupposes an extension of the territory in which the guerrilla action takes place, but an excessive increase of this territory should be avoided. A strong base of operations must always be preserved and continuously strengthened during the course of the war. Within this territory, the indoctrination* of local residents should take place; the irreconcilable enemies of the revolution should be quarantined; all simple defensive measures, such as trenches, mines, and communications, should be perfected.

When the guerrilla band has achieved a respectable level of armed power in terms of arms and the number of combatants, it should proceed to the formation of new columns. This is similar to the beehive that at a particular moment releases a new queen, who goes off to another region with a part of the swarm. The mother hive with the most outstanding guerrilla chief will stay in the less dangerous places, while the new columns will penetrate other enemy territories following the cycle already described.

The time will come when the territory occupied by the columns is too small for them; and in advancing toward regions strongly defended by the enemy, it will be necessary to confront powerful forces. At that moment the columns combine to offer a compact fighting front, and a war of positions commences, a war conducted by regular armies. Nevertheless, the former guerrilla army cannot cut itself off from its base, and should create new guerrilla bands operating behind the enemy lines as the original bands did, proceeding in this way to penetrate enemy territory until it is controlled.

This is how the guerrillas reach the stage of attack, of the encirclement of fortified bases, of the defeat of reinforcements, of mass action, ever more committed, throughout the entire national territory, finally accomplishing the objective of the war: victory.

3 Guerrilla Tactics

In military terms, tactics are the practical methods of achieving great strategic objectives.

In one sense, they complement strategy, and in another they are more specific rules within it. As a means to an end, tactics are much more variable, much more flexible than the final objectives, and they should be adjusted continually during the struggle. There are tactical objectives that remain constant throughout a war and others that vary. The first thing to be considered is the adjusting of guerrilla operations to the enemy's actions.

The fundamental characteristic of a guerrilla band is mobility. Within a few minutes it can move away from a specific theatre and in a few hours farther still from the region, if that becomes necessary; this mobility allows the guerrillas to constantly change fronts and avoid any kind of encirclement. As circumstances of the war permit, the guerrilla band can dedicate itself exclusively to fleeing from an encirclement, which is the enemy's only way of forcing it into a decisive encounter that might be unfavorable; it can also change the battle into a counter-encirclement (small groups of guerrillas are assumed to be surrounded by the enemy when suddenly the enemy itself is surrounded by stronger contingents; or men positioned in a safe place serve as a lure, leading to the encirclement and annihilation of the entire troop and supply of the attacking force). A feature of this mobile war is the "minuet," so named for its similarity to the dance: the guerrilla bands encircle an enemy position, such as an advancing column; it is surrounded completely from the four points of the compass,14 with five or six guerrillas at each point, far enough away to avoid being encircled themselves; the battle is started at any one of the points, and the army moves toward it; the guerrillas then retreat, always maintaining visual contact, and initiate an attack from another point. The army will repeat its action and the guerrilla band the same. Thus, successively, an enemy column can be kept immobilized, and forced to expend large quantities of ammunition, weakening the morale of its troops at no great risk to the guerrillas.

This same tactic can be applied at night, but closing in more and showing greater aggression, because in these conditions counter-encirclement is much more difficult. Movement by night is another important trait of the guerrilla band, enabling it to advance into an attack position and to organize in a new territory where the danger of betrayal might exist. The numerical inferiority of the guerrilla band makes it necessary that attacks are always carried out by surprise; this is the great advantage that allows the guerrillas to inflict losses on the enemy without suffering losses themselves. In a battle between 100 men on one side and 10 on the other, the losses are not equal if there is one casualty on each side. The enemy loss can always be overcome, representing only one percent of their effective forces. A loss for the guerrilla band requires more time to be replaced as it involves a highly specialized soldier and represents 10 percent of the operating forces.

Dead guerrilla soldiers should never be left with their arms and ammunition. The duty of every guerrilla fighter, whenever a compañero falls, is to recover immediately these extremely precious elements of struggle. Specifically, the care that must be taken of ammunition and the method of using it are other characteristics of guerrilla warfare. In any combat between a regular force and a guerrilla band it is always possible to distinguish one from the other by their different manner of fire: a regular army will use a great deal of firepower, the guerrillas' shots will be sporadic and accurate.

At one time, one of our heroes, now dead, had to employ his machine gun for nearly five minutes, burst after burst, in order to slow the advance of enemy soldiers. This caused considerable confusion in our forces, because they assumed from the rhythm of fire that that key position must have been taken by the enemy; this was one of the rare occasions where a departure from the rule of saving fire had been necessary because of the importance of the position being defended.

Another elementary characteristic of the guerrilla soldier is flexibility, an ability to adapt to any circumstance, and to convert all accidents of the action to advantage. Contrary to the rigidity of classical methods of war, guerrilla fighters invent their own tactics at every minute of the battle and constantly surprise the enemy.

In the first case, there are only elastic positions, specific places that the enemy cannot pass, and places of diverting him. After easily overcoming difficulties in a gradual advance, the enemy is frequently surprised to find himself suddenly and solidly caught with no possibility of moving forward. This is because, when they have been selected on the basis of a careful study of the terrain, the guerrilla-defended positions are almost invulnerable. It is not the number of attacking soldiers that counts, but the number of defending soldiers. Once that number is in position, it can nearly always successfully hold off a battalion. It is a major task of the chiefs to choose carefully the timing and the place for defending a position without retreat.

The form of attack of a guerrilla army is also different; starting with surprise and ferocity, implacable, it suddenly converts itself into total passivity. The surviving enemy, resting, believes that the attacker has left; he begins to relax, to return to routine life within the besieged position, when suddenly a new attack bursts forth in another place, with the same characteristics, while the main body of the guerrilla band lies in wait to intercept reinforcements. At other times the guerrillas will suddenly attack an outpost defending the camp, overwhelm and capture it. The fundamental thing is surprise and rapidity of attack.

Acts of sabotage are very important. A clear distinction must be made between sabotage, a revolutionary and highly effective method of warfare, and terrorism, a measure that is generally ineffective and indiscriminate in its results, since it often makes victims of innocent people and destroys many lives that would be valuable to the revolution. Terrorism should be considered a valuable tactic when it is used to put to death some noted leader of the oppressive forces who is known for his cruelty, his efficiency in repression, or for another reason that makes his elimination useful. But the killing of insignificant individuals is never advisable, since it results in increased reprisals, and inevitable deaths.

There is one very controversial point about terrorism. Many consider that by provoking police oppression, it hinders all more or less legal or semi-clandestine contact with the masses and makes impossible the united action that might be necessary at a critical moment. This is true; but in a civil war the repression by the governmental power in certain towns might already be so great that, in fact, all forms of legal action are suppressed, and any mass action that is not supported by arms is ruled out. Therefore, it is necessary to be circumspect in adopting methods of this nature and to assess the general favorable consequences for the revolution. At any rate, well- managed sabotage is always a very effective weapon. It should not be used to immobilize means of production, which would paralyze a sector of the population (in other words, leave them unemployed), unless this also affects the normal life of the society. Sabotage against a soft-drink factory is ridiculous, but sabotage against a power plant is absolutely correct and advisable. In the first instance, a certain number of workers are put out of work without disrupting the rhythm of industry; in

11

the second case, there will also be displaced workers, but this is entirely justified by the paralysis of regional life. We will return to the technique of sabotage later.

Aviation is one of the favorite weapons of a conventional army in modern times, supposedly a decisive one. Nevertheless, it is useless during the early phases of a guerrilla war, when there are only small concentrations of guerrillas in rugged places. The effectiveness of aviation is in its systematic destruction of visible and organized defenses; and for this there must be large concentrations of men who construct these defenses, something nonexistent in warfare of this nature. Planes are also potent against marches by columns on level ground or places without cover; however, this vulnerability can be easily avoided by conducting marches at night.

One of the enemy's weakest points is road and rail trans- portation. It is virtually impossible to guard every meter of a transport route, a road, or a railroad. At any point a considerable amount of explosives can be planted that will make the road impassable; and by detonating explosives when a vehicle passes by, besides cutting off the road, considerable loss of life and materiel can be inflicted on the enemy.

The sources of explosives are varied: they can be brought from other zones; or unexploded bombs dropped from enemy planes can be used, although these do not always work; or they can be manufactured in secret laboratories within the guerrilla zone. The techniques of detonation are quite varied; their manufacture also depends on the conditions of the guerrilla band.

In our laboratory we made powder that we used as a cap, and we invented various devices for exploding mines at the desired moment. Those that produced the best results were electric, but the first mine we exploded was a bomb which had been dropped from a plane by the dictatorship but had not exploded; we adapted it by inserting various caps and adding a gun with the trigger pulled by a cord. At the moment an enemy truck passed, the weapon was fired to set off the explosion.

These techniques can be developed to a high degree. For example, we have learned that in Algeria today, in the struggle against the French colonial power, they are using tele-explosive mines, that is, mines exploded by radio at great distances from the point where they are located.

The tactic of setting up ambushes along roads in order to explode mines and annihilate survivors is one of the most profitable for obtaining arms and ammunition. The surprised enemy cannot use their ammunition and has no time to flee; so with a small expenditure of ammunition significant results are achieved. As the enemy receives blows, they also change their tactics, and instead of isolated trucks, moves in veritable motorized columns. However, by choosing the terrain well, the same result can be produced by breaking up the column and concentrating forces on one vehicle. In these cases the essential elements of guerrilla tactics must always be kept in mind. These are: perfect knowledge of the area; surveillance and foresight as to the lines of escape; vigilance over all the secondary roads that might bring in reinforcements to the point of attack; intimacy with people in the zone so as to ensure their help in regard to supplies, transport, and temporary or permanent hiding places if it becomes necessary to leave wounded compañeros behind; numerical superiority at a chosen point of action; total mobility; and the possibility of counting on reserves.

If all these basic tactics are employed, surprise attacks along the enemy's lines of communication yield important dividends.

A fundamental part of guerrilla tactics is the treatment of the people in the zone. The treatment of the enemy is similarly important; the norm should be absolute inflexibility during attack, an absolute inflexibility toward all the contemptible elements that resort to informing and assassination, and the greatest clemency possible toward the enemy soldiers who go into battle performing — or believing that they are performing — their military duty. It is a good policy to take no prisoners while there are no significant operational bases and no unassailable positions. Survivors should be set free. The wounded should be cared for with all available resources at the time of the action. Conduct toward the civil population should be governed by great respect for all the customs and traditions of the people of the zone, in order to demonstrate effectively, through deeds, the moral superiority of the guerrilla fighter over the oppressing soldier.

4 Warfare On Favorable Terrain

As already stated, the guerrilla struggle will not always take place on the most favorable terrain for the employment of its tactics; but when it does, that is, when the guerrilla band is located in zones difficult to reach, either because of dense forests, steep mountains, or impassable deserts or marshes, the general tactics, based on the fundamental postulates of guerrilla warfare, must always be the same.

An important point to consider is the way to engage the enemy. If the zone is so dense, so difficult that an organized army can never reach it, the guerrilla band should advance to the regions where the army can get to and where there are possibilities for combat.

The guerrilla band should fight as soon as its survival has been assured; it must constantly leave its refuge to fight; it does not have to be as mobile as in those cases where the terrain is unfavorable; it must adjust itself to the conditions of the enemy, but is not required to move as quickly as in those areas where the enemy can concentrate a large number of men in a few minutes. Neither is the nocturnal character of this warfare so important; it will be possible in many cases to carry out daytime operations, especially mobilizations by day, though subjected to enemy observation by land and air. It is also possible to pursue military action for a much longer time, above all in the mountains; it is possible to undertake battles of long duration with very few guerrillas, and it is very probable that the arrival of enemy reinforcements at the field of battle can be prevented.

A close vigilance over the access points is, however, an axiom never to be forgotten by guerrilla fighters. Their aggressiveness (on account of the difficulties that the enemy faces in bringing in reinforcements) can be greater, they can get closer to the enemy, fight much more directly, more frontally and for a longer time, although all of this may be qualified by various factors, for example, such as the amount of ammunition.

Fighting on favorable terrain, particularly in the mountains, presents many advantages but also the inconvenience that it is difficult in a single operation to capture a large quantity of arms and ammunition, owing to the precautions that the enemy takes in these areas. (The guerrilla soldier must never forget the fact that the enemy must serve as the source of arms and ammunition.) The guerrilla band will be able to "dig in" here much more rapidly than on unfavorable ground, that is, to form a base from which to engage in a war of positions, where small industries may be established as they are needed, as well as hospitals, centers for education and training, storage facilities, propaganda organs, etc., adequately protected from aircraft or from long-range artillery.

In these conditions the guerrilla band can expand its numbers, including noncombatants and perhaps even a system of training in the use of those arms that eventually fall into the hands of the guerrilla army.

The size of a guerrilla band is extremely flexible, depending on the territory, the means available of obtaining supplies, the flight of oppressed people from other zones, the arms available, and the necessities of organization. But, in any case, it is far more practicable to establish a base and expand with the support of new combatant elements.

This type of guerrilla band's radius of action will be as wide as conditions or the operations of other bands in adjacent territory permit. The range will be restricted by the time it takes to reach a secure zone from the zone of operations; assuming that marches must be made at night, it is not possible to operate more than five or six hours away from a point of maximum security. Small guerrilla bands that work constantly at weakening a territory can go farther from the security zone.

For this type of warfare the preferred arms are long-range weapons requiring minimal expenditure of bullets, supported by a group of automatic or semi-automatic arms. Of the rifles and machine guns available in the US markets, one of the best is the M-1 rifle, called the Garand.19 This should, however, only be used by those with some experience, since it has the disadvantage of expending too much ammunition. Medium- heavy arms, such as tripod machine guns, can be used on favorable ground, affording a greater margin of security for the weapon and its personnel, but they will always be weapons of defense and not of attack.

An ideal composition for a guerrilla band of 25 fighters would be 10 to 15 single-shot rifles and about 10 automatic arms between Garands and hand machine guns, including light and easily portable automatic arms such as the Browning or the more modern Belgian FAL and M-14 automatic rifles. Nine-millimeter weapons are among the best hand machine guns because they carry more ammunition. The simpler the construction, the better, because this increases the chance of being able to replace parts. All this depends on the armaments the enemy uses, since the ammunition they have is what we will use when their arms fall into our hands. Heavy arms are practically impossible to use. Aircraft cannot see anything in these zones and cease to operate; tanks and cannons cannot do much because of the difficulties of advancing.

A very important consideration is supply. For this very reason, remote zones generally present special problems, since there are few peasants and thus animal and food supplies are scarce. Stable lines of communication must be maintained in order to be able always to rely on a minimum of stockpiled food in the event of any unfortunate contingency.

In this kind of operational zone there are generally no possibilities for sabotage on a large scale; the inaccessibility brings a lack of installations, telephone lines, aqueducts, etc., that could be damaged by direct action.

Animals are important for supply purposes, the mule being the best option in rough country. Adequate pasturage permitting good nutrition is essential. The mule can manage in extremely hilly country impossible for other animals. In the most difficult situations it may be necessary to resort to transport by men. Each individual can carry 25 kilograms for many hours a day and for many days.

The lines of communication with the outside should include a series of intermediate points staffed by totally reliable people, where goods can be stored and where contacts can be hidden at critical times. Internal lines of communication can also be created; their extension will depend on the stage of development reached by the guerrilla band. In some zones of operations in the recent Cuban [revolutionary] war, many kilometers of telephone lines were established, roads were built, and a messenger service was maintained sufficient to cover all areas in minimal time.

There are also other possible means of outside communication we did not use in the Cuban war but which are perfectly appropriate, such as smoke signals, signals using light reflected by mirrors, and carrier pigeons.

For the guerrillas it is of vital importance to maintain their arms in good condition, to capture ammunition, and, above all, to have adequate shoes. The first efforts to create industries should therefore be directed toward meeting these needs. Shoe factories can initially be cobbler workshops that can replace half-soles on old shoes, later developing into various organized factories producing a good daily average of shoes. The manufacture of powder is fairly simple, and a lot can be accomplished with a small laboratory, bringing in the necessary materials from outside. Mines constitute a grave danger for the enemy; large areas can be mined for simultaneous explosion, destroying up to hundreds of men.

5 Warfare On Unfavorable Terrain

To conduct warfare in territory that is not particularly hilly, lacks forests, and has many roads, all the fundamentals of guerrilla warfare must be observed, with only the forms altered. The quantity, not the quality, of guerrilla warfare will change. For example, following the same order as before, the mobility of this type of guerrilla band should be extraordinary; strikes should preferably be made at night; they should be extremely rapid, almost explosive, and the guerrillas should then withdraw to a different place from their starting point, as far as possible from the scene of the action, assuming that there is no place secure from the repressive forces that the guerrillas can use as their garrison.

A guerrilla can walk between 30 and 50 kilometers during the night; marching is also possible during the first hours of daylight, unless the operational zones are closely watched or there is a danger that local people will see the passing troops and notify the pursuing army of the guerrilla band's location and route. In these cases it is always preferable to operate at night, keeping as quiet as possible both before and after the action; the first hours of darkness are best. Here too there are exceptions to the general rule, as sometimes dawn might be preferable. It is never wise to let the enemy get used to a certain form of warfare; it is necessary to vary constantly the places, the hours, and the forms of operation.

We have already said that the action cannot be for long, but must be rapid; it must be highly effective, last a few minutes, and be followed by an immediate withdrawal. The arms employed here will not be the same as in the case of actions on favorable ground; a large quantity of automatic weapons is preferable. In night attacks marksmanship is not the determining factor, but rather concentrated fire; the more automatic arms firing at short distance, the more possibilities there are of annihilating the enemy.

Furthermore, the mining of roads and the destruction of bridges are tactics of great importance. Guerrilla attacks will be less aggressive so far as perseverance and duration are concerned, but they can be very violent, and they can utilize different arms, such as mines and the shotgun. Against open vehicles heavily loaded with soldiers — the usual method of transporting troops — and even against closed vehicles that do not have special defenses or against buses, for example, the shotgun is a tremendous weapon. A shotgun loaded with large shot is the most effective. This is not a secret of guerrilla warfare but is used also in major wars; the US used shotgun platoons armed with high-quality weapons and bayonets for assaulting machine-gun nests.

An important problem to explain is that of ammunition; this will almost always be taken from the enemy. It is therefore necessary to strike where there is the absolute guarantee of replacing whatever ammunition is expended, unless there are large reserves in secure places. In other words, a devastating attack against a group of men should not be undertaken at the risk of expending all ammunition without being able to replace it. In guerrilla tactics it is always necessary to keep in mind the grave problem of procuring the war materiel required to continue the fight. For this reason guerrilla arms should be the same as those of the enemy, except for weapons such as revolvers and shotguns, for which ammunition can be obtained locally or in the cities.

The number of people in a guerrilla band of this type should not exceed 10 to 15. In establishing a single combat unit it is of utmost importance to always consider the limitations on numbers: 10, 12, 15 guerrillas can hide anywhere and at the same time can help each other in putting up a powerful resistance to the enemy. Four or five would perhaps be too small a number, but when the number exceeds 10 there is a greater possibility that the enemy will discover them in their camp or on the march.

Remember that the pace of the guerrilla band on the march is equal to the pace of its slowest person. It is more difficult to achieve a uniform marching speed with 20, 30, or 40 guerrillas than with 10. And on the plains, the guerrilla fighter must essentially be a runner. There the practice of hitting and running is most useful. The guerrilla bands on the plains suffer the enormous disadvantage of being subject to rapid encirclement and of not having secure places where they can set up a firm resistance; they must therefore live in conditions of absolute secrecy for a long time, since it would be dangerous to trust any local person whose fidelity is not perfectly established. Enemy reprisals are so violent, usually so brutal, inflicted not only on the head of the family but frequently on the women and children as well, that pressure on individuals lacking firmness may result at any moment in capitulation and their revealing information as to where the guerrilla band is located and how it is operating.

This would immediately result in encirclement, with the inevitable unfortunate consequences, although not necessarily fatal ones. When conditions, the quantity of arms, and the rebelliousness of the people demand an increase in the number of fighters, the guerrilla band should be divided. If necessary, all can regroup at a particular moment to deal a blow, but in such a way that immediately afterwards they can disperse toward separate zones, again divided into small groups of 10, 12, or 15.

Entire armies can be organized under a single command and respect and obedience assured to this command without the necessity of being in a single group. Therefore, the election of the guerrilla chiefs and the certainty that they coordinate ideologically and personally with the overall chief of the zone are very important.

The bazooka is a heavy weapon that can be used by the guerrilla band because of its easy portability and operation. Today the rifle-fired anti-tank grenade can replace it. Naturally, it will be a weapon taken from the enemy. The bazooka is ideal for firing on armored vehicles, and even on unarmored vehicles that are loaded with troops, and for seizing small military bases of just a few men in a short time; but it is important to point out that a man can only carry three shells, and even this requires considerable exertion.

As for using the heavy arms taken from the enemy, naturally nothing should be scorned; but there are weapons such as the tripod machine gun, the heavy 50-millimeter machine gun, etc., which, when captured, should be utilized on the understanding that they might be lost again. In other words, in the unfavorable conditions that we are now considering, a battle to defend a heavy machine gun or other weapon of this type cannot be allowed; they should simply be used until the tactical moment when they must be abandoned. In our Cuban war of liberation, to abandon a weapon constituted a grave offense, and there was never a case where it was necessary. Nevertheless, we mention this in order to explain clearly the only situation in which abandonment would not be such a critical offense. On unfavorable ground, the guerrilla's weapon is the personal weapon of rapid fire.

A peasant population will usually inhabit an accessible zone, and this enormously facilitates supply. Having trustworthy people and making contact with establishments that provide supplies to the population, it is possible to maintain a guerrilla band perfectly without having to devote time or money to long and dangerous lines of communication. Furthermore, it is well to reiterate that the smaller the number of guerrillas, the easier it will be to provide them with food. Essential supplies such as bedding, waterproof material, mosquito nets, shoes, medicines, and food can be found within the zone, since these are items used daily by the local population.

Communications will be much easier in the sense of being able to count on a larger number of guerrillas and more roads; but they will be more difficult in regard to the security necessary for sending messages between distant points, since it will be necessary to rely on a range of contacts that have to be trustworthy. There will be the danger of an eventual capture of one of the messengers, who are constantly crossing enemy lines. If the messages are not so important, they should be verbal; if of great importance, writing in code should be used. Experience shows that transmission by word of mouth greatly distorts any communication.

For these same reasons industry will have much less importance, as well as being much more difficult to carry out. It will not be possible to have factories making shoes or arms. Practically speaking, industry will have to be limited to small workshops, carefully hidden, where shotgun shells can be recharged and mines, simple grenades, and other bare necessities of the moment manufactured. On the other hand, it is possible to make use of all the friendly local workshops to make whatever is necessary.

This brings us to two consequences that flow logically from what has been said. First, the favorable conditions for establishing a permanent camp in guerrilla warfare determine the degree of productive development of a particular location. All favorable conditions, all the comforts of life usually induce people to settle down; but the opposite is the case for the guerrilla band. The more facilities there are for social life, the more nomadic, the less certain the life of the guerrilla fighter. In reality, this is the result of one and the same principle. The title of this section is "Warfare on Unfavorable Terrain," because everything that is favorable to human life: communications, urban and semi- urban concentrations of large numbers of people, land easily worked by machine — all these place the guerrilla fighter in a disadvantaged position.

The second conclusion is that as guerrilla warfare must necessarily include the extremely important factor of work among the masses, this task is even more important in the unfavorable zones, where a single enemy attack can produce a catastrophe. Indoctrination should be constant, as should be the struggle for unity of the workers, the peasants, and other social classes that live in the zone, in order to achieve the greatest homogeneous attitude toward the guerrillas. This task with the masses, this continuous attention to the huge problem of relations between the guerrilla band and the local residents, must also govern the attitude taken toward the case of an individual recalcitrant enemy soldier: he should be eliminated without hesitation if he is a danger. In this respect the guerrilla band must be severe. Enemies cannot be permitted to exist within the operational zone that offers no security.

6 Urban Warfare

If, during the war, the guerrilla bands move in on the cities and penetrate the surrounding countryside in such a way as to be able to create conditions of some security, it will be necessary to give these urban bands special education, or rather, a special organization.

It is essential to recognize that an urban guerrilla band can never emerge of its own accord. It will be born only after certain conditions necessary for its survival have been created. Therefore, the urban guerrilla will always be under the direct command of chiefs located in another zone. The function of this guerrilla band will not be to carry out independent actions but to coordinate its activities with the overall strategic plans in such a way as to support the action of larger groups situated in another area, contributing specifically to the success of a particular tactical objective, without the operational freedom of other types of guerrilla bands. For example, an urban band will not be able to choose the nature of its operation: whether to destroy telephone lines, to make attacks in another locality, or to surprise a patrol of soldiers on a distant road; it will do exactly what it is told. If its function is to cut down telephone poles or electric wires, to destroy sewers, railroads, or water mains, it will limit itself to carrying out these tasks efficiently.

It should not number more than four or five. The limitation on numbers is important, because the urban guerrilla must be considered as operating on exceptionally unfavorable terrain, where the enemy's vigilance will be much greater and the possibilities of reprisals as well as of betrayal are increased enormously. Another aggravating factor is that the urban guerrilla band cannot go far from the places where it is going to operate; added to speedy action and withdrawal there is also a limit on the distance of withdrawal from the scene of action and the need to remain totally hidden during the daytime. This is a nocturnal guerrilla band in the extreme, with no possibility to change its mode of operation until it can take part as an active combatant in the siege of the city when the insurrection is very advanced.

The essential qualities of the guerrilla fighter in this unfavorable situation are discipline — perhaps to the highest degree — and discretion. No more than two or three friendly houses can be relied on to provide food; it is almost certain that an encirclement under these conditions equals death. Besides, weapons will not be the same kind as those used by other groups. They will be for personal defense, only those that do not hinder a rapid flight or betray a secure hiding place. The group should have not more than one or two sawed-off automatic weapons, with pistols for the other members.

Preferably they will concentrate on prescribed sabotage actions and never carry out armed attacks, except by surprising one or two members or agents of the enemy troops.

For this they need a broad range of equipment. The guerrilla fighter must have good saws, large quantities of dynamite, picks and shovels, apparatus for lifting rails, and, in general, adequate tools for the work to be carried out. These should be hidden in places that are secure but easily accessible to those who will need them.

If there is more than one guerrilla band, they will depend on a single chief to give orders as to the necessary tasks through contacts of proven trustworthiness that live openly as ordinary citizens. In certain cases guerrilla fighters will be able to maintain their peacetime work, but this is very difficult; practically speaking, the urban guerrilla band is constituted by a group of individuals who are already outside the law, in a situation of war, in unfavorable conditions as already described.

The importance of the urban struggle is extraordinary. A good operation of this nature extended over a wide area can almost completely paralyze the commercial and industrial life of the sector and place the entire population in a situation of unrest, of anguish, almost of impatience for the development of violent events that will relieve the suspense. If, from the moment war is initiated, the future possibility of such a struggle is anticipated and organization of specialists in this field commences, much more rapid action can be guaranteed and lives and precious time will be saved.

CHAPTER II: THE GUERRILLA BAND

1 The Guerrilla Fighter: Social Reformer

We have already identified the guerrilla fighter as one who shares the longing of the people for liberation and who, after peaceful means are exhausted, initiates the struggle and con- verts himself into an armed vanguard of the fighting people. From the commencement of the struggle the guerrilla is com- mitted to destroying an unjust order and has the intention, more or less hidden, to replace the old with something new.

We have also said that in the current conditions, at least in America and in almost all countries with little economic development, the countryside offers ideal conditions for the struggle, thus the basic social claims that the guerrilla fighter will raise begin with changes in the structure of agrarian property.

In this period the banner of the struggle will be agrarian reform. At first this goal may or may not be completely defined in its extent and limits; it may simply refer to the age-old hunger of the peasant for the land he or she works or wishes to work.

The conditions in which the agrarian reform will be realized depend on the conditions that existed before the struggle began, and on the social depth of the struggle. But the guerrilla fighter, as the conscious element of the vanguard of the people, must display the moral conduct of a true priest of the desired reform. To the stoicism forced by the difficult conditions of warfare should be added an austerity born of rigid self-control that prevents a single excess, a single slip, whatever the circumstances. The guerrilla soldier should be an ascetic.

Social relations will vary according to the development of the war. At the beginning it will not be possible to attempt any changes in the social order of the area.

Goods that cannot be paid for in cash will be paid for with bonds; and these should be redeemed at the first opportunity.

The peasant must always be given technical, economic, moral, and cultural assistance. The guerrilla fighter will be a kind of guardian angel who has dropped into the zone, always helping the poor and harassing the rich as little as possible in the first phases of the war. But as the war develops, contradictions will become sharper; the time will arrive when many of those who regarded the revolution sympathetically at the start will place themselves in a position diametrically opposed to it; and they will make the first move into battle against the popular forces. At that moment the guerrilla fighter should act to become the standard-bearer of the people's cause, punishing every betrayal with justice. Private property should acquire a social function in the war zones. In other words, excess land and livestock not essential for the maintenance of a wealthy family should pass into the hands of the people and be distributed equitably and fairly.

The right of the owners to receive payment for possessions used for the social good should always be respected; but this payment will be made in bonds ("bonds of hope," as they were called by our teacher, General Bayo, referring to the common interest that is thus established between debtor and creditor).

The land and property of notorious and active enemies of the revolution should pass immediately into the hands of the revolutionary forces. Furthermore, taking advantage of the heat of the war — those moments in which human fraternity reaches its greatest intensity — all kinds of cooperative work should be stimulated, as far as the mentality of the local people will allow.

As social reformers, guerrilla fighters should not only pro- vide an example in their own lives, but should also constantly give an orientation on ideological issues, explaining what they know and what they wish to do at the right time. They should also make use of what they learn as the months or years of the war strengthen their revolutionary convictions, making them more radical as the potency of arms is demonstrated, as the outlook of the local people becomes a part of their spirit and of their own life, and as they understand the justice and the vital necessity of many changes, the theoretical importance of which they understood before, but perhaps not the practical urgency.

Very often this occurs because the initiators of a guerrilla war, or rather the directors of guerrilla warfare, are not those who have bent their backs day after day over the furrow. They understand the necessity for change in the social treatment of the peasants, but have never suffered this bitter treatment personally. What happens then (here, I am drawing on the Cuban experience and expanding on it) is a genuine interaction between those leaders, who by their actions teach the people the fundamental importance of the armed struggle, and the people themselves, who rise in rebellion and teach the leaders these practical necessities we are discussing. In this way, as a product of the interaction between the guerrilla fighters and their people, a progressive radicalization appears, further accentuating the revolutionary nature of the movement and giving it a national scope.

2 The Guerrilla Fighter As Combatant

The life and activities of the guerrilla fighter, as sketched in general outline, demand a range of physical, mental, and moral qualities needed for adapting oneself to prevailing conditions and for completely fulfilling any assigned mission.

To the question of what the guerrilla soldier should be like, the first answer is that he or she should preferably be an inhabitant of the zone. If this is the case, they will have friends who will help; as a local resident, they will know the area (and this knowledge of the terrain is one of the most important factors in guerrilla warfare); and since they will be accustomed to local peculiarities they will be able to work better, not to mention that they will add to all this the enthusiasm that arises from defending one's own people and struggling to change a social regime that affects one's own world.

The guerrilla combatant is a night combatant; to say this also means to say that he or she must have all the special qualities that such a struggle requires. The guerrilla must be cunning and able to march to combat across plains or mountains without detection, and then to fall on the enemy, taking advantage of the factor of surprise, which must be emphasized again as so important in this type of fight. After causing panic by this surprise, the guerrillas should launch themselves implacably into the fight, never permitting a single weakness in their compañeros and taking advantage of every sign of weakness on the enemy's part. Striking like a tornado, destroying everything, giving no quarter unless the tactical circumstances demand it, judging those who must be judged, sowing panic among the enemy combatants, the guerrillas nevertheless treat defenseless prisoners benevolently and show respect for the dead.

A wounded enemy should be treated with care and respect unless his former life warrants the death penalty, in which case he will be treated according to his deserts. Prisoners can never be kept unless a secure base of operations, invulnerable to the enemy, has been established. Otherwise, the prisoner will become a dangerous menace to the security of the local people of the region or to the guerrilla band itself, because of the information that he can give on rejoining the enemy army. If he has not been a notorious criminal, he should be set free after receiving a lecture.2

Guerrilla fighters should risk their lives whenever necessary and be ready to die without hesitation; but, at the same time, they should be cautious and never expose themselves un- necessarily. All possible precautions should be taken to avoid defeat or annihilation. For this reason it is extremely important in every battle to maintain vigilance over all the points from which enemy reinforcements may arrive and to take precautions against encirclement, the consequences of which are usually not physically disastrous but which damage morale by causing a loss of faith in the prospects of the struggle.

Nevertheless, the guerrilla fighter should be audacious, and after carefully analyzing the dangers and possibilities in an action, always ready to take an optimistic attitude toward circumstances and to see reasons for a favorable outcome even at times when the analysis of the adverse and favorable conditions does not appear to be positive.

To be able to survive in the midst of these conditions of struggle and enemy action, guerrilla fighters must have a degree of adaptability that allows them to identify themselves with the environment in which they live, to become a part of it, and to take advantage of it as an ally to the greatest possible extent. The guerrilla fighter also needs rapid comprehension and an instantaneous ingenuity in order to be able to change tactics depending on the course of the action.

These faculties of adaptability and inventiveness in popular armies are what ruin the statistics of the warmongers and bring them to a halt.

The guerrilla fighter must never for any reason leave a wounded compañero at the mercy of the enemy troops, because this means abandoning him or her to an almost certain death. At whatever cost, the wounded must be removed from the combat zone to a secure place. The greatest exertions and the greatest risks must be taken in this task. The guerrilla soldier must be an extraordinary compañero.

At the same time they should be close-lipped. Everything that is said and done in their presence should be kept strictly to themselves. A single unnecessary word should never be let slip, even with one's own comrades-in-arms, since the enemy will always try to plant spies among the ranks of the guerrillas in order to discover their plans, location, and livelihood.

Besides the moral qualities mentioned, the guerrilla fighter should possess a range of very important physical qualities. The guerrilla fighter must be indefatigable, able to produce an additional effort even when exhaustion seems unbearable. Profound conviction, evident in every facial expression, forces the guerrilla to take another step, and this not the last one, since it will be followed by another and another and another until the place designated by his chiefs is reached.

The guerrilla should be able to endure extremities, to with- stand not only the privations of food, water, clothing, and shelter to which they are frequently subjected, but also the sickness and wounds that often must be cured by nature without much help from the surgeon. This must be so, because many times the individual who leaves the guerrilla zone to recover from sickness or wounds will be assassinated by the enemy.

To meet these conditions guerrilla fighters need an iron constitution that will enable them to resist all these adversities without becoming ill, and to use their lives as hunted animals as another way of strengthening themselves. Assisted by a natural adaptability, guerrillas become part of the landscape in which they fight.

All these considerations raise the question: what is the ideal age for a guerrilla fighter? These limits are always very difficult to define precisely, because individual and social characteristics differ. A peasant, for example, will be much more resistant than a person from the city. A city dweller who is accustomed to physical exercise and a healthy life will be much more effective than a person who has spent their life behind a desk. But generally the maximum age of combatants in the totally nomadic stage of the guerrilla struggle should not exceed 40, although there can be exceptional cases, above all among the peasants. One of the heroes of our struggle, Commander Crescencio Pérez, came to the Sierra Maestra when he was 65 and he immediately became one of the most useful members of the troop.

We could also ask if the members of the guerrilla band should be drawn from a certain social class. It has already been stated that the social composition should be adjusted to that of the particular operational zone; in other words, the combatant nucleus of the guerrilla army should be composed of peasants.

The peasant is evidently the best soldier; but the other strata of the population should by no means be excluded or deprived of the opportunity to fight for a just cause. In this respect there are important individual exceptions.

We have not yet set a lower age limit. We believe that minors less than 16 years old should not be accepted, except in very special circumstances. In general these young kids, virtually children, do not have sufficient development to cope with the tasks, the weather, and the suffering to which they will be subjected.

The best age for a guerrilla fighter varies between 25 and 35 years, a stage at which life has assumed a definite shape for most people. An individual who sets out at that age, abandoning their home, children, and entire world, must have thought through their responsibility and made a firm decision not to retreat a step. There are extraordinary cases of children who have reached the highest ranks of our Rebel Army as combatants, but this is not common. For every one of them who displayed great fighting qualities, there were dozens who should have been returned to their homes and who frequently became a dangerous burden for the guerrilla band.

As we have said, the guerrilla fighter is a soldier who carries their house on their back like a snail; therefore, one's back- pack must be arranged in such a way that the least number of utensils will be of the greatest possible service. Only what is indispensable will be carried, and will be guarded at all times as something fundamental and never to be lost other than in extreme adversity.

By the same token, armaments will be only what can be carried. Reprovisioning is very difficult, especially bullets; to keep them dry, always to keep them clean, to count them one by one so that none is lost — these are the watchwords. And the gun should always be kept clean, well greased, and with the barrel shining. It is advisable for the leader of each group to impose some penalty or punishment on those who do not maintain their arms in these conditions.

People with such outstanding commitment and firmness must have an ideal that sustains them in the adverse conditions we have described. This idea is simple, straightforward, without any great pretension, and in general does not go very far; but it is so firm, so clear, that a person will give their life for it without the least hesitation. For almost all peasants this ideal is the right to have and work a piece of their own land and to enjoy just social treatment. For workers it is to have a job, to receive an adequate wage, as well as just social treatment. For students and professionals more abstract ideas such as liberty might be their motives for the struggle.

This raises the question: what is the life of the guerrilla fighter like? The normal routine is the long hike. Let us take the example of a mountain guerrilla fighter located in wooded regions under constant enemy harassment. In these conditions the guerrilla band moves during daylight hours, without eating, in order to alter its position; when night arrives, camp is set up in a clearing near a water supply according to a routine, each group assembling in order to eat together; at dusk the fires are lighted with whatever is around.

Guerrilla fighters eat when they can and anything they can. Sometimes fabulous meals disappear down combatants' gullets; at other times they fast for two or three days without reducing their capacity for work. Home will be the open sky; between it and a hammock a sheet of waterproof nylon is placed, and beneath the sheet and hammock are the guerrilla fighter's treasures: their backpack, gun, and ammunition. There are places where it is not wise to remove one's shoes, because of the possibility of a surprise at- tack by the enemy. Shoes are another precious treasure. Who- ever has a pair has the security of a happy existence within the limits of the prevailing circumstances.

The guerrilla fighter might go for days in a place, avoiding all contact that has not been previously arranged, staying in the roughest zones, familiar with hunger, at times thirst, cold, heat; sweating during the continuous marches, letting the sweat dry on one's body and adding to it new sweat with no possibility of regular cleanliness (although this also depends somewhat on the individual's disposition, as does everything else).

During the recent war, on entering the town of El Uvero after a 16-kilometer march and a battle that lasted two hours and 45 minutes in a hot sun (after several days spent in very adverse conditions along the sea with high temperatures and a boiling sun), our bodies gave off a peculiar and offensive odor that repelled anyone who came near. Our noses were completely accustomed to this lifestyle; guerrilla fighters' hammocks are known for their characteristic, individual odor.

In such circumstances breaking camp should be done rapidly, leaving no traces behind; vigilance must be extreme. For every 10 guerrillas asleep, one or two should be on watch, with the guards changed continually and a sharp vigil maintained over all the entrances to the camp.

Campaign life teaches several tricks for preparing meals, some to help speed preparation; others to add seasoning with little things found in the forest; still others for inventing new dishes that give a more varied character to the guerrillas' menu, which is mainly roots, grains, salt, a little oil or lard, and very sporadically, pieces of the meat of some animal that has been slaughtered. This is the scenario for a group operating in tropical regions.

Within the framework of the combatant's life, the most interesting event — the one that brings convulsions of joy and puts new vigor in everyone's steps — is the battle. As the climax of the guerrilla's life, combat is sought at the opportune moment, either when an enemy camp sufficiently weak to be annihilated has been located and investigated; or when an enemy column is advancing directly toward the territory occupied by the liberating force. The two cases are different.

Against an enemy camp the action will be a thin encirclement and essentially will become a hunt for the members of the columns that try to break the encirclement. An entrenched enemy is never the favorite prey of the guerrilla fighter; he prefers the enemy to be on the move, nervous, not knowing the terrain, fearful of everything and without natural protections for defense. To be behind a parapet with powerful arms for repelling an offensive, however bad the situation, will never be the same as being in a long column that is suddenly attacked in two or three places and cut. If the attackers are not able to encircle the column and destroy it totally, they will withdraw prior to any counterattack.

If there is no possibility of defeating those entrenched in a camp by means of hunger or thirst or by a direct assault, the guerrilla band should retire after the encirclement has yielded its destructive fruits in the relieving columns. In cases where the guerrilla column is too weak and the invading column too strong, the action should be concentrated against the vanguard. There should be a special preference for this tactic, whatever the hoped-for result, since after the leading ranks have been hit several times, thereby spreading the news among the soldiers that those in the front are constantly dying, the reluctance to occupy those positions will provoke nothing less than mutiny. Therefore, attacks should target that point even if they also target other points of the column.

The guerrilla fighters' ability to perform their functions and adapt themselves to the environment will largely depend on their equipment. Even though united in small groups, they will have individual needs. Besides their regular shelter, they should have in their backpack everything needed for survival in case they find themselves alone for some time.

In presenting this list of equipment we refer essentially to what should be carried by an individual in rough country at the beginning of a war, with frequent rainfall, some cold weather, and harassment by the enemy; in other words, in a similar situation to that we faced at the beginning of the Cuban war of liberation.

The equipment of the guerrilla fighter is divided into the essential and the accessory. Among the first is a hammock. This provides adequate rest; it is easy to find two trees from which to hang it; and, in cases where one sleeps on the ground, it can serve as a mattress. Whenever it is raining or the ground is wet, a frequent occurrence in tropical mountain zones, the hammock is indispensable for sleeping. A piece of waterproof nylon cloth complements

Hammock with a nylon roof.

it. The nylon should be large enough to cover the hammock when tied from its four corners, and with a line strung through the center to the same trees from which the hammock hangs. This last line serves to make the nylon into a kind of tent by raising a center ridge and allowing water to run off.

It is cold in the mountains at night so a blanket is indispensable. It is also necessary to carry a garment such as a jacket that will help one to bear the extreme changes of temperature. Clothing should be rough work trousers and shirt, which may or may not be a uniform. Shoes should be of the best possible construction and, moreover, since without good shoes marches are very difficult, they should be one of the first articles laid up in reserve.

Since guerrilla fighters carry their homes in their backpacks, the latter is very important. The more primitive types may be made from any kind of sack carried by two ropes; but canvas backpacks found in the market or made by a harness maker are preferable.

The guerrilla fighter should always carry some personal food besides that which the troop carries or consumes in its camps. Indispensable articles are: lard or oil, which is necessary for fat consumption; canned goods, which should not be consumed except in circumstances where food for cooking cannot be found or when there are too many cans and their weight impedes the march; preserved fish, which has great nutritional value; condensed milk, which is also nourishing, particularly because of the large quantity of sugar it contains; something for the sweet tooth; powdered milk, which is also easily transportable; sugar is another essential part of the supplies, as is salt, without which life becomes sheer martyrdom; and something with which to season the meals, such as onion, garlic, etc., depending on what can be found locally. This completes the list of the essentials.

The guerrilla fighter should carry a plate, knife, and fork, which will serve all the various necessary functions. The plate can be a camping or military type or a pan that can be used for cooking anything from a piece of meat to yams or a potato, or for brewing tea or coffee.

Special greases are necessary for maintaining rifles; and these must be carefully administered — sewing machine oil is very good if there is no special oil available. Cloths that can be used for cleaning weapons are needed as well as a rod for cleaning the inside of the gun, something that should be done often. The ammunition belt can be a commercial type or homemade, depending on the circumstances, but it should be constructed so that not a single bullet will be lost. Ammunition is the key to the battle, without which everything else is in vain; it must be guarded like gold.

A canteen or a bottle for water is essential, since it will often be needed where water is not available. Medicines that have a general use should be carried: for example, penicillin or some other type of antibiotic, preferably the types taken orally, well sealed; medicines for lowering fever, such as aspirin; and others to treat the endemic diseases of the area. These may be malaria tablets, sulfas for diarrhea, medicine against all types of parasites; in other words, the medicine should be appropriate for the region. Where there are poisonous animals it is recommended to carry the right serums. Surgical instruments will complete the medical equipment. Small personal items for attending to less important injuries should also be included.

A common and extremely important comfort in the guerrilla fighter's life is a smoke, such as cigars, cigarettes, or pipe tobacco; a smoke is a great friend to the solitary soldier in moments of rest. Pipes are useful, because they mean every last piece of tobacco from the butts of cigars and cigarettes can be used at times of scarcity. Matches are extremely important, not only for lighting a smoke, but also for starting fires; this is one of the great problems in the woods in rainy periods. It is preferable to carry both matches and a lighter, so that if the lighter runs out of fuel, matches can be a substitute.

Soap should be carried, not just for personal hygiene, but also for cleaning eating utensils, because intestinal infections or irritations are frequent and can be caused by spoiled food left on dirty cookware. With this set of equipment, the guerrilla fighter can be assured that he will be able to live in the woods under adverse conditions, no matter how bad, for as long as required to overcome the situation.

There are accessories that are sometimes useful and others that are a bother but are very useful. The compass is one of these; at the start this will be used a lot for orientation, but little by little knowledge of the countryside will make it unnecessary. In mountainous regions a compass is not much use, because impassable obstacles will probably cut off the route indicated. Another useful article is an extra nylon cloth for covering all the equipment when it rains. Remember that in tropical countries it rains continuously during certain months, and that water is the enemy of everything the guerrilla fighter must carry: food, ammunition, medicine, paper, and clothing.

A change of clothing can be carried, but this usually shows inexperience. The usual practice is to carry no more than an extra pair of pants, eliminating underwear and other articles, such as towels. The guerrilla fighter learns through experience to conserve their energy in carrying a backpack from one place to another, and will, little by little, get rid of everything that has no essential value.

Along with a piece of soap that is equally useful for washing utensils as for personal hygiene, a toothbrush and paste should be carried. It is worthwhile also to bring a book, which can be exchanged for others among members of the guerrilla band. These books can be good biographies of past heroes, histories, or economic geographies, preferably of the country, and works of general character that can raise the cultural level of the soldiers and discourage the tendency toward gambling or other undesirable pastimes. There are periods of boredom in the life of the guerrilla fighter.

Whenever there is extra space in the backpack, it should be used for food, except in those zones where food is easily and readily available. Sweets or food of lesser importance that supplement the basic items can be carried. Crackers can be one of these, although they take up a lot of space and break up into crumbs. In thick forests a machete is useful; in very wet places a small bottle of gasoline or some light resinous wood for kindling, such as pine, will make fire building easier when the wood is wet.

A small notebook and pen or pencil for taking notes and for letters to the outside or communication with other guerrilla bands should always be part of the guerrilla fighter's equipment. Pieces of string or rope should be kept as these have many uses, along with needles, thread, and buttons for clothing. The guerrilla fighter who carries this equipment will have a solid house on their back, rather heavy but furnished to ensure a comfortable life during the hardships of the campaign.

3 Organization Of A Guerrilla Band

There is no rigid scheme for the organization of a guerrilla band; there will be innumerable differences depending on the environment in which it operates. For the sake of argument, we will suppose that our experience has a universal application, but it should always be kept in mind that there will possibly be new forms that better match the particular characteristics of a given armed group. The size of the component units of the guerrilla force is one of the most difficult problems to deal with: the number and composition of the troops will differ, as we have already explained. Let us suppose a force is situated on favorable terrain, mountainous, with conditions not so bad as to necessitate perpetual flight, but not good enough for establishing a base of operations. The combat units of an armed force in this situation should not be more than 150 guerrillas, and even this number is rather high; ideally, the unit would be about 100. This constitutes a column and would be led by a commander.

according to Cuban organizational practice, remembering that in our war we abolished the ranks of corporal and sergeant because of their identification with the dictatorship.

On this basis, the commander heads this entire force of 100 to 150 guerrillas; and there will be as many captains as there are groups of 30 to 40. The captain's role is to direct and unify their platoon, making it fight almost always as a unit and looking after the distribution of guerrillas and the general organization. In guerrilla warfare, the squad is the functional unit. Each squad, made up of approximately eight to 12 fighters, is commanded by a lieutenant, who leads the group as the captain leads the platoon, but is always subordinate to the captain.

The operational tendency of the guerrilla band to function in small groups makes the squad the true unit. Eight to 10 guerrillas are the maximum that can act as a unit in a battle in these conditions; therefore, the squad, which will often be separated from the captain even though they fight on the same front, will operate under the orders of its lieutenant; there are exceptions, of course. A squad should not be broken up or kept dispersed at times when there is no fighting. Each squad and platoon should know who the immediate successor is in case the chief is killed, and such a person should be sufficiently trained to be able to take over their new responsibilities immediately.

One of the fundamental problems of the troop is food supply; in this everyone from the last person to the chief must be treated alike. This acquires great importance, not only because of the chronic shortage of supplies, but also because meals are the only daily events. The troops, who have a keen sense of justice, measure the rations with a sharp eye; the least favoritism toward anyone should never be allowed. If, in certain circumstances, the meal is served to the entire column, a regular order should be established and strictly observed, and at the same time the quantity and quality of food given to each person should be carefully checked. In the distribution of clothing the problem is different, these being articles of individual use. Here there are two considerations: first, the demand for necessities, which will almost always be greater than the supply; and second, the length of service and merits of each of the applicants. The length of service and merits, something very difficult to calculate precisely, should be noted in a special book by someone as- signed this responsibility under the direct supervision of the column chief. The same should occur with other articles that become available and are of individual rather than collective use. Tobacco and cigarettes should be distributed according to the general rule of equal treatment for all.

This task of distribution should be a specifically assigned responsibility. It is preferable that the designated persons be attached directly to the command. The command performs, therefore, the very important administrative tasks of liaison, as well as all the other special tasks that are necessary. Officers of the greatest intelligence should be included in the command, and soldiers attached to the command should be the brightest and most dedicated, since they will usually bear a greater burden than the rest of the troop. Nevertheless, they can have no special treatment at mealtime.

Each guerrilla fighter carries their full equipment; there is also a range of implements for general use that should be equitably distributed within the column. For this, too, rules can be established, depending on the number of unarmed persons in the troop. One method is to distribute all extra materiel, such as medicines, medical, dental or surgical instruments, extra food, clothing, general supplies, and heavy weapons equally among all platoons, which will then be responsible for their safekeeping. Each captain will distribute these supplies among the squads, and each chief of squad will distribute them among the guerrillas. Another solution, which can be used when not all the troop is armed, is to create special squads or platoons assigned to transport; this works out well, since it leaves the soldier who already has the weight and responsibility of their rifle free of an extra load. In this way the danger of losing materiel is reduced, since it is concentrated; and at the same time there is an incentive for the porter to carry more and to carry better and to demonstrate more enthusiasm, since this is the way he will win his right to a weapon in the future. These platoons will march in the rear and will have the same duties and the same treatment as the rest of the troop.

The tasks to be performed by a column will vary depending on its activities. While encamped, there will be special teams for keeping watch. These should be experienced, specially trained, and they should receive some special reward for this duty. This might be increased independence, or, if there is an excess of sweets or tobacco after proportional distribution to each column, something extra for the members of those units that carry out special tasks. For example, if there are 100 guerrillas and 115 packets of cigarettes, the 15 extra packs of cigarettes can be distributed among the members of these units. The vanguard and the rearguard units, separated from the rest, will have special vigilance duties; but each platoon should also have its own sentries. The farther from camp the watch is maintained, the greater the security of the group, especially when it is in open country.

The places chosen should be high, overlooking a wide area by day and difficult to approach by night. If the plan is to stay several days, it is worthwhile to construct defenses that will permit sustained fire in case of an attack. These defenses can be demolished when the guerrilla band moves on, or they can be left if circumstances no longer make it necessary to hide the column's tracks.

Where permanent camps are established, the defenses should be constantly improved. Remember that in a mountainous zone on carefully chosen ground, the only heavy weapon that is effective is the mortar. Using roofs reinforced with local materials, such as wood, rocks, etc., it is possible to make good refuges that are difficult for the enemy forces to approach and which will offer protection for the guerrilla forces from mortar shells.

It is very important to maintain discipline in the camp, and this should have an educational function. The guerrilla fighters should be required to go to bed and get up at set times. Games that have no social function and that damage troop morale and the consumption of alcohol should be prohibited. All these tasks are performed by a commission of internal order elected from those combatants with the greatest revolutionary merit. Their other role is to prevent the lighting of fires in places where they might be visible from a distance, or that raise columns of smoke before nightfall; they must also see that the camp is kept tidy and that when the column leaves no traces remain, if this is necessary.

Great care must be taken with fires that can leave traces for a long time; they should be covered over with earth; papers, cans, and scraps of food should also be burned. Total silence must prevail when the column is on the march. Orders are passed by gestures or by whispers passed from mouth to mouth until they reach the last person. If the guerrilla band is marching through unknown places, cutting a road, or being led by a guide, the vanguard will be approximately 100 or 200 meters or even more ahead, depending on the nature of the terrain. Wherever confusion might occur about the route, a guerrilla will be left at each turning to wait for those following, and this will be repeated until the last person in the rearguard has passed. The rearguard will also be somewhat separated from the rest of the column, keeping a watch on the roads in the rear and trying to erase the troop's tracks as far as possible. If there is a side road that might present a danger, it is always necessary to have a group keeping watch on it until the last person has passed. It is more practical that each platoon utilize its own members for this special duty, with each having the obligation to pass the guard to the following platoon and then to rejoin their own unit; this process will be continued until the entire troop has passed.

The march should be uniform and in an established order, always the same. Thus it will always be known that platoon number one is the vanguard, followed by platoon number two, and then platoon number three, which may be the command; then number four, followed by the rearguard or platoon num- ber five or other platoons that make up the column, always in the same order. During night marches silence should be even stricter and the distance between each combatant shorter, so that no one will get lost and make it necessary to shout or shine any light. Light is the enemy of the guerrilla fighter at night.

If all this marching is directed toward an attack, then on reaching a given point — the place to which everyone will return after the objective is accomplished — the extra weight of such things as backpacks and cooking utensils will be deposited, and each platoon will proceed with nothing more than its arms and fighting equipment. Trustworthy people, who have checked out the terrain and have observed the location of the enemy guards, should have reviewed the point of attack. The leaders, knowing the orientation of the base, the number of men that defend it, etc., will make the final plan for the attack and send combatants to their positions, always keeping in mind that a good number of the troops should be assigned to intercept reinforcements. In cases where the attack against a base is merely a diversion designed to provoke the sending of reinforcements along roads that can be easily ambushed, someone should communicate the result to the command as soon as the attack has been carried out, in order to break the encirclement, if necessary to prevent being attacked from the rear. In any case there must always be a sentry on the roads that lead to the combat area while the encirclement or direct attack is being carried out.

By night a direct attack is always preferable. It is possible to capture a camp if there is enough momentum and the required presence of mind and if the risks are not excessive.

An encirclement necessitates waiting and taking cover, closing in on the enemy steadily, trying to harass him in every way, and above all, trying to force him by fire to come out. When the circle has been closed to short range, the "Molotov cocktail" is a weapon of extraordinary effectiveness. Before coming within range for the "cocktail," shotguns with a special charge can be employed. These weapons, which in our war we christened "M-16s," consist of a 16-caliber sawed-off shotgun with a pair of legs added in such a way that with the butt of the gun they form a tripod. The weapon will thus be mounted at an angle of about 45 degrees; this can be adjusted by moving the legs back and forth. It is loaded with an open shell from which all the shot has been removed. A cylindrical stick extending from the muzzle of the gun is used as the projectile. A bottle of gasoline resting on a rubber base is placed on the end of the stick. This apparatus will fire the burning bottles 100 meters or more with a fairly high degree of accuracy. This is an ideal weapon for encirclements when the enemy has a lot of constructions made of wood or other inflammable material; it is also good for firing against tanks in hilly country.

Once the encirclement ends in victory, or, having achieved its objective, is withdrawn, all platoons retire in an orderly manner to the place where the backpacks were left, and the normal routine is resumed.

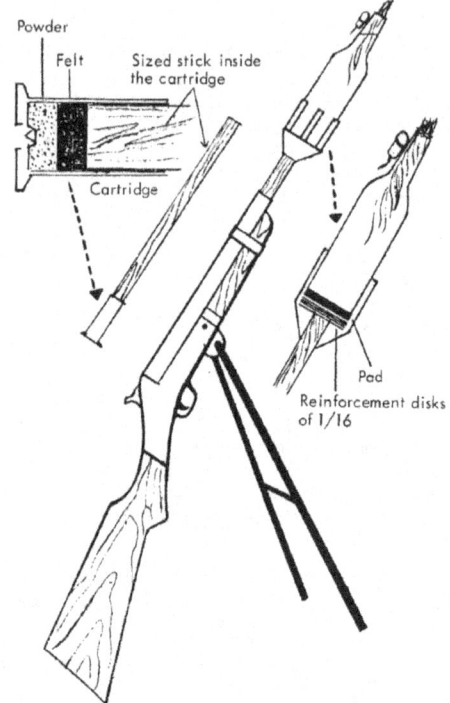

Adapting a 16 gauge shotgun to launch a molotov cocktail

In this phase, the guerrilla fighter's nomadic life produces not only a deep sense of fraternity among the guerrillas but at times also dangerous rivalries between groups or platoons. If these are not channeled to create a beneficial spirit of emulation, there is the risk that the unity of the column will be damaged. The education of the guerrilla fighters is crucial from the very beginning of the struggle; the social purpose of the fight and their duties should be explained in order to deepen their understanding, and to give them lessons in morale that will help forge their characters. Each experience should be a new source of strength and not simply one more episode in the fight for survival.

One of the greatest elements of education is example. The chiefs must therefore constantly present an example of a pure and committed life. A soldier's promotion should be based on courage, ability, and a spirit of sacrifice; those who do not possess these qualities to the highest degree should not be given responsible assignments, because this may lead to unfortunate incidents at any time.

The conduct of the guerrilla fighter will be judged whenever he approaches a house to ask for something. The residents will draw favorable or unfavorable conclusions about the guerrilla band depending on the manner in which any service, food, or other necessity is solicited, and the methods used to get what is wanted. The chief should discuss this problem in detail, emphasizing its importance; he should also teach by example. On entering a town, all drinking of alcohol should be prohibited and the troops should be exhorted beforehand to display the best possible discipline. The entrances and exits to the town should be constantly watched.

The organization, combat ability, heroism, and spirit of the guerrilla band will undergo a test of fire during encirclement by the enemy — the most dangerous situation of the war. In our guerrilla jargon in the recent war, the phrase "encirclement face" described the fear evident in someone who was frightened. The hierarchy of the deposed [Batista] regime arrogantly called its campaigns "encirclement and annihilation." Nevertheless, for a guerrilla band that knows the countryside and that is bound ideologically and emotionally to its chief, this is not a particularly serious problem. There is only a need to take cover, to try to slow up the enemy's advance, impede their action with heavy equipment, and wait for nightfall, the guerrilla fighter's natural ally. Then, with the greatest possible stealth, after exploring and choosing the best route, the band will depart, utilizing the most adequate means of escape and maintaining absolute silence. It is extremely difficult in the conditions of darkness to prevent a group of guerrillas from escaping encirclement.

4 Combat

Combat is the most important drama in the guerrilla life. It occupies only a short time; but nevertheless these stellar moments acquire an extraordinary importance, since each small encounter is a battle of a fundamental kind for the combatants.

We have already pointed out that an attack should be carried out in such a way as to guarantee victory. Besides general observations regarding the tactical role of attack in guerrilla warfare, the different features of each action should be noted. In the first instance, we will describe the type of battle conducted on favorable terrain, because this is the original model of guerrilla warfare; and it is in this aspect that certain principles must be examined before dealing with other problems through a study of practical experience. Warfare on the plain is always the result of an advance by the guerrilla bands consequent on their being strengthened and on changed conditions; this implies an increase in the guerrillas' experience and their ability to use this experience to their possible advantage.

In the first stage of guerrilla warfare, enemy columns will deeply penetrate insurgent territory; depending on the strength of these columns two different types of guerrilla attacks can be made. In chronological order, the first is to inflict systematic losses on the enemy's offensive capacity over a certain number of months. This action is concentrated on the vanguards; un- favorable terrain restricts the advancing columns' ability to defend their flanks; therefore, there will always be one point of the vanguard that is exposed as it penetrates and offers security to the rest of the column. When there are not sufficient combatants and reserves and the enemy is strong, the guerrillas should always aim for the destruction of this vanguard point. The system is simple; only a certain level of coordination is necessary. At the moment when the vanguard appears at the selected place — the steepest possible — a deadly fire is let loose on them, after a certain number of men have been allowed to penetrate. A small group must stall the rest of the column for some time while arms, munitions, and equipment are being collected. The guerrilla soldier should always remember that their source of arms is the enemy and that, except in special circumstances, they should not engage in any battle that will not lead to the capture of weapons.

When the guerrilla band is strong enough, a complete encirclement of the column can be carried out; or at least this impression should be given. In this case, the guerrilla front line must be sufficiently strong and well covered to resist the enemy's frontal assaults, naturally taking into account both offensive power and combat morale. At the moment when the enemy is detained in some chosen place, the rearguard guerrilla forces attack the enemy's rear. In such a selected place, flanking maneuvers will be difficult; snipers, outnumbered, perhaps by eight or 10 times, will have the entire enemy column within the circle of fire. Whenever there are sufficient forces in these cases, all roads should be protected with ambushes in order to block reinforcements. The encirclement will be closed gradually, particularly at night. Guerrilla fighters know the places where they fight, the invading column does not; guerrilla fighters grow at night, and the enemy sees his fear growing in the darkness.

In this way, without too much difficulty, a column can be totally destroyed; or at least will suffer such losses as to prevent its return to battle and to force it to take some time to regroup.

When the guerrilla band has a small force that aims to halt or slow the advance of the invading column, groups of between two and 10 snipers should spread out around the column at each of the four cardinal points. In this situation an assault can begin, for example, on the right flank; when the enemy focuses their response on that flank and fires that way, shooting begins at that moment from the left flank; at another moment from the rearguard or from the vanguard; and so forth.

With a very small expenditure of ammunition it is possible to hold the enemy in check indefinitely.

The tactics for attacking an enemy convoy or position must be adapted to the conditions of the area selected for the combat. In general, the first attack on an encircled place should be made at night against an advance post, with surprise assured. An attack carried out by skillful commandos can easily liquidate a position, thanks to the advantage of surprise. For a regular encirclement the escape routes can be controlled with a few guerrillas and access roads defended with ambushes; these should be spread out in such a way that if one is unsuccessful, falls back, or simply withdraws, a second remains, and so on successively. In cases where there is no surprise factor, the success of an attempt to overrun a camp will depend on the capacity of the encircling force to block the attempts of the rescue columns. In these cases, artillery, mortars, airplanes, and tanks will usually support the enemy. On favorable terrain, the tank is not so dangerous; it must travel by narrow roads and is an easy victim of mines. The offensive capacity of a formation of these vehicles is generally absent or reduced in these circumstances, since they must proceed in single file or at most two abreast. The best and surest weapon against the tank is the mine; but in a close fight, which may easily take place in steep places, the "Molotov cocktail" has an extraordinary value. We will not yet talk about the bazooka, which for the guerrilla force is a decisive weapon but difficult to acquire, at least in the first stages.

A trench with a roof is a defense against the mortar, which is a formidably potent weapon when used against encirclement; but on the other hand, against mobile attackers it loses its effectiveness unless it is used in large batteries. Artillery does not play a big role in this type of fight, since it has to be located where there is suitable access and it does not see the targets, which are constantly shifting. Aviation constitutes the principal arm of the oppressor forces, but its power of attack is also much reduced by the fact that its only targets are small trenches, generally hidden. Aircraft can drop high-explosive or napalm bombs, both of which constitute inconveniences rather than real dangers. Moreover, as the guerrillas draw as close as possible to the enemy's defensive lines, it becomes very difficult for planes to effectively attack the positions of the vanguard.

Anti-tank trap.

For attacks against camps made of wood or inflammable constructions the "Molotov cocktail" is a very important arm at a short distance. At longer distances, bottles containing inflammable material with the fuse lighted can be launched from a 16-caliber shotgun, as described earlier.

Of all the different types of mines, the most effective is the remotely exploded mine, although it requires the most technical know-how; but contact, fuse, and above all, electric mines with their lengths of cord are also extremely useful, and on mountainous roads constitute defenses for the popular forces that are virtually invulnerable.

A good defense against armored cars traveling along roads is to dig sloping ditches in such a way that the tank enters them easily and afterwards cannot get out, as the picture shows. These can easily be hidden from the enemy, especially at night or when infantry cannot advance in front of the tanks because of resistance by the guerrilla forces.

Another common form of enemy advance in zones that are not too steep is in trucks that are more or less open. Armored vehicles head the columns and the infantry follows behind in trucks. Depending on the strength of the guerrilla band it may be possible to encircle the entire column, following the general rules; or attacking some of the trucks and simultaneously exploding mines can split it. Swift action is necessary in order to seize the arms of the fallen enemy and withdraw.

For an attack on open trucks, a weapon of great importance, which should be used to its full potential, is the shotgun. A 16- caliber shotgun with large shot can sweep 10 meters, nearly the whole area of the truck, killing some of the occupants, wounding others, and provoking an enormous confusion. Grenades, if available, are also excellent weapons for these situations.

For all these attacks, surprise is fundamental because, at least at the moment of firing the first shot, it is one of the basic requirements of guerrilla warfare. Surprise is not possible if the peasants of the zone know the insurgent army is present. For this reason all movements of attack should be made at night. Only guerrillas of proven discretion and loyalty should know about these movements and establish the contacts. The march should be made with backpacks full of food, in order to be able to live two, three, or four days in the ambush areas.

The discretion of the peasants can never really be trusted, first because there is a natural tendency to talk and to discuss events with other family members or with friends; and also because of the enemy soldiers' inevitably cruel treatment of the population after a defeat. Terror can be sown, and this terror leads to someone's talking too much, revealing important information, in the effort to save their life.

In general, the site selected for an ambush should be at least one day's march from the guerrilla band's regular camp, since the enemy will almost always know more or less accurately where it is.

We said before that the nature of fire in a battle indicates the location of the opposing forces: on one side, violent and rapid firing by the soldier of the line, who has plenty of ammunition as a rule; on the other side, the methodical, sporadic fire of the guerrilla fighter who knows the value of every bullet and who endeavors to expend it with the greatest economy, never firing one shot more than necessary. It is not reasonable to allow an enemy to escape or to fail to use an ambush to the full in order to save ammunition, but the amount to be expended in particular circumstances should be calculated in advance and the action carried out according to these calculations.

Ammunition is the big problem for a guerrilla fighter. Arms can always be obtained, and those acquired are not used up by the guerrillas as ammunition is; moreover, arms are generally captured with their ammunition, but ammunition is never or rarely obtained alone. Each confiscated weapon will have its load, but does not contribute to others because there is no extra. The tactical principle of saving firepower is fundamental in this type of warfare.

A guerrilla chief who takes pride in their role takes great care in withdrawal. This should be timely, rapid, and managed in order to save all the wounded and their equipment, the back-packs, ammunition, etc. The rebels should never be surprised while withdrawing, neither should they allow themselves to become surrounded. Therefore, guards must be posted along the selected route at every point where the enemy army is expected to bring forward its troops in an attempt to close a circle; and there must be a communications system that will permit rapid reports when a force tries to surround the rebels.

In combat there must always be some unarmed guerrillas. These people recover the guns of compañeros who are wounded or dead, guns seized in battle or belonging to prisoners; they will take charge of the prisoners, of evacuating the wounded, and of transmitting messages. There should also be a good corps of messengers with legs of steel and a proven sense of responsibility who will relay the necessary warnings in the least possible time.

The number of people needed in addition to the armed combatants varies; but a general rule is two or three for each 10, including those who will be present at the battle scene and those who will fulfill the necessary tasks in the rearguard, keeping watch on the route of withdrawal and performing the messenger services mentioned above.

When a defensive type of war is being fought, that is to say, when the guerrilla band is attempting to block the passage of an invading column beyond a certain point, the action becomes a war of positions; but at the outset it should always have the element of surprise. In this case, since trenches as well as other defensive systems that will be easily noticed by the peasants will be used, it is necessary that these remain in the friendly zone. In this type of warfare the government generally establishes a blockade of the region, and the peasants who have not fled must go to buy their basic provisions outside the zones of guerrilla action. If these people leave the region at a critical moment, as we are now describing, this can constitute a serious danger because of the information they might pass on to the enemy army. The policy of complete isolation must serve as the strategic principle of the guerrilla army in these circumstances. The defenses and the entire defensive apparatus should be arranged in such a way that the enemy vanguard will always fall into an ambush. It is very important as a psychological factor that the man in the vanguard will die with no chance of escape in every battle, because this creates a growing consciousness of the risk within the enemy army, until eventually no one wants to be in the vanguard; and it is obvious that a column with no vanguard cannot move, since someone has to assume that responsibility. Furthermore, encirclements can be carried out if they are expedient; or diversionary maneuvers such as flank attacks; or the enemy can simply be detained frontally. In every case, places that might be utilized by the enemy for flank attacks should be fortified.

We are now assuming that more guerrillas and arms are available than in the battles described above. Clearly, a large personnel is required to blockade all possible roads leading into a zone, as there might be very many roads. The variety of traps and attacks against armored vehicles will be increased here, in order to give the greatest security possible to the systems of fixed trenches which can be located by the enemy. Generally, in this type of battle, the command will be to defend the positions to the death if necessary; and it is essential to assure every combatant the greatest chance of survival.

The more a trench is hidden from distant view, the better; above all, it is important to provide cover so that mortar fire will be ineffective. Mortars of 60 or 81 millimeters, the usual campaign calibers, cannot penetrate a good roof made with simple local materials, such as one made from a base of wood, earth, and rocks covered with some camouflage material. An escape route for an emergency must always be constructed, so that the defender can get away with less risk.

The sketch below shows the way these defenses were constructed in the Sierra Maestra,

Refuge against mortar fire.

Fixed lines of fire do not exist. The lines of fire are something more or less theoretical; they are established at certain critical moments, but they are extremely elastic and permeable on both sides. What does exist is a broad "no man's land." But a civilian population inhabits "no man's land" in a guerrilla war, and this civilian population collaborates to some extent with both sides, even though the overwhelming majority supports the insurrectionary band. These people cannot be removed en masse from the zone because of their numbers and because this would create problems of supply for whichever of the contenders tried to provide them with food. This "no man's land" experiences periodic incursions (generally during the daytime) by the repressive forces and at night by the guerrilla forces. This zone becomes a very important maintenance base for the guerrillas and should be nurtured in a political way, always establishing the best possible relations with the peasants and merchants.

In this type of warfare the tasks of indirect combatants — in other words, those who do not carry arms — are extremely important. We have already outlined some of the features of contact work in combat zones; but this is an institution within the guerrilla organization. Contacts to the most distant command post or to the most distant guerrilla group should be linked in such a way that messages can be passed along by the most rapid system in the region — whether it is an area easily defended, in other words, favorable terrain, or unfavorable. A guerrilla band operating on unfavorable terrain will be unable to use modern communications systems, such as the telegraph, roads, etc., apart from radios located in military garrisons that can be defended. If these fall into enemy hands, the codes and frequencies must be changed, a rather annoying task.

We are speaking here from memory about experiences in our war of liberation. The daily and accurate report on all the enemy's activities was complemented by our contacts. The espionage system should be carefully studied, well worked out, and personnel selected with great care. A counter-spy can do enormous damage, but even in less extreme cases, the harm that can result from exaggerated information that misjudges the danger is very great. It is not likely that a risk will be underrated; the tendency of people in the countryside is to overrate and exaggerate. The same mystical mentality that imagines phantasms and all kinds of supernatural beings also creates monstrous armies where there is scarcely a platoon or an enemy patrol. The spy should appear to be as neutral as possible, not known by the enemy to have any connection with the forces of liberation. This is not as difficult a task as it seems; many such individuals are encountered in the course of the war: merchants, professionals, and even clergy can assist in this task and give timely information.

One of the most important features of a guerrilla war is the significant difference between the information that reaches the rebel forces and the information possessed by the enemy. While the latter must operate in regions that are absolutely hostile, encountering sullen silence on the part of the peasants, in nearly every house the rebels have a friend or even a relative, and news is constantly conveyed through the contact system until it reaches the central command of the guerrilla force or of the guerrilla group in the zone.

A serious problem is created when an enemy incursion takes place in territory that has become openly pro-guerrilla, where all the peasants respond to the popular cause. The majority of peasants try to escape with the popular army, abandoning their children and their work; others might even flee with the whole family; some wait to see what will happen. The most serious problem caused by an enemy penetration into guerrilla territory is that of a group of families finding themselves in a tight, possibly desperate situation. They should receive the utmost help, but should also be warned of the consequences of flight into an inhospitable area away from their usual place of work, exposed to the hardships of such an existence.

It is not possible to talk of a "pattern of repression" by the people's enemies. Although the general methods of repression are always the same, they usually commit crimes of greater or lesser intensity depending on specific social, historical, and economic circumstances. There are places where the departure of a person for the guerrilla zone, leaving their family and home, provokes no great reaction. Elsewhere, this is enough to provoke the burning or seizure of their belongings, or even the death of all members of their family. Adequate distribution and organization of the peasants who are going to be affected by an enemy advance must be arranged, depending on the local customs in the war zone or country concerned.

Obviously, preparations must be made to expel the enemy from the territory by taking action against their supplies, completely cutting their lines of communication, destroying their attempts to supply themselves through the actions of small guerrilla bands, and in general forcing them to devote large numbers of men to the problem of supplies.

In all combat situations a very important element is the correct utilization of reserves wherever battle begins. Because of its character, a guerrilla army can rarely count on reserves, since it always strikes so that every individual's efforts are regulated and focused. Nevertheless, some guerrillas should be positioned ready to respond to an unforeseen development, to block a counter-offensive, or to handle a situation at any moment. Within the organization of the guerrilla band, assuming that the conditions and possibilities allow, a utility platoon can be kept ready to be dispatched to the place of greatest danger. It might be christened the "suicide platoon" or something along those lines, a title that truthfully reflects its role. This "suicide platoon" should be everywhere a battle is decided: in the surprise attacks against the vanguard, in the defense of the most vulnerable and dangerous positions, in other words, wherever the enemy threatens to break the steadiness of the line of fire. It should be made up strictly of volunteers. Admittance to this platoon should be regarded as a great honor. Over time it becomes the darling within the guerrilla column, and the guerrilla fighter who wears its insignia enjoys the admiration and respect of all their compañeros.

5 Beginning, Development, And End Of A Guerrilla War

We have now thoroughly defined the nature of guerrilla war- fare. Next, we will discuss the ideal development of this war, from the emergence of a single nucleus on favorable terrain and beyond.

In other words, we will theorize further on the basis of the Cuban experience. At the start, there is a more or less armed, more or less homogeneous group that devotes itself almost exclusively to hiding in the roughest and most inaccessible places, making little contact with the peasants. It strikes a lucky blow and its fame grows; a few peasants, dispossessed of their land or engaged in a struggle to preserve it, and young idealistic members of other classes join the nucleus; acquiring greater audacity, it starts to operate in populated areas, making more contact with the local people; it repeats attacks, always fleeing afterwards; suddenly it engages in combat with some column or other and destroys its vanguard. People continue to join; it has increased in number, but its organization remains exactly the same; less cautious, it ventures into more populous zones.

Later it sets up temporary camps for several days; it abandons these on receiving news of an approaching enemy army, after being bombed, or when it becomes aware of such dangers. The numbers in the guerrilla band increase as work among the masses converts every peasant into an enthusiast for the war of liberation. Finally, an inaccessible place is chosen, a settled life begins, and the first small industries are established: a shoe factory, a cigar and cigarette factory, a clothing factory, an arms factory, a bakery, hospitals, possibly a radio transmitter, a printing press, etc.

The guerrilla band now has an organization, a new structure. It is the head of a large movement with all the characteristics of a small government. A court is established for the administration of justice, laws might be promulgated, and the work of indoctrination of the peasant masses continues, now also extended to any local workers, drawing them into the cause. An enemy action is launched and defeated; the number of rifles increases; with these the number of people fighting with the guerrilla band increases. The time arrives when its radius of action will not have increased in the same proportion as its personnel; at that moment a force of appropriate size is separated, a column or a platoon, perhaps, and this moves to another combat zone.

The work of this second group will begin with rather different characteristics because of the experience that it brings and because of the influence of the troops of liberation on the war zone. The original nucleus also continues to grow; it has now received substantial support in food, sometimes in guns, from various places; recruits continue to arrive; the administration of government, with the promulgation of laws, continues; schools are established, permitting the indoctrination and training of recruits. The leaders learn steadily as the war develops, and their ability to command develops with the added responsibilities of the qualitative and quantitative growth of their forces.

If there are remote territories, a group sets out in that direction at a certain moment, in order to confirm the advances that have been made and to continue the cycle.

But there is also an enemy territory, unfavorable for guerrilla warfare. Small groups begin to penetrate there, assaulting the roads, destroying bridges, planting mines, and sowing alarm. With the ups and downs characteristic of warfare the movement continues to grow; by this time the extensive work among the masses makes easy movement of the forces possible in unfavorable territory, and so opens the final stage, which is urban guerrilla warfare.

Sabotage increases significantly throughout the entire zone. Life is paralyzed; the zone is conquered. The guerrillas then move on to other zones, where they confront the enemy army along defined fronts; by now heavy arms have been captured, perhaps even some tanks; the fight is more equal. The enemy falls when a series of partial victories becomes transformed into definitive victories, in other words, when the enemy has to accept

41

battle in conditions imposed by the guerrilla band; there he is annihilated and compelled to surrender.

This is a sketch of what occurred in the different stages of the Cuban war of liberation; but it has a universal significance. Nevertheless, it will not always be possible to count on the same degree of intimacy with the people, the conditions, and the leadership that existed in our war. It is unnecessary to state:

Fidel Castro epitomizes the ultimate qualities of a combatant and statesman; our journey, our struggle, and our triumph we owe to his vision. We cannot say that without him the victory of the people would not have been achieved; but that victory would certainly have cost much more and would have been less complete.

CHAPTER III: ORGANIZATION OF THE GUERRILLA FRONT

1 supply

A good supply system is of primary importance to the guerrilla band. A group of guerrillas in contact with the soil must live from the products of this soil and at the same time ensure that the livelihood of those who provide the supplies, the peasants, is maintained. Especially at the beginning of the difficult guerrilla struggle, it is not possible for the group to dedicate its efforts to producing its own food, apart from the fact that these supplies could be easily discovered and destroyed by the enemy in territory likely to be completely penetrated by the repressive forces. In the first stages, therefore, supply is always internal.

As the guerrilla struggle develops, it will be necessary to arrange supply from outside the combat territory. At first, the band survives only on whatever the peasants have; occasionally it might be possible to reach a store to buy something, but it is never possible to have supply lines because there is no territory in which to establish them. The supply line and the food store are conditional on the development of the guerrilla struggle.

The first task is to gain the absolute confidence of the residents of the zone; and this confidence is won by a positive approach to their problems, by help and a constant orientation program, by defending their interests and punishing anyone who attempts to take advantage of the instability in which they live in order to pressure, dispossess the peasants, seize their harvests, etc. The approach should be simultaneously soft and hard: soft and spontaneously helpful to all those who honestly sympathize with the revolutionary movement, hard against anyone attacking it outright, fomenting dissension, or simply passing on important information to the enemy army.

Little by little the territory will be consolidated, and then actions will be easier. The fundamental principle must be always to pay for all merchandise taken from a friend. This might be crops or items from commercial establishments. They will often be donated, but at other times the economic conditions of the peasantry prevent such donations, and there are cases in which necessity forces the guerrilla band to take food from stores without paying for it, simply because there is no money. In this instance, the merchant should always be given a bond, a promissory note, something that acknowledges the debt, "the bonds of hope" discussed previously. It is better to use this method only with people who are outside the liberated territory, and in such cases to pay as soon as possible all or at least part of the debt. When conditions have improved sufficiently to maintain a territory permanently free from the rule of the opposing army, it is possible to establish collective plantings, whereby the peasants work the land for the benefit of the guerrilla army. In this way an adequate, permanent food supply is guaranteed.

If the volunteers for the guerrilla army greatly outnumber its arms, and political circumstances prevent these people from entering zones dominated by the enemy, the rebel army can put them to work directly on the land, harvesting crops; this guarantees supply and adds something to the service record of combatants and their prospects for future promotion. Nevertheless, it is preferable that the peasants sow their own crops, so that this work is performed more efficiently, with more enthusiasm and skill. When conditions have further

43

ripened, depending on the crops involved it is possible to arrange the purchase of entire harvests, which can be left in the field or in a warehouse for the use of the troops.

When mechanisms are established to supply the peasant population as well, all food supplies can be concentrated in order to facilitate a system of barter among the peasants, with the guerrilla army acting as intermediary.

If conditions continue to improve, taxes can be established; these should be as low as possible, especially for the small producer. It is important to pay attention to every detail of relations between the peasant class and the guerrilla army, which emanates from that class.

Taxes might be collected sometimes in money or in the form of a part of the harvest, which will increase the food supplies. Meat is one of the primary necessities. Its production and preservation must be guaranteed. If the zone is not secure, peasants with no apparent connection to the guerrillas can establish farms that can produce chickens, eggs, goats, and pigs, starting with stock that has been bought or confiscated from the large landowners. In the areas with big estates there are usually large numbers of cattle. These can be killed and salted and the meat maintained in condition for consumption for a long period of time.

This can also be a source of hides. A more or less primitive leather industry can be developed to provide leather for shoes, one of the fundamental accessories in the struggle. In general, essential foods are the following (depending on the zone): meat, salt, vegetables, starches, or grains. The basic food is always produced by the peasants; it might be yams as it was in the mountainous regions of Oriente province in Cuba; it might be corn, as in the mountainous regions of Mexico, Central America, and Peru; potatoes, in Peru; in other zones, such as Argentina, beef; wheat in others; but it is always necessary to guarantee a basic food supply for the troops as well as some kind of fat to assist in food preparation; these might be animal or vegetable fats.

Salt is one of the essential supplies. When close to the sea and able to get to it, small dryers should be established immediately; these will guarantee some production in order to always have a reserve and the ability to supply the troops. Remember that in rough places such as these, where only some of the foods are produced, it is easy for the enemy to establish an encirclement that can greatly damage the flow of supplies into the zone. It is better to plan for such eventualities through peasant groups and civil organizations in general. The residents of the zone should have a minimum food supply on hand that will permit them at least to survive, even though poorly, during the hardest phases of the struggle. An attempt should be made to collect rapidly a good provision of foods that do not decompose — grains, corn, wheat, rice, etc., which can last quite a long time; also flour, salt, sugar, and all types of canned goods; furthermore, the essential seeds should be sown.

The time will arrive when all the food problems of the troops in the zone are solved, but large quantities of other items will be needed: leather for shoes, if it has not been possible to create an industry for supplying the zone; cloth and all the additional things necessary for clothing; paper, a press or mimeograph machine for newspapers, ink, and various other implements. In other words, the need for products from the outside world will increase as the guerrilla bands become more organized and the organization becomes more complex. In order to meet these needs adequately the organized supply lines must function perfectly. These organizations are composed basically of friendly peasants; they should have two poles, one in the guerrilla zone and one in a city. Departing and radiating out from the guerrilla zones, supply lines will spread throughout the whole territory, permitting the passage of materials. Little by little the peasants become accustomed to the danger (in small groups they can work marvels) and come to deposit whatever is required in the indicated spot without running extreme risks. These errands can be run at night with mules or other similar transport animals or with trucks, depending on the area. In this way, a very good supply can be achieved. This type of supply line is for areas near the operational zone.

It is also necessary to organize a supply line to remote areas, which should provide the money needed for making purchases and also the implements that cannot be produced in small towns or provincial cities. The organization will be assisted with direct donations from sectors sympathetic to the struggle, exchanged for secret "bonds," which should be delivered. The personnel charged with the management of this operation should always be strictly controlled. Serious consequences should follow any neglect of the indispensable moral duties involved in this responsibility. Purchases can be made with cash and also with "bonds of hope" when the guerrilla army, having departed from its base of operations, threatens a new zone. In these circumstances, there is no way to avoid seizing merchandise from its owner, who will have to rely on the good faith and capacity of the guerrillas to pay their debts.

For all supply lines that pass through the countryside, it is necessary to have a series of houses, terminals, or way-stations, where supplies can be hidden during the day while waiting to be moved by night. Only those directly in charge of the food supplies should know these houses. The least possible number of local residents should know about this transport operation, and these must be individuals in whom the organization has the greatest confidence.

The mule is one of the most useful animals for these tasks. With an incredible resistance to fatigue and an ability to walk in the steepest terrain, the mule can carry more than 100 kilograms for days. Requiring only simple food, mules are an ideal means of transport. The mule train should be properly provided for with horseshoes, and the muleteers should understand their animals and take great care of them. In this way it is possible to have regular four-footed armies of unbelievable effectiveness. But often, despite the strength of the animal and its ability to bear up through the hardest days, difficulty of passage will make it necessary to leave the cargo in particular spots. To avoid this necessity, a special team should be assigned to cut trails for this kind of animal. If all these conditions are met, if an adequate organization is created, and if the rebel army maintains the required excellent relationship with the peasants, an effective and lasting supply for the entire troop is guaranteed.

2 Civil Organization

The civil organization of the insurrectional movement is very important on both the external and the internal fronts. Naturally, their features are as different as their functions, although they both perform tasks that fall under the same name. The collections that can be carried out on the external front, for example, are not the same as those that can take place on the internal front; the same goes for the tasks of propaganda and supply. Let us begin with the tasks on the internal front.

On considering the "internal front," we are referring to a territory largely dominated by the forces of liberation. Further, it is assumed that the zone is suitable for guerrilla warfare, because when the right conditions do not exist, in other words, when the guerrilla struggle is taking place in unsuitable terrain, the guerrillas can extend their reach but not deepen it; they can create channels through new areas, but cannot establish an internal mechanism because the entire zone is penetrated by the enemy. On the internal front we might have a range of organizations that perform specific roles to improve administrative efficiency. Generally, propaganda belongs directly to the guerrilla army, but it also can be separated if kept under its control. (This matter is so important that we will discuss it elsewhere.) Collections are a function of the civil organization, as are the general tasks of organizing the peasants and also the workers, if they are present. Both these should be run by one council.

Collections, as we explained in the previous chapter, can be conducted in various ways: through direct or indirect taxes, through direct or indirect donations, and through confiscations; all this goes to complete the large part of the guerrilla army's supplies.

Bear in mind that the zone should never be impoverished by the direct actions of the rebel army, even though it will be indirectly responsible for the impoverishment that results from enemy encirclement, a fact that the adversary's propaganda will repeatedly point out. Precisely for this reason, conflicts should not be created by direct causes. For example, there should be no regulations that prevent the farmers in a liberated zone from selling their produce outside that territory, except in extreme and transitory circumstances and with a full explanation of these exceptions to the peasantry. Every act of the guerrilla army should always be accompanied by propaganda explaining the reasons for it. These reasons will generally be well understood by peasants who have sons, fathers, brothers, or relations within this army, which has become their own.

Considering the importance of relations with the peasantry, organizations should be created to make regulations for them, organizations that exist not only within the liberated zone, but also have connections with the adjacent areas. Precisely through these connections it is possible to penetrate a zone for a future expansion of the guerrilla front. The peasants will sow the seed with oral and written propaganda, with accounts of life in the other zone, of the laws that have already been issued to protect the small peasant, of the spirit of sacrifice of the rebel army; in a word, they create the necessary atmosphere for helping the rebel troops.

The peasant organizations should also have connections of some kind that will permit the transport and sale of crops by the rebel army networks in enemy territory through more or less benevolent intermediaries, more or less friendly to the peasant class. The merchant can be devoted to a cause for which he might take risks, but he is also devoted to money and this means he will take advantage of any opportunity to gain a profit.

We have already mentioned the importance of the department of road construction in connection to supply problems. When the guerrilla band has attained a certain level of development, it might have four or five more or less permanent centers, and will no longer wander about through various regions without a camp. Routes should be established ranging from small trails allowing the passage of a mule to good roads for trucks. In all this, the capacity of the rebel army's organization must be kept in mind, as well as the offensive capacity of the enemy, who might destroy these constructions and even make use of roads built by their opponent to reach the camps more easily. The essential rule is that roads are for assisting supply in places where no other solution is possible; they should not be constructed except in circumstances where there is virtual certainty that the position can be maintained against an adversary's attack.

Another exception would be roads built without great risk to facilitate communication between points that are not of vital importance.

Furthermore, other means of communication can be established. One of these that is extremely important is the telephone, the lines of which can be strung in the forest conveniently using trees for posts. There is the advantage that they are not visible to the enemy from above. The telephone also presupposes a zone that the enemy cannot penetrate.

The council — or central department of justice, revolutionary laws, and administration — is one of the vital features of a guerrilla army fully constituted and with its own territory. The council should be directed by an individual who knows the laws of the country; all the better if he or she understands the necessities of the zone from a juridical point of view; he or she can proceed to draft a series of decrees and regulations that help the peasants normalize and institutionalize life in the rebel zone.

For example, during our experience in the Cuban war we issued a penal code, a civil code, rules for supplying the people, and guidelines for the agrarian reform. Subsequently, the laws were passed setting qualifications for candidates in the national elections that were to be held later; the Agrarian Reform Law of the Sierra Maestra was also established. The council is likewise in charge of accounting operations for the guerrilla column or columns; it is responsible for handling financial problems and at times intervenes directly in issues of supply.

All these recommendations are flexible, based on an experience in a particular place, conditional on its geography and history; they will be modified in different geographical, historical, and social situations. Besides the council, the general health of the zone must be considered; this might involve central military hospitals that can provide the most complete care to all the peasants. Whether adequate medical treatment can be given will depend on the stage reached by the revolution. Civil hospitals and civil health administration, where officers and army personnel have the dual function of caring for the people and showing them how to improve their health, are linked directly to the guerrilla army. The major health problems among people in these conditions are rooted in their total ignorance of elementary principles of hygiene and this aggravates their already precarious situation.

The collection of taxes, as mentioned above, is also a function of the general council.

Warehouses are very important. As soon as a place is taken that can serve as a base for the guerrilla band, warehouses should be established in the most orderly fashion possible. These will guarantee minimal care of goods and, most importantly, will provide the control needed for equalizing distribution and keeping it equitable in the future.

On the external front, the functions are different both in terms of quantity and quality. For example, propaganda should provide a national orientation, explaining the victories won by the guerrillas, calling on workers and peasants to take effective mass action, and providing news of any victories on its own front. Soliciting funds is completely secret and should be carried out with the greatest possible care, isolating small collectors in the chain completely from the treasurer of the organization.

This organization should be spread throughout areas that complement one another so as to form a whole: zones that might be provinces, states, cities, villages, depending on the size of the movement. In each organization there must be a finance commission that takes charge of the disposal of funds collected. Money can be collected by selling bonds or through direct donations. When the struggle is more advanced, taxes can be collected; when industries come to recognize the great force that the insurrectional army possesses, they will agree to pay. Procurement of supplies should be adjusted to the needs of the guerrilla bands; a supply chain is organized so that the more common items are procured nearby, and things that are scarce or impossible to find locally are sought in larger centers. An effort should be made to keep the chain as limited as possible, known to the smallest number of people; in that way it can carry out its mission for a longer time.

Sabotage should be directed by the civil organization in the external sector, in coordination with the central command. In special circumstances, after careful analysis, assaults on individuals will be initiated. In general, we do not consider this desirable, except for the purpose of eliminating some figure notorious for his villainies against the people and the virulence of his repression. Our experience in the Cuban struggle shows that it would have been possible to save the lives of many fine comrades who were sacrificed in the performance of missions of little value. Sometimes these can result in the reprisal of enemy bullets and a loss of combatants that does not match the results obtained. Indiscriminate assaults and terrorism should not be employed. Much more preferable is an effort directed at large concentrations of people in whom the revolutionary idea can be planted and nurtured, so that at a critical moment they can be mobilized and, with the help of the force of arms, tip the balance in the revolution's favor.

This also requires the popular organizations of workers, professionals, and peasants, who work at sowing the seed of the revolution among their respective masses, explaining issues, providing revolutionary publications for reading, teaching the truth. Truth must be one of the features of revolutionary propaganda. In this way, the masses will slowly be won over. Those who do the best work can be selected for incorporation into the rebel army or assigned to other big responsibilities.

This is the outline of civil organization within and outside the guerrillas' territory during a popular struggle. There are possibilities of improving all these features significantly; I repeat again, I am discussing here our experience in Cuba; new experiences can vary and improve these concepts. We offer an outline, not a bible.

3 The Role Of Women

The part women can play in the development of a revolutionary process is of extraordinary importance. This needs to be stressed because in all our countries, with their colonial mentality, there is a certain underestimation of women that becomes real discrimination.

Women are capable of performing the most difficult tasks, of fighting beside men; and despite current belief, do not create conflicts of a sexual nature among the troops if a sufficient ideological and organizational base exists.

In the rigors of a combatant's life a woman is a compañero who brings the qualities appropriate to her sex, but she can work the same as a man and she can fight; she is weaker, but no less resistant than a man. She can perform every combat task that a man can at a given moment, and on certain occasions in the Cuban struggle she performed a relief role. Naturally, female combatants are a minority. When the internal front is being consolidated, as many combatants as possible who do not possess indispensable physical characteristics should be removed; women then can be assigned a considerable number of specific occupations, of which one of the most important — perhaps the most important — is communication between different combatant forces, especially between those in enemy territory. The transport of things, messages, or money, small items and those of great importance, should be assigned to women in whom the guerrilla army has total confidence; women can act as couriers using a thousand tricks; it is a fact that however brutal the repression, however thorough the search, a woman receives less harsh treatment than a man and can carry her message or an important or confidential object to its destination.

As a simple messenger, either by word of mouth or in writing, a woman can always perform her task with more freedom than a man, attracting less attention and at the same time inspiring less fear of danger in the enemy soldier; a man who commits brutalities will often act on the impulse of fear or apprehension that he himself will be attacked, since this is one form of action in guerrilla warfare.

Women can wear special belts beneath their skirts to carry messages between separated forces, messages to outside the lines, even to outside the country, and also objects of considerable size, such as bullets. In this phase, women can also perform their traditional role in peacetime: a soldier living in the extremely harsh conditions of guerrilla warfare is happy to be able to look forward to a seasoned meal which actually tastes like something. (One of the great tortures of the war was eating a cold, sticky, tasteless mess.) A female cook can greatly improve the diet and, moreover, is easier to assign to these domestic tasks; one of the problems in guerrilla bands is that those who perform it scorn all labor of a civilian character; they are constantly trying to get out of these tasks in order to participate in the active combat forces.

A task of great importance for women is to teach the basics of reading, including revolutionary theory, primarily to the peasants of the zone, but also to the revolutionary soldiers. The organization of schools, which is a part of the civil organization, should be done principally through women, who arouse more enthusiasm among children and enjoy more affection from the school community. Similarly, when the fronts have been consolidated and a rear exists, the role of social worker also falls to women who investigate the various economic and social evils of the zone with a view to changing them as far as possible.

A woman plays an important part in medical matters as nurse, and even as doctor, with an infinitely superior gentleness to that of her rough compañero-in-arms, a gentleness that is so much appreciated at times when a man is helpless, without comfort, perhaps suffering severe pain and exposed to the many kinds of risks that are part of a war of this nature.

Once the stage of creating small war industries has begun, women can also contribute, especially in the manufacture of uniforms, a traditional employment of women in Latin American countries. With a simple sewing machine and a few patterns she can work miracles. Women can take part in all levels of civil organization; they can replace men perfectly well and should do, even to carry weapons if there is a shortage of couriers, although this is a rare occurrence in guerrilla life.

Men and women should be properly educated, in order to avoid all kinds of misconduct that can damage troop morale; but unattached individuals who love each other should be allowed to marry and live as man and wife in the mountains, in compliance with the simple requirements of the guerrilla band.

4 Health

One of the serious problems a guerrilla fighter confronts is exposure to the accidents of life, especially to wounds and illness, which are very common in guerrilla warfare. The doctor performs an extraordinarily important role in the guerrilla band, not just in saving lives, where their scientific intervention might have little impact because of the limited resources available; the doctor also reinforces the patient morally and makes him or her feel that there is someone nearby who is dedicating all their efforts to minimizing their pain; the doctor gives the sick or wounded the comfort of knowing that someone will remain at their side until they are cured or out of danger.

The organization of hospitals depends largely on the stage of development of the guerrilla band. Three fundamental types of hospital organization corresponding to various stages can be mentioned.

In the first, nomadic phase, the doctor — if there is one — always travels with their compañeros, as just one more person; he or she will probably have to perform all the other functions of the guerrilla fighter, including that of fighting, and will suffer at times the depressing and desperate task of treating patients when the means of saving lives are not available. This is the stage in which the doctor has the most influence over the troops, the greatest impact on their morale. During this period of the guerrilla band's development the doctor achieves their full potential as a true priest, who seems to carry in their poorly equipped backpack the consolation needed by the guerrillas. The value of a simple aspirin to someone who is suffering badly is beyond calculation, when administered by the friendly hand of someone who makes that suffering his own. Therefore, in the first stage, the doctor must be a person who totally identifies with the ideas of the revolution, because their words will affect the troops much more deeply than those of any other member.

In the normal course of events in guerrilla warfare a further stage is reached that might be called "semi-nomadic." There will be camps, occasionally visited by the guerrilla troops; friendly, secure houses where it is possible to store objects and even leave the wounded; and a growing tendency for the guerrilla troop to become settled. At this stage, the task of the doctor is less frustrating; he or she might have emergency surgical equipment in their backpack and another more complete kit for less urgent operations in a friendly house. It will be possible to leave the sick and wounded in the care of peasants who will offer their help with great dedication. The doctor can also count on a greater number of medicines kept in convenient places; as far as possible, these should be completely catalogued, depending on the circumstances. In this same semi-nomadic state, if the guerrilla band operates in places that are absolutely inaccessible, hospitals can be established to which the sick and wounded can go to recover.

In the third stage, when there are zones invulnerable to the enemy, a true hospital network can be constructed. In its most developed form, this might consist of three different types of center. In the combat category there should be a doctor, the combatant most loved by the troops, a fighter, who does not need very extensive knowledge. I say this because their task is principally one of giving relief and preparing the sick or wounded, while the real medical work is performed in hospitals situated more securely. A good surgeon should not be sacrificed in the line of fire.

When soldiers fall in the front line, stretcher-bearers — if available, which will depend on the level of organization of the guerrilla band — will carry the wounded to the first post; if they are not available, the compañeros themselves will perform this duty. Transport of the wounded in rough zones is one of the most delicate of all tasks and one of the most painful experiences in a soldier's life. The transport of the wounded is probably harder on the compañeros than the wounded soldier, however grave the injury, because of the guerrillas' spirit of sacrifice. The transport can be carried out in different ways depending on the nature of the terrain. In rough and wooded places, which are typical in this kind of warfare, it is necessary to walk in single file. Here the best system is to use a long pole, with the patient carried in a hammock that hangs from it.

The guerrillas take turns carrying the weight, one in front and one behind. They should swap positions with two other compañeros frequently, since the shoulders suffer severely and the individual is gradually worn out carrying this delicate and heavy burden.

After having been checked at the first hospital, the wounded soldier then proceeds with the information of this initial treatment to a second center, where there might be surgeons and specialists, depending on the possibilities of the guerrilla troop, and where more serious life-saving operations are performed and individuals can be relieved from danger — this is the second stage. At a third level, there are much better hospitals that can conduct investigations into causes and effects of illnesses that prevail among the residents of the zone. These hospitals, which correspond to a sedentary life, are not only centers for convalescence and for less urgent operations, but also serve the civil population, where the hygiene specialists have an educational role. Dispensaries that can monitor individual treatment should also be established. If the supply capability of the civil organization is sufficient, the hospitals of this third group can have a range of facilities that provide diagnosis, possibly even laboratory and x-ray facilities.

Other useful individuals are the doctor's assistants. They are generally youths with some vocation and some knowledge, with fairly strong physiques; they do not bear arms, sometimes because their vocation is medicine, but usually because there are insufficient arms for everyone who wants them. These assistants will be in charge of carrying most of the medicines, an extra stretcher, or a hammock, if circumstances make this possible. They must take charge of the wounded in any battle that is fought.

The essential medicines should be obtained through contacts with health organizations that exist in the enemy's territory. Sometimes they can be obtained from such organizations as the International Red Cross, but this should not be counted on, especially in the initial period of the struggle. It is necessary to organize an apparatus that will ensure the rapid transport of the medicines required in case of danger and that will gradually provide all the hospitals with the supplies necessary for their work, military as well as civil. Moreover, contacts should be made in the surrounding areas with doctors who would be able to help the wounded whose cases are beyond the capacities or the facilities of the guerrilla band.

The doctors required for this type of warfare have different characteristics. The combatant doctor, the compañero of the guerrillas, is needed in the first stage; their functions develop as the action of the guerrilla band becomes more complicated and a range of connected organizations is created. General surgeons are the best acquisition for an army of this type. If an anesthetist is available, so much the better; although almost all operations are performed not with gas anesthesia but using largactil and sodium pentothal, which are much easier to administer and easier to procure and preserve. Besides general surgeons, bone specialists are very useful, because fractures often occur from accidents in the zone; they are also often the result of bullet wounds in limbs. The clinic primarily serves the peasant masses, since sicknesses in the guerrilla armies are generally easy for anyone to diagnose. The most difficult task is to cure problems caused by nutritional deficiencies.

In a more advanced stage, if there are good hospitals, there might even be laboratory technicians, in order to have complete facilities. Appeals should be made to all sectors of the profession whose services are needed; it is quite likely that many will respond to this call and come to offer their help. All kinds of medical professionals are needed; surgeons are very useful, so are dentists. Dentists should be advised to come with simple field instruments and a campaign-type drill, with which they can do practically everything necessary.

5 Sabotage

Sabotage is one of the invaluable weapons of a people fighting a guerrilla war. Its organization falls under the civil or clandestine branch, since naturally sabotage should only be carried out outside the territories dominated by the revolutionary army; but this organization should be directly commanded and oriented by the guerrillas' general staff, which will be responsible for deciding the industries, communications, or other objectives that are to be attacked.

Sabotage has nothing to do with terrorism; terrorism and personal assaults are entirely different tactics. We sincerely believe that terrorism is negative, that in no way does it produce the desired effects, that it can turn people against a particular revolutionary movement, and that it brings a loss of life to its agents far greater than any benefit. On the other hand, attempts on the lives of particular individuals are acceptable, but only in very special circumstances, where it will eliminate a key opponent. But specially trained, heroic, self-sacrificing human beings should never be used to eliminate a minor assassin whose death might provoke a reprisal and the annihilation of all the revolutionaries involved in the action, and others besides.

Sabotage can be of two types: sabotage on a national scale against particular targets, and local sabotage against combat lines. Sabotage on a national scale should be aimed principally at destroying communications. Each type of communication can be destroyed in a different way; all of them are vulnerable. For example, telegraph and telephone poles are easily destroyed by sawing them almost all the way through, so that at night they appear to be in normal condition, until a sudden kick brings one pole down and this drags along with it all those that are weak, producing a major power failure.

Bridges can be attacked with dynamite; if there is no dynamite, steel bridges can be destroyed very easily with an oxyacetylene blowtorch. A steel truss bridge should be cut in its main beam and in the upper beam from which the bridge hangs. When these two beams have been cut at one end with the torch, they are then cut at the opposite end. The bridge will fall completely on one side and will be twisted and destroyed. This is the most effective way to bring down a steel bridge without dynamite. Railroads can also be destroyed, as can roads and culverts; sometimes trains can be blown up, if the guerrilla band is strong enough.

At certain times, utilizing the right equipment will also destroy the vital industries of each region. In these cases, an overview of the problem is necessary to ensure that a workplace is not destroyed unless such action will be decisive, since the consequences will be massive unemployment and hunger. The enterprises belonging to the potentates of the regime should be eliminated (and attempts made to convince the workers of the need for doing so), unless this will bring very grave social consequences.

We stress the key factor of sabotage against communications. The great strength of the enemy army against the rebels in the less mountainous zones is rapid communication; we must, therefore, constantly undermine that strength by knocking out railroad bridges, culverts, electric lights, telephones; also aqueducts, and in general everything that is necessary for a normal and modern life.

Around the combat lines sabotage should be performed in the same way but with much more audacity, with much more dedication and frequency. Here it is possible to rely on the invaluable assistance of the flying patrols of the guerrilla army, which can descend into these zones and help the members of the civil organization perform the task. Again, sabotage should be aimed principally at communications, but with greater persistence. All factories, all centers of production that are capable of providing the enemy with something necessary to maintain their offensive against the popular forces, should also to be liquidated.

Emphasis should be placed on seizing merchandise, cutting off supplies as far as possible, if necessary intimidating the large landowners who want to sell their farm produce, burning vehicles that travel along the roads, and using them as blockades. It is advantageous in every act of sabotage that frequent contact be made with the enemy army at points not too far away, always following the hit and run tactic. It is not necessary to put up a serious resistance, but simply to show the adversary that in the area where the sabotage has been carried out there are guerrilla forces ready to fight. This forces him to take a large number of troops, to move cautiously, or not to move at all.

In this way, little by little, all the cities in the zone surrounding guerrilla operations will be paralyzed.

6 War Industry

War industries within the guerrilla army sector are the product of a rather long evolution; they also depend on the guerrillas' control of geographically favorable territory. From the time that liberated zones are created and the enemy establishes a strict blockade over all supply chains, different departments can be organized as necessary, as already described. There are two fundamental industries, of which one is the manufacture of shoes and leather goods. It is not possible for a troop to walk without shoes in steep, wooded zones, with rocks and thorns. It is very difficult to march without shoes in such conditions; only the locals, and not all of them, can do it. The rest must have shoes. The industry is divided into two parts, one for putting on half-soles and repairing damaged shoes; the other will be dedicated to the manufacture of rough shoes; there should be a small but complete toolkit for making shoes; because this is a simple industry practiced by many people in such regions it is very easy to create. Connected with the shoe workshop there should always be a workshop making all kinds of canvas and leather goods for the troop's use, such as cartridge belts and backpacks. Although these items are not essential, they contribute to comfort and give a feeling of autonomy, of adequate supply, and of self-reliance to the troop.

An armory is the other vital industry for the small internal organization of the guerrilla band. This also has different functions: the simple repair of damaged weapons, rifles, and other available arms; the manufacture of certain types of combat arms that the inventiveness of the people will create; and the preparation of mines with various devices. When conditions permit, equipment for the manufacture of powder can be added. If it is possible to manufacture the explosives as well as the detonation devices in the liberated territory, brilliant achievements can be made in this area, which is very important because communications over land can be completely paralyzed by the appropriate utilization of mines.

Another important group of industries makes iron and tin products. All the blacksmith's labor will be focused on making equipment for the mules, such as shoes. The tinsmith will fabricate plates and canteens, which are especially important. A foundry can be combined with the tinsmith's shop. By melting soft metals it is possible to make grenades, which with a special type of charge will contribute in an important way to the armament of the troop. There should be a technical team for general repair and construction work of various kinds, the "service battery," as it is called in regular armies. With the guerrillas it would operate in the same way, taking care of all necessities, but with nothing of the bureaucratic spirit.

A person must be in charge of communications. They will have as their responsibility not only propaganda communications, such as radio directed toward the outside, but also telephones and all kinds of routes. They will use the civil organization as necessary in order to perform their duties effectively. Remember that we might be in a war in which we are subject to attack by the enemy and that often many lives depend on rapid communication.

For the troop's pleasure, it is a good idea to have cigarette and cigar factories; the tobacco can be bought in selected places and transported to the liberated territory where the items to be consumed by the soldiers can be manufactured. An industry for preparing leather from hides is also of great importance. All these are simple enterprises that can operate quite well anywhere, and are easy to establish in the guerrilla situation. The industry for making leather requires a small cement construction; also it uses large amounts of salt; but it will be an enormous advantage to the shoe industry to have its own supply of material. Salt should be made in revolutionary territory and collected in large quantities. It is made by evaporating water of a high saline content. The sea is the best source, though there might be others. It need not be purified of other ingredients for the purpose of consumption, although it might have a nasty taste at first.

Meat should be preserved in the form of beef jerky, which is easy to prepare. This can save many lives among the troop in extreme situations. It can be preserved with salt in large barrels for a fairly long time, and it can then be used anywhere.

7 Propaganda

The ideas of the revolution should be disseminated through whatever media is available, as broadly as possible. This re- quires full equipment and an organization, which will consist of two complementary wings covering the entire national area: one for propaganda originating outside the liberated territory, that is, from the national civil organization; and the other for propaganda originating within, that is, from the base of the guerrilla army. In order to coordinate these two forms of propaganda, the functions of which are strictly interlinked, there should be a single director.

Propaganda of the national type from civil organizations outside the liberated territory should be disseminated through newspapers, bulletins, and proclamations. The most important newspapers will be devoted to national affairs in general and will give the public exact information about the state of the guerrilla forces, always observing the fundamental principle that in the long run truth is the best policy. Apart from these publications of general interest, there must be others, more specialized, for different sectors of the population. A publication for the countryside should bring to the peasantry a message from their compañeros in all the free zones who have already felt the beneficial effects of the revolution; this strengthens their aspirations. A workers' newspaper will have similar features, with the sole difference that it cannot always offer a message from the combatant sector of that class, since it is likely that workers' organizations will not operate within the framework of guerrilla warfare until the final stages.

The great watchwords of the revolutionary movement, the watchwords of a general strike (which at an opportune moment will assist the rebel forces) and the need for unity should be explained. Other periodicals can be published; for example, one explaining the tasks of those sectors throughout the entire island that are not combatants but which nevertheless carry out various acts of sabotage, attacks, etc. Within the organization there can be periodicals directed at the enemy's soldiers; these will explain facts of which they are otherwise kept ignorant. News bulletins and movement proclamations are very useful.

The most effective propaganda is that prepared within the guerrilla zone. Priority will be given to the dissemination of ideas among the local people of the zone, explaining the theoretical significance of the insurrection, already understood by them as fact. In this zone there will also be peasant periodicals, the general organ of all the guerrilla forces, and bulletins and proclamations; and besides all these, the radio.

All problems should be discussed by radio — for example, the way to defend oneself from air attacks, and the location of the enemy forces, even citing familiar names among them. National propaganda will use newspapers similar to those pre- pared outside the liberated territory, but it can include more up to date and more precise news, reporting facts and battles that are extremely interesting to the reader. Information on international affairs will be confined almost exclusively to commentaries on facts that are directly related to the liberation struggle.

The spoken word on the radio, above all, is the most effective propaganda, which can be most easily spread over the entire national area and can appeal to the reason and the sentiments of the people. The radio is a factor of extraordinary importance. At times when war fever is more or less palpitating in every person in a region or a country, the inspiring, burning word enhances this fever and imparts it to all future combatants. It explains, teaches, inflames, and confirms the future positions of both friends and enemies. Nevertheless, the radio should be ruled by the fundamental principle of popular propaganda, which is the truth; it is preferable to tell the truth, however insignificant, than a large lie, artfully embellished. The radio should give reports, especially of battles, of all kinds of en- counters, and assassinations committed by the repression; it should also give doctrinal orientations and practical lessons to the civil population; and, from time to time, speeches by the revolutionary leaders.

We consider it useful that the main newspaper of the movement bears a name recalling a great and unifying image, perhaps a national hero or something similar; in-depth articles should also explain where the movement is headed and create a consciousness about the great national problems, besides offering sections of more lively interest to the reader.

8 Intelligence

"Know yourself and your adversary and you will be able to fight a hundred battles without a single disaster."

This Chinese proverb is as valuable for guerrilla warfare as a biblical psalm. Nothing helps the combatant forces more than correct intelligence. This comes spontaneously from the local residents, who will tell their friendly army, their allies, what is happening in various places; but in addition, this should be completely systematized. As we have seen, there should be a postal organization with the right contacts both within and outside guerrilla zones for carrying messages and goods. An intelligence service should also be in direct contact with enemy fronts, which men and women, especially women, should infiltrate; they should be in permanent contact with soldiers and gradually discover what there is to be discovered. The system must be coordinated in such a way that crossing the enemy lines into the guerrilla camp can be achieved without mishap.

If this is done well by competent agents the insurgent camp will be able to sleep more peacefully.

The essential aspect of this intelligence will be concerned, as already stated, with the front line of fire or the forward enemy camps that border the no man's land; but it should also develop along with the guerrilla band, increasing its depth of operation and its potential to predict larger troop movements in the enemy's rear. All local people are intelligence agents for the guerrilla band in the places where it is dominant or makes incursions, but it is a good idea to have persons especially assigned to this task. The peasants, not accustomed to precise military language, have a strong tendency to exaggerate, so their reports must be checked. As the spontaneous forms of popular collaboration are established and organized, it is possible to use the intelligence apparatus not only as an extremely important backup but also as a weapon of attack by using its personnel, for example, as "sowers of fear." Pretending to be on the side of the enemy soldiers, they sow fear and instability by spreading discouraging news. Mobility, the basic tactic, can be developed to the maximum. By knowing exactly the places where the enemy troops are going to attack, it is easy to avoid them, or when the time is ripe, to attack them where they least expect it.

9 Training And Indoctrination

The liberation soldier is trained essentially through life in the guerrilla band, and no leader can exist without having learned their difficult role in daily armed exercises. Some compañeros can be taught how to handle arms, concepts of orientation, the correct approach to the civil population, methods of fighting, etc.; but precious time of the guerrilla band should not be consumed in methodical teaching. This only begins when there is a large liberated area and a large number of persons are needed to perform a combat role. Schools for recruits will then be established.

These schools then have a very important function; they will train new soldiers from those who have not passed through that excellent sieve of formidable privations — the life of a guerrilla combatant. From the beginning, other privations must be suffered to convert them into the truly chosen. After having passed through very difficult tests, they will be able to be incorporated into the kingdom of a beggar army that moves without leaving a single trace. They should have two types of physical exercises: gymnastics as part of commando training, which demands agility in attack and withdrawal; and hard and exhausting hikes that will toughen the recruit for this kind of existence. Above all, they should live in the open air. They should suffer all inclemency of the weather in close contact with nature, as the guerrilla band does.

The school for recruits must have workers who will take care of its supply needs; for this there should be cattle sheds, grain sheds, gardens, a dairy, everything necessary, so that the school will not be a burden on the general finances of the guerrilla army. The students can rotate the tasks of supply, either as punishment for bad conduct or simply as volunteers.

Target Practice.

This will depend on the nature of the zone where the school is being held. We believe that a good principle is to assign volunteers and to cover the remaining work quotas with those who have the poorest conduct and show the poorest disposition for learning warfare.

The school should have a small medical facility with a doctor or nurse, depending on possibilities; this will give the recruits the best possible attention.

Shooting is the basic training. The guerrilla fighter should be properly trained in this respect, so he or she will try to use the least possible amount of ammunition. Practice begins with what is called dry shooting. This consists of placing the rifle firmly on a kind of wooden apparatus as shown in the picture. Without moving or firing the rifle, the recruits point at a movable board with a hole in the center, until they think they have the center exactly in their sight. The board allows each "shot" to be marked on the fixed board behind, which remains stationary. If the mark for several tries gives a single point, this is excellent. When circumstances allow, practice with 22-caliber rifles can start; this is very useful. If there is an excess of ammunition or an urgent need for preparing soldiers, they will have the chance to use bullets.

One of the most important courses in the school for recruits, one considered basic and which can be given anywhere in the world, is what to do in an air attack. Our school had been positively identified from the air and was attacked once or twice a day. The way in which the students responded to this continuous bombardment on their regular place of instruction showed which of the young guerrillas had the potential to be useful soldiers in battle.

The important aspect in such a school for recruits — something that must never be neglected — is indoctrination; this is crucial because individuals enter without a clear conception of why they have come, with only a few vague concepts of liberty, freedom of the press, etc., without any great ideological foundation. Because of this, indoctrination should be carried out with the greatest dedication and for as long as possible. These courses should teach the basic history of the country, explained with a clear grasp of the economic factors behind each historical event; the classes should include accounts of the national heroes and their response to certain injustices; and finally an analysis of the national situation or the situation in the zone. A short primer should be studied thoroughly by all members of the rebel army, as it might become a draft for future use in popular education.

There should also be a teacher training school, where texts to be used can be selected, considering the contribution that each book can make to the educational process.

Reading should be encouraged at all times, with an effort to promote books that are worthwhile and that develop the recruit's ability to encounter the world of letters and great national questions. The desire for further reading will follow; their circumstances will awaken a new aspiration for under- standing in the soldiers; little by little, the recruits will observe in their routine tasks the enormous advantages of those who have passed through the school over the rest of the troop, their capacity for analyzing problems and their superior discipline, which is another of the fundamental things that the school must teach.

This discipline that is internal, not mechanical, and based on a rational understanding, can produce formidable results in moments of combat.

10   The Organizational Structure Of The Army Of A Revolutionary Movement

As we have seen, a revolutionary guerrilla army, whatever its zone of operations, must rely on a noncombatant organization for a range of extremely important auxiliary missions. Later, we will see that this whole organization converges to lend the army the utmost help, since the armed struggle is obviously the key factor in the triumph.

The military organization is headed by a commander, in the Cuban case by a commander-in-chief, who names the leaders of the different regions or zones; these leaders have authority to govern their respective territories of activity, to name column commanders — the chiefs of each column — and the other lower officers.

Under the commander-in-chief there will be the zone commanders; under them several columns of varying size, each with a column commander; under the column commanders there will be captains and lieutenants, which, in our guerrilla organization, was the lowest grade. In other words, the lieutenant was the first rank above a soldier.

This is not a model but a description of one case, of how the organization worked in one country where it proved possible to triumph over an army that was fairly well organized and well armed. In other respects, too, our experience is not a model. It simply shows how, as events unfold, it is possible to organize an armed force. The ranks certainly have no importance, but what is important is that no rank should be conferred that does not correspond to the effective battle force commanded; no one should be promoted who has not passed through the sieve of sacrifice and struggle, for that would conflict with morality and justice.

This description applies to a well-developed army, already capable of waging serious combat. In the first stage of the guerrilla band, the chief can take whatever rank he likes, but he will still command only a small group.

One of the most important features of military organization is disciplinary punishment. Discipline must be one of the bases of action of the guerrilla force (this must be stressed again and again), which, as we have also said, should emanate from a carefully reasoned internal conviction; this produces an individual with inner discipline. When discipline is violated, the offender must be punished, whatever their rank, and punished severely, in a way that hurts.

This is important, because a guerrilla soldier feels pain differently from a soldier in a regular army. The punishment of spending 10 days in jail is a magnificent period of rest for the guerrilla fighter: 10 days with nothing to do but eat, no marching, no work, no standing the customary guards, sleeping at will, resting, reading, etc. From this it is obvious that the deprivation of liberty is not recommended for a guerrilla situation.

Where the combat morale of the individual is very high and self-respect strong, deprivation of their right to be armed can constitute a real punishment and provoke a positive reaction. In such cases, this is an effective punishment.

The following painful incident is an example. During the battle for one of the cities in Las Villas province in the final days of the war, we found an individual asleep in a chair while others were attacking positions in the town center. When questioned, the man responded that he was sleeping because he had been deprived of his weapon for firing accidentally; he was told that this was not the way to react to punishment and that he should regain his weapon, not in this way, but in the front line of combat.

A few days went by, and as the final assault on the city of Santa Clara began, we visited the first-aid hospital. A dying man there extended his hand, recalling the episode I described, stating that he had been able to recover his weapon and had earned the right to carry it. He died a short time afterwards.

This was the level of revolutionary morale that our troops attained through the continual exercise of armed struggle. It is not possible to achieve this at the beginning, when many are still frightened and subjective tendencies put a brake on revolutionary influences; but it is reached eventually through work and through the force of continual example.

Long night watches and forced marches can also serve as punishments; but such marches are not really practical, as they exhaust the individual for no reason other than that of punishment, and they require guards who also wear themselves out; the guards suffer the further inconvenience of having to keep a watch on the persons being punished, who are soldiers of limited revolutionary mentality.

In the forces directly under my command I imposed the punishment of arrest with privation of sweets and cigarettes for slight offenses and a total deprivation of food for worse offenses. The result was magnificent, even though the punishment was terrible; it is recommended only in very special circumstances.

Appendices

Appendix A - Underground Organization of the First Guerrilla Band

Guerrilla warfare develops in accordance with a series of laws, some derived from the general laws of warfare and others related to its special characteristics; if it is really intended to begin the struggle from some foreign country or from different and remote regions within the country, it is obvious that underground work must start with a small nucleus of initiates, acting apart from the mass action of the people. If the guerrilla movement arises spontaneously from the reaction of a group of individuals to some form of coercion, it is possible that the organization of this guerrilla nucleus to defend itself from annihilation will be enough. Nevertheless, in general, guerrilla warfare starts from a well-considered act of will: a chief with some prestige starts an uprising for the salvation of his people, beginning this work in difficult conditions in a foreign country.

Almost all popular movements against dictators in recent times have suffered from the same fundamental weakness of inadequate preparation; the rules of conspiracy, which demand extreme secrecy and caution, have not generally been observed. The governmental power of the country often knows about the intentions of the group or groups in advance, either through its secret services or from imprudent revelations, or in some cases, from outright declarations, as occurred in our case, for example, when the invasion was announced and encapsulated in Fidel Castro's phrase: *"In the year 1956 we will be free or we will be martyrs."*

Absolute secrecy, a total denial of information to the enemy, should be the primary base of the movement; secondly — and also very important — is the selection of the human material. At times this selection can be made easily, but at other times it will be extremely difficult because it is necessary to rely on those elements that are available, long-time exiles or persons who offer themselves when the call goes out simply because they understand that it is their duty to enroll in the battle to liberate their country, etc. There might not be the appropriate facilities for thoroughly checking these individuals. Nevertheless, even though elements of the enemy regime might be able to infiltrate the movement, it is inexcusable that they might later be able to pass on information; in the period just prior to an action all participants should be concentrated in secret places known only to one or two individuals, they should be under the strict vigilance of their chiefs and without the slightest contact with the outside world. Whenever there are concentrations, whether in preparation for departure or in order to carry out preliminary training or simply to hide from the police, it is always necessary to keep all new personnel, about whom there is no clear knowledge available, away from the key places.

In underground conditions no one, absolutely no one, should know anything more than what is strictly indispensable; and there should not be any talk in front of those who do not need to know something. When certain concentrations are formed, it is necessary even to control letters that leave and arrive in order to have total knowledge of the contacts that the individuals maintain; no one should be permitted to live alone, or to go out alone; personal contacts of the future member of the liberating army, contacts of any type, should be prevented by every means. However positive the role of women in the struggle, it must be emphasized that they can also play a destructive part. The temptation women present to young men living away from their usual lifestyle in a special psychological situation is well known. As dictators are well aware of this weakness, they will try to use it for infiltrating spies. Occasionally, the relationship of these women with their superiors is clear and even notorious; at other times, it is extremely difficult to prove even the slightest evidence of contact; therefore, it is necessary also to prohibit relations with women.

The revolutionary in a clandestine situation preparing for war should be a complete ascetic; this is also a test of one of the qualities that will later be the basis of their authority: discipline. If an individual repeatedly disobeys the orders of his superiors and makes contact with women, develops friendships that are not permitted, etc., he should be separated immediately, not merely because of the potential danger in the contacts, but simply because of the violation of revolutionary discipline.

Unconditional help should not be expected from a government, whether friendly or simply negligent, that allows its territory to be used as a base of operations; the situation should always be regarded as though the movement was operating in a totally hostile environment. The few exceptions to this only confirm the general rule.

We cannot speak here of the number of persons that should be trained. This depends on so many and such varied conditions that it is practically impossible to specify; it is only possible to discuss the minimum number with which a guerrilla war might be initiated. In my opinion, considering the inevitable desertions and limitations, in spite of the rigorous process of selection, there should be a nucleus of 30 to 50 guerrillas; this figure is sufficient to initiate an armed struggle in any country of Latin America within a favorable territory for operations, where there is hunger for land, persistent injustice, etc.

As already stated, weapons should be the same as those used by the enemy. Always considering that every government is in principle hostile to a guerrilla action being undertaken from its territory, the bands that prepare themselves should not be greater than approximately 50 to 100 guerrillas per unit. In other words, although there is no objection to 500 guerrillas initiating a war, all 500 should not be concentrated in one place. Firstly, because so many will attract attention and in the case of a betrayal or raid, the whole group is destroyed; secondly, it is much more difficult to raid several places at once.

The central headquarters for meetings might be more or less known, and the exiles will go there for all kinds of meetings; but the leaders should be present only occasionally, and there should be no compromising documents. The leaders should use as many different houses as possible, those least likely to be under surveillance. Arms deposits should be distributed in several places, if possible; these locations should be an absolute secret, known to only one or two people.

Weapons should be handed to those who are going to use them only when the war is about to be initiated; this is to avoid punitive action against trainees that might lead to their imprisonment and the loss of arms that are very difficult to procure, when the popular forces are in no state to suffer such a loss.

Another important factor requiring attention is the preparation of the forces for the extremely hard fight that is coming; these forces should have strict discipline, high morale, and a clear comprehension of the task to be performed, without bravado, without illusions, without false hopes of an easy triumph. The struggle will be bitter and long, setbacks will be suffered, almost to the brink of annihilation; only high morale, discipline, faith in final victory, and exceptional leadership can save the situation. This was our experience in Cuba, even though at one time just

12 men formed the nucleus of the future army, because all these conditions were met and because the person who led us was named Fidel Castro.

Besides ideological and moral preparation, thorough physical training is necessary. The guerrillas will, of course, select a mountainous or very rough zone for their operations; at any rate, in whatever situation they find themselves, the basic tactic of the guerrilla army is the march, and neither slow nor tired men can be tolerated. Adequate training therefore includes exhausting hikes day and night, day after day, gradually in- creasing, always carried on to the brink of exhaustion, using emulation to increase speed. Resistance and speed will be the key to the first guerrilla nucleus; also, a range of theoretical principles should be taught, such as orienteering, reading, and forms of sabotage. If possible, there should be training with military rifles, frequent firing, especially at distant targets, and a lot of instruction about ways to economize bullets.

To the guerrilla fighter, economy and utilization of ammunition down to the last bullet should almost become a religious tenet. If all these warnings are followed, the guerrilla forces can attain their final destiny.

Appendix 2 - Holding Power

Naturally, ultimate victory cannot be won until the army that sustained the former regime has been systematically and totally destroyed. Furthermore, all the institutions that protected the former regime should be smashed; but as this is solely a guerrilla manual we will limit ourselves to considering the problem of national defense in the case of a war or aggression against the new power.

The first thing we will encounter is world public opinion; "the serious press," the "truthful" news agencies belonging to the monopolies of the United States and other countries will begin an attack on the liberated country, an attack as aggressive and systematic as the laws of popular reform. For this reason not a single element from the former army can be retained. Militarism, blind obedience, traditional concepts of military duty, discipline and morale cannot be eradicated with one blow. Neither can the victors — who are good fighters, decent and kind-hearted, but at the same time generally lacking education — coexist with the vanquished, who are proud of their specialized military knowledge in combat weaponry, mathematics, fortifications, logistics, etc., and who totally despise the uncultured guerrilla fighters.

Of course, there are cases of individual military men who break with the past and join the new organization with a spirit of absolute cooperation. These people are doubly useful, because they combine their love of the people's cause with the knowledge necessary for creating the new popular army. As a consequence of smashing the old army and dismembering it as an institution, and the occupation of its former posts by the new army, it will be necessary to reorganize the new force. Its former guerrilla character, operating to a certain extent under separate chiefs, without any kind of plan, must be restructured; it is very important to emphasize that the operational concepts of the guerrilla band should still serve as the guidelines. In other words, these concepts will determine the organic formation and the best framework for the popular army. Care should be taken to avoid the error we made during the first months of trying to fit the new popular army into the old vestments of military discipline and organization. This error can lead to serious disorder and can lead to a complete lack of organization. The transition must be made with caution.

Preparation should begin immediately for the new defensive war that will have to be fought by the people's army, which has been accustomed to independence of command within the common struggle and initiative in the management of each armed group. This army will have two immediate problems: one will be the incorporation of thousands of last-minute revolutionaries, good and bad, whom it is necessary to train in the rigors of guerrilla life and who must be provided with revolutionary indoctrination in accelerated and intensive courses. Revolutionary education, providing the essential ideological unity to the popular army, is the basis of national security both in the long and short runs. The other problem is the difficulty of adaptation to the new organizational structure.

A group should immediately be created to take charge of disseminating the new revolutionary truths among all units of the army. It should explain to the soldiers, peasants, and workers, who have all come from the people, the justice and the truth of each revolutionary act, the goals of the revolution, why there is a struggle, why so many compañeros have died without being able to see the victory. Combined with this intensive indoctrination, accelerated courses of primary instruction that begin to overcome illiteracy should also be organized, in order to raise the level of the rebel army gradually so that it becomes an instrument of high technical qualifications, solid ideological structure, and magnificent combative power.

Time will bring these three qualities. The military apparatus will continue to be perfected over time; the former combatants can be given special courses to prepare them as professional military men who will then give annual courses of instruction to the people joining voluntarily or through conscription. This process will depend on national characteristics and cannot be presented as a model.

From this point on, we are expressing the opinion of the command of [Cuba's] Rebel Army in regard to the policy followed in the concrete situation of Cuba, considering the threat of foreign invasion, the situation in the modern world at the end of 1959 and the beginning of 1960, with the enemy in our sights, analyzed, evaluated, and awaited without fear. In other words, we are no longer theorizing for the instruction of others about what has already been done; rather we are analyzing what others have done in order to apply it ourselves in our own national defense.

As our problem is to theorize about the Cuban case, to locate and test our hypothesis on the map of Latin American realities, we present the following as an epilogue.

Appendix 3 - Epilogue: Analysis of the Situation in Cuba, Present and Future.

It is now one year since the dictator fled, the culmination of a long civil and armed struggle by the Cuban people. The government's achievements in the social, economic, and political fields are enormous; nevertheless, they should be analyzed, and every conclusion reviewed in order to demonstrate to the people the precise dimensions of our Cuban revolution. This national revolution, fundamentally agrarian in nature, having the enthusiastic support of workers, the middle class, and today even the owners of industry, has acquired a continental and international transcendence, protected as it is by the unshakable determination of the people and by the particular features that animate it.

This will not try to summarize, however briefly, all the laws passed, all of them of undeniable benefit to the people. It should suffice to focus on a few of these laws in order to demonstrate the logical course of development, from beginning to end, in the progressive and necessary order that responds to the interests of the Cuban people.

The first alarm bell for the parasitic classes of the country sounds with the rent law, the reduction of electricity rates, and the nationalization of the telephone company and the subsequent reduction in call charges — all decreed in rapid succession.

Those who saw Fidel Castro and those who made this revolution as nothing more than old-style politicians, easily manipulated idiots with beards as their only distinction, now began to suspect that something more profound was emerging from the bosom of the Cuban people and that their privileges were about to disappear. The word "communism" began to surround the leading figures and the triumphant guerrilla fighters; consequently, the word "anticommunism," the dialectical opposite, attracted all those resentful or dispossessed of their unjust sinecures.

The law on vacant plots and the law on time payments accentuated this sense of malaise among usurious capitalists. But these were minor skirmishes with the reactionaries; every- thing was still all right and possible. "This crazy kid," Fidel Castro, could be counseled and set on the right path, the good "democratic" path, by a Dubois or a Porter. One had to have faith in the future.

The Agrarian Reform Law came as a tremendous jolt; most of those affected by it now saw clearly. Gaston Baquero, the voice of reaction, was one of the first; he had accurately predicted what would happen and had retired to the more tranquil waters of the Spanish dictatorship. There were still some who believed that "the law is the law," and that other governments had previously promulgated similar laws, theoretically designed to help the people; but carrying out these laws was something else. This unruly and complicated child nicknamed INRA (National Institute of Agrarian Reform) was regarded at first with peevish and touching paternalism from the ivory towers of learning, pervaded as they were by social doctrines and respectable theories of public finance that were considered beyond the uncultured and absurd mentality of a guerrilla fighter. But INRA advanced like a tractor or a war tank, because it is both a tractor and tank, breaking down the walls of the great estates as it passed, and creating new social relations in land ownership. This Cuban agrarian reform arose with several important characteristics for Latin America. Yes, it was anti-feudal in the sense that it eliminated the Cuban-style latifundia, annulled all contracts that demanded payment of land rent in crops, and liquidated the servitude existing mainly in coffee and tobacco production, among our most important agricultural products. But it also was an agrarian reform in a capitalist framework to destroy the monopoly pressure that limits the possibilities of human beings, isolated or as collectives; our reform helped peasants work their land honorably and produce without fear of the creditor or the master. From the beginning, it was exemplified by offering the peasants and agricultural workers — those who give themselves to the soil — what they needed in terms of technical assistance from competent personnel; machinery; financial help provided through credits from INRA or parastate banks; and significant help from the "Association of People's Stores" that has been developed on a large scale in Oriente [province] and is in the process of being developed in other provinces. The state stores, replacing the old usurers, provide fair financing and pay a fair price for crops.

In contrast to the three other great agrarian reforms in America (Mexico, Guatemala, and Bolivia) the most important unique feature of the Cuban reform has been the decision to go all the way, without concessions or exceptions of any kind. This thorough-going agrarian reform respects no rights other than the rights of the people, with no discrimination based on class or nationality: the weight of the law falls equally on the United Fruit Company and on the King Ranch, as on the big Creole latifundistas.

Under these conditions land is prepared for the production of crops that the country needs — rice, oil-producing grains, and cotton — which are being developed intensively. But the nation is not satisfied and is going to recover all its stolen wealth. Its rich sub-soil, which has been a battlefield of monopolist voracity and struggle, is effectively recovered through the petroleum law. This law, like the agrarian reform and all the measures taken by the revolution, meets Cuba's undeniable needs, responding to the urgent demands of a people that wishes to be free, that wishes to be master of its economy, that wishes to prosper and to reach ever higher goals of social development. But for this very reason it represents an example for the continent that the oil monopolies fear. It is not that Cuba directly hurts the petroleum monopoly substantially; there is no reason to believe the country to be rich in reserves of the precious fuel, even though there are reasonable hopes of obtaining a supply that will satisfy our domestic needs. But Cuba's law provides a palpable example to the brother peoples of America, many of them ravaged by these monopolies or pushed into internal wars in order to satisfy the needs or appetites of competing trusts, while Cuba, on the other hand, shows what is possible in Latin America, and the exact moment when action should be considered. The great monopolies also turn their uneasy gaze on Cuba; not only has the little Caribbean island dared to eliminate the interests of the omnipotent United Fruit Company (the legacy of Mr. Foster Dulles to his heirs), but the empires of Mr. Rockefeller and the Deutsch group have also suffered under the lash of intervention by the popular Cuban revolution.

This law, like the mining law, is the response of the people to those who try to hold them back with threats of force, with air attacks, with all kinds of punishment. Some say that the mining law is as important as the agrarian reform. In general, we do not consider this is so important for the country's economy, but it introduces another new feature: a 25 percent tax on exported products, to be paid by companies that sell our minerals abroad (so that they now leave behind them something other than a hole in the ground). This not only contributes to Cuba's well- being, but also increases the relative strength of the Canadian monopolies in their struggle with the current exploiters of our nickel. Thus the Cuban revolution eliminates the latifundia, limits the profits of the foreign monopolies, limits the profits of the foreign intermediaries that dedicate their parasitic capital to the commerce of importation — launching into the world a new policy in Latin America, which dares to break the mining giants' monopoly, and leaves at least one of them in difficulty. This represents a potent new call to the neighbors of one of the great monopolistic nations, and has repercussions throughout the continent. The Cuban revolution breaks through the barriers of the news media and spreads its truth like a shower of dust among the Latin American masses anxious for a better life. Cuba is the symbol of new nationhood and Fidel Castro the symbol of liberation.

By a simple law of gravity the little island of 114,000 square kilometers and six-and-a-half million inhabitants assumes the leadership of the anticolonial struggle in America, where the serious wavering by other countries has allowed Cuba to take the heroic, glorious, and dangerous advanced post. The economically stronger nations of colonial America, those in which national capitalism develops stunted by the continuous, relentless, and at times violent struggle with the foreign mon- opolies, now cede their place gradually to this small, new champion of liberty, since their governments do not have sufficient force to carry the fight forward. This is not a simple task, nor is it free from danger and difficulties, but requires the backing of an entire people and an enormous charge of idealism and spirit of sacrifice, especially in the virtual isolation we face in Latin America. Small countries have tried to maintain this post before Guatemala: the Guatemala of the quetzal, that dies when imprisoned in a cage, the Guatemala of the Indian Tecum Uman, felled by the direct aggression of the colonialists. Bolivia, the country of Murillo, the proto-martyr of Latin American independence, yielded to the terrible hardships of the struggle after providing three examples that served as the foundation of the Cuban revolution: the suppression of the army, agrarian reform, and nationalization of mines — the greatest source of wealth, and at the same time, the greatest source of tragedy.

Cuba is aware of these previous examples, knows the failures and the difficulties, but also knows that we are at the dawning of a new era in the world; the pillars of colonialism have been swept aside by the force of the national and popular struggle in Asia and Africa. Now the unity of the peoples is not based on religion, customs, desires, racial identification or discrimination, it comes from similar economic and social conditions and from a similar yearning for progress and recovery. Asia and Africa joined hands in Bandung*; Asia and Africa have come to join hands with colonial and indigenous America through Cuba, here in Havana.

On the other hand, the great colonial powers have lost ground in the face of the struggle of the peoples. Belgium and Holland are two caricatures of empires; Germany and Italy have lost their colonies. France is bitterly fighting a war that is lost. England, diplomatic and skillful, slowly withdraws politically while maintaining the economic connections.

US capitalism replaced some of the old colonial capitalisms in the countries that began their independent life, but it knows that this is transitory and that there is no real security for its financial speculations in these new territories; the octopus cannot attach its suckers firmly there. The claws of the imperial eagle are trimmed. Colonialism has died or is dying a natural death in all these places.

Latin America is something else. The English lion with its voracious appetite departed from our America some time ago, when the charming, young Yankee capitalists installed the "democratic" version of the English clubs, imposing their sovereign authority over every one of the 20 republics.

This is the colonial realm of US monopoly, its reason for being and its last chance, its own "backyard." If all the Latin American peoples should raise the flag of dignity, as Cuba has done, monopoly would tremble; it would have to accommodate itself to a new political-economic situation and to a substantial pruning of its profits. Monopoly does not like profits to be pruned, and the Cuban example, this "bad example" of national and international dignity, is gaining strength in the countries of Latin America. Every time an impudent people cries out for liberation, Cuba is blamed; and Cuba is guilty in the sense that Cuba has shown the way, the way of a popular armed struggle against supposedly invincible armies, how to fight in rough places and wear down and destroy the enemy away from his bases, in a word, the way of dignity.

This Cuban example is bad, a very bad example, and monopoly cannot sleep peacefully while this bad example remains at its feet, confronting danger, advancing toward the future. It must be destroyed, they declare. Intervene against this bastion of "communism," cry the servants of monopoly disguised as representatives in Congress. "The Cuban situation is very disturbing," say the artful defenders of the trusts. We all know what they mean: "It must be destroyed."

So, what are the options for aggressive action to destroy the bad example? One could be described as purely economic. This starts with a restriction on credit by US banks and suppliers to all businessmen, national banks, and even the National Bank of Cuba. US credit is thus restricted, and through the medium of associates an attempt is made to have the same policy adopted in all Western European countries; but this alone is not sufficient.

The denial of credits strikes a first, hard blow at the economy, but recovery is rapid and the balance of trade evens out, since the victimized country is accustomed to living day to day. More pressure is required. The sugar quota becomes part of the dance: yes, no, no, yes. Frantically the calculating machines of the agents of monopoly tally up all their accounts and reach the final conclusion: it is very dangerous to reduce the Cuban quota and impossible to cancel it. Why very dangerous? Because besides being bad politics, it would stimulate the appetite of

10 or 15 other supplier countries, causing them tremendous unease as they would all consider they had a right to more. It is impossible to cancel the quota, because Cuba is the largest, most efficient, and cheapest provider of sugar to the United States, and because 60 percent of the interests that profit directly from the production and commerce in sugar are US interests. Moreover, the balance of trade is favorable to the United States; whoever does not sell cannot buy; and it would set a bad example to break a treaty. Besides, the supposed US gift of paying nearly three cents above the market price is only the result of US incapacity to produce sugar cheaply. The high wages and the low productivity of the soil prevent the Great Power from producing sugar at Cuban prices; and by paying this higher price for a product, they are able to impose onerous treaties on all beneficiaries, not only Cuba. It is impossible to eliminate the Cuban quota.

We do not consider it likely that monopolists will employ a variant of the economic approach in bombing and burning cane fields, hoping to cause a sugar shortage. This appears instead to be a measure calculated to undermine confidence in the revolutionary government's authority. (The corpse of the US mercenary stains more than a Cuban house with blood; it also stains a policy. And what can be said of the gigantic explosion of arms destined for the Rebel Army?)

Another vulnerable place where the Cuban economy can be pressured is the supply of raw materials, such as cotton. However, it is well known that there is an over-production of cotton in the world, and any difficulty of this nature would be transitory. Fuel? This requires some attention; it is possible to paralyze a country by depriving it of fuel, and Cuba produces very little petroleum; it has some heavy fuel that can be used to operate its steam-driven machinery and some alcohol that can be used in vehicles; besides, there are large amounts of petroleum in the world. Egypt can sell it, the Soviet Union can sell it, possibly Iraq will be able to sell it soon. It is not possible to develop a purely economic strategy.

Another possible form of aggression to add to this economic variant is an intervention by some puppet power, the Dominican Republic, for example, which would be something of a nuisance; but the United Nations would undoubtedly step in, with nothing concrete having been achieved.

Incidentally, the new course followed by the Organization of American States creates a dangerous precedent for intervention. Under the cover of the Trujillo pretext, monopoly consoles itself by constructing a means of aggression. It is sad that Venezuelan democracy has put us in the difficult position of having to oppose an intervention against Trujillo. What a good turn it has done the pirates of the continent!

Among the new possibilities for aggression is the assassination of the same old "crazy kid," Fidel Castro, who has now become the focus of the monopolies' fury. Naturally, matters must be arranged so that the other two dangerous "international agents," Raúl Castro and the author, are also eliminated. This solution is appealing; if simultaneous assaults on all three or at least on the principal leader succeeded, it would be a bonus for the reactionaries. (But never forget the people, Messrs. Monopolists and agents, the omnipotent people who in their fury at such a crime would crush and erase all those who had anything to do directly or indirectly with an assault on any of the leaders of the revolution; it would be impossible to restrain them.)

Another feature of the Guatemalan variant is to pressure the arms suppliers so as to force Cuba to buy from communist countries, and to then use this as an occasion to let loose another torrent of insults. This could produce results. Someone in our government has said, "It might be that they will attack us as communists, but they are not going to eliminate us as imbeciles."

Therefore a direct aggression by the monopolies begins to seem necessary; all the variants are being shuffled and studied in the IBM machines, calculating every result. It occurs to us right now that the Spanish variant could be used. The Spanish variant would be one in which some initial pretext is seized on for an attack by exiles with the help of volunteers, volunteers who would be mercenaries, of course, or simply the troops of a foreign power, well supported by sea and air — we should say, with enough support to be successful.

It could also begin as a direct aggression by some state like the Dominican Republic, which would send some men, our brothers, and many mercenaries to die on these beaches in order to provoke a war; this would prompt the "pure-intentioned" monopolists to say that they did not want to intervene in this "disastrous" struggle between brothers; they would merely restrict the war by maintaining vigilance over the skies and seas of this part of America with cruisers, battleships, destroyers, aircraft carriers, submarines, minesweepers, torpedo boats, and airplanes. And it might happen that while these zealous guardians of continental peace were not allowing a single boat to pass carrying anything for Cuba, some, many, or all of the boats headed for the unhappy country of Trujillo would escape the iron vigilance. Also they might intervene through some "reputable" inter-American body, to put an end to the "crazy war" that "communism" had unleashed in our island; or, if this device of the "reputable" Latin American organization was no use, they might intervene directly, as they did in Korea, using the name of the international body in order to restore peace and protect the interests of their citizens.

Perhaps the first step in the aggression will not be against us, but against the constitutional government of Venezuela, in order to remove our last point of support on the continent. If this happens, it is possible that the center of the struggle against colonialism will move from Cuba to the great country of [Simón] Bolívar. The people of Venezuela will rise to defend their liberties with all the enthusiasm of those who know they are fighting a decisive battle, that behind defeat lies the darkest tyranny and behind victory the certain future of America. A wave of popular struggles might disturb the peace of the monopolist cemeteries that our subjugated sister republics have become.

Many factors operate against the enemy's chance of victory, but there are two key ones. First, the external: this is 1960, the year of the underdeveloped peoples, the year of the free people, the year in which there will finally be respect for the voices of the millions of beings who are lucky enough not to be governed by the possessors of the means of death and payment. Furthermore — and this is an even more powerful argument — an army of six million Cubans will grab hold of weapons as a single person in order to defend their territory and their revolution. Cuba will be a battlefield where the army will be nothing other than part of the people in arms; if destroyed in a frontal war, hundreds of guerrilla bands under a dynamic command and a single center of orientation would fight the battle all over the country. In cities the workers will die in their factories or workplaces, and in the countryside the peasants will deal out death to the invader from behind every palm tree and from every furrow of the new mechanically plowed field that the revolution has given them.

And around the world international solidarity will create a barricade of hundreds of millions of people protesting against the aggression. Monopoly will see how its pillars are shaken and how the web of its newspaper lies is blown away by a single puff. But suppose they dare to defy the popular outrage of the world, what will happen then?

The first thing to recognize, considering our position as a very vulnerable island without heavy weapons, with a very weak air force and navy, is the need to apply the concept of guerrilla warfare to the fight for national defense.

Our ground units will fight with all the fervor, commitment, and enthusiasm of which the sons of the Cuban revolution are capable in these glorious years of our history; but in the worst case, we are prepared to continue fighting even after the destruction of our army organization in a frontal combat. In other words, in confronting large concentrations of enemy forces that have succeeded in destroying our army, we would immediately transform ourselves into a guerrilla army with broad mobility, with our column commanders having unlimited authority, although a central command situated somewhere in the country would provide the necessary orders and establish the overall strategy.

The mountains would be the last line of defense of the organized armed vanguard of the people, which is the Rebel Army; but in every house of the people, on every road, in every forest, in every part of our national territory the struggle would be fought by the great army of the rearguard, the entire people trained and armed in the manner now to be described.

As our infantry units will not have heavy arms, they will concentrate on anti-tank and anti-aircraft defense. Mines in very large numbers, bazookas or anti-tank grenades, highly mobile anti-aircraft cannons and mortar batteries will be our only significant powerful weapons. The veteran infantry soldier, although equipped with automatic weapons, will know the value of ammunition and will guard it with loving care. Special installations for reloading shells will accompany each army unit, maintaining reserves of ammunition, even in these precarious conditions.

The air force will probably be badly hit in the first moments of an invasion of this type. We are basing our calculations upon an invasion by a great foreign power or by a mercenary army of some small power, helped either openly or covertly by this great power. The national air force, as I said, will be destroyed, or almost destroyed; only reconnaissance or liaison planes will remain, especially helicopters for minor functions.

The navy will also be organized for this mobile strategy; small launches will offer the enemy the most minimal target and maintain maximum mobility. In such circumstances, the enemy army becomes desperate to find something to receive its blows; all he will find is a mobile, impenetrable, gelatinous mass that retreats and never presents a solid front, while inflicting wounds from every side.

It is not easy to overpower an army of the people that is prepared to continue as an army in spite of defeat in a frontal battle. Two great masses of the people are united around it: the peasants and the workers. The peasants have already shown their effectiveness in capturing the small band that was marauding in Pinar del Rio. These peasants will be trained principally in their own regions; but the platoon commanders and the superior officers will be trained, as is now already being done, in our military bases. From there they will be distributed throughout the 30 zones of agrarian development that form the new geographical division of the country. These will constitute 30 more centers of peasant struggle, charged with defending to the last their lands, their social conquests, their new houses, their canals, their dams, their ripening harvests, their independence, in a word, their right to live.

At the beginning they will present a firm resistance to any enemy advance, but if this proves too strong for them, they will disperse, each peasant becoming a peaceful cultivator of the soil during the day and a fearsome guerrilla fighter at night, the scourge of the enemy forces. Something similar will take place among the workers; the best among them will also be trained to act as leaders of their compañeros, teaching them the principles of defense. Each social class, however, will have different tasks. The peasant will fight the typical battle of the guerrilla fighter; he should learn to be a good shot, to take advantage of all the difficulties of the terrain and to disappear without ever showing his face. The workers, on the other hand, have the advantage of operating in a modern city, a large and efficient fortress; at the same time their lack of mobility is a handicap. The worker will learn first to blockade the streets with barricades of any available vehicle, furniture, or utensil; to use every block as a fortress with communications through holes made in interior walls; to use that terrible defense weapon, the "Molotov cocktail"; and to coordinate gunfire from the innumerable crevices provided by the houses of a modern city.

From the worker masses assisted by the national police and those armed forces charged with the defense of the city, a powerful block of the army will be created; but it should expect to suffer great losses. The struggle in the cities in these conditions cannot achieve the facility and flexibility of the struggle in the countryside: many will fall, including many leaders, in this popular struggle. The enemy will use tanks that will be destroyed rapidly as soon as the people learn their weaknesses and not to fear them; but before that the tanks will leave their tally of victims.

There will be other organizations linked to those of the workers and peasants: first, the student militias, which will contain the flower of the student youth, directed and coordinated by the Rebel Army; organizations of youth in general, who will participate in the same way; and organizations of women, who will provide great encouragement by their presence and who will do such auxiliary tasks for their compañeros in the struggle as cooking, taking care of the wounded, giving final comfort to those who are dying, doing laundry, in a word, showing their compañeros-in-arms that they will never be absent in the difficult times of the revolution. All this is achieved by the widespread organization of the masses supplemented by patient and careful education, an education that begins and is confirmed in knowledge acquired from their own experience; it must focus on rational and truthful explanations of the facts of the revolution.

The revolutionary laws should be discussed, explained, studied in every meeting, in every assembly, wherever the leaders of the revolution are present for any purpose. Furthermore, the speeches of the leaders, and in our case particularly of the undisputed leader, should constantly be read, commented on, and discussed. People should gather together in the country- side to listen to the radio, and where there is more advanced technology, to watch on television those magnificent popular lessons given by our prime minister [Fidel Castro].

The participation of the people in politics, that is to say, in the expression of their own desires made into laws, decrees, and resolutions, must be constant. Vigilance against any manifestation of opposition to the revolution should also be constant; and vigilance over morale within the revolutionary masses should be stricter, if this is possible, than vigilance against the non-revolutionary or the disaffected. If the revolution is not to take the dangerous path of opportunism, it can never be allowed that a revolutionary of any level should be excused for grave offenses of conduct or morality simply because they are a revolutionary. Their former service record might provide extenuating circumstances and can always be taken into consideration in deciding on the punishment, but the act itself must always be punished.

Respect for work, especially for collective work and work for collective ends, should be cultivated. Volunteer brigades to construct roads, bridges, docks or dams, and school cities should receive a strong impulse; this helps to forge unity among those people who demonstrate their love for the revolution by their works.

An army connected in this way to the people, that feels an intimacy with the peasants and the workers from which it has arisen, that knows all the special techniques of warfare and is psychologically prepared for the worst contingencies, knows itself to be invincible. And it will become even more invincible as the army and the citizenry embody the righteous phrase of our immortal Camilo: "The army is the people in uniform." For this, for all this, in spite of the monopolists' desire to suppress Cuba's "bad example," our future shines brighter than ever

PART 4

ONE HUNDRED FIFTY QUESTIONS TO A GUERRILLA

ALBERTO BAYO GIROUD

1955

ONE HUNDRED FIFTY QUESTIONS TO A GUERRILLA

By Alberto Bayo Giroud

1. In order for a guerrilla to succeed, exactly what preconditions should exist?

To be right in your struggle against the injustices which a people suffer, whether from foreign invasion, the imposition of a dictatorship, the existence of a government which is an enemy to the people, an oligarchic regime, etc. If these conditions do not exist, the guerrilla war will always be defeated. Whoever revolts unrighteously reaps nothing but a crushing defeat.

2. Who should take part in a guerrilla unit?

Primarily only young men and women who are firm in their convictions, cautious in their dealings, have proven their spirit of self-sacrifice, personal courage, patriotism, and great dedication to the cause of the people should take part in a guerrilla war.

3. In addition to these moral qualifications what else must one who intends to join our guerrilla organization do?

He must truthfully and in detail answer questions on a questionnaire which includes such information as the applicant's full name; place and date of birth; marital status; names of parents; names of spouse, children, etc.; places of work since the age of eighteen; names of friends in the Revolutionary Movement; whether he has ever been arrested; and many other questions which our Movement has worked out. The applicant must give a history of his political position. After completion of the questionnaire and our obtaining a favorable impression from the investigation of the data supplied, he will be admitted to the appropriate guerrilla unit.

4. If the results of the investigation of his questionnaire reveal the applicant to be an informer or spy who intends to enter our ranks to betray zips, what shall we do with him?

He will be judged by the Summary Court Martial as a traitor to the revolution.

5. If in spite of all steps we take, a despicable spy infiltrates the organization, what shall we do with him?

Once his status has been verified as such, he will be judged by Court Martial and without pity sentenced to death. We can pardon a political enemy who fights for an ideal which in our estimation is wrong. but never a spy. Such a man deserves no consideration even though to the enemy he may be a hero or martyr. The accused should be given every right which his situation warrants, especially since he may really be an agent working for us who was ordered by his supervisors to engage in counterespionage.

6. How many guerrillas work in a guerrilla unit?

The ideal number is between ten and twenty. The fewer the men, the greater the mobility.

7. How fast does a guerrilla unit make an amphibious landing and how is it achieved?

The unit is only as fast as the slowest of its members. To effect a landing everything must be planned and rehearsed in advance so that as soon as the unit hits the beach every member moves quickly, silently, well disciplined and well briefed in his particular task. Those who are assigned to take the hills commanding the beach move off to the left flank; those who are to take and hold the center run forward and assume their positions, then rapidly unload the material from the boat as quickly as possible, maintaining discipline and absolute silence as though they were a group of deaf mutes, not even being able to signal to one another.

8. What is done with guerrillas who cannot withstand long marches?

They are brought together to form slower units within which, however, everyone has to keep up.

9. Who should captain a guerrilla unit?

The captain should be the one who because of his special qualifications of command ability, character, intelligence, caution, zest for combat, etc. is nominated for the position.

10. Should a guerrilla be informed of the higher command organization?

Yes, he should know it and abide by it so that when there are casualties there will be no disagreement as to who is to command a unit. Vacated positions are taken over by the person with the next highest authority and who will be respected and obeyed by all subordinates.

11. What weapons should a guerrilla unit carry?

The unit should be equipped with the same type of rifles to facilitate the supply of ammunition, and in addition, it is good to have a light machine gun which is always useful in our operations. Each guerrilla should always carry his own first-aid kit, canteen, a watch synchronized with the unit leader's, and many need field glasses. A guerrilla should also wear as a belt a rope some six feet long which can be used at night by a companion who holds on to one end thus not losing contact with his unit. This "tail" is worn wound around the waist. The part left over is what his companion, following behind, holds on to. No one is ever lost this way, no matter how dark the night is. It can be used in scaling peaks, crossing rivers, and for tying up bundles of firewood.

12. How should the guerrilla unit be equipped?

Its men should have good heavy shoes with thick soles and count on one good compass per unit. These are indispensable. Maps of the sector should always be available in order not to have to ask directions of any peasant. But if necessary he should only be used to confirm data already on the map.

13. How should a guerrilla unit be organized?

Exactly like an army corps, with its staff, its different positions and responsibilities filled by guerrillas so all the work does not fall on one man. Therefore the guerrilla unit is composed of the following sections: intelligence, operations, sabotage, recruiting, training, armament, munitions, quartermaster, sanitation, and propaganda.

14. What are the duties of each of these sections?

Intelligence should compile all the information it can on all members of the guerrilla unit, all enemies, those indifferent to the movement; on the location of water, springs and rivers; on roads, highways, trails, bridges; on the conduct of the guerrilla members; on sympathizers who wish to join the unit; on soldiers, informers, spies, etc. At the same time it will obtain or make maps of the terrain and the principal targets in the sector assigned to the unit. It will conduct espionage and counterespionage activities, keep records on unit personnel regarding all combat performance whether outstanding or unimpressive; and carry on cryptographic work (coding and deciphering messages, documents of courts martial, etc.).

> The **Intelligence Section** should be under the direction of the second in command of the guerrilla unit, who should himself possess a high degree of intelligence, wisdom, and caution.

> The **Operations Section** will supervise all attacks and other missions the unit undertakes and will evaluate the results of these endeavors. It consults with the comrades responsible for carrying out the missions, keeps the commander posted on the development of projects so he can make the final decision as to whether the operation will be put into effect. When the captain is unable to command a unit because of wounds, severe illness, or necessary absence, the head of Operations takes over his command, filing all data required for operations, both proposed and ready for accomplishment, along with different scale maps of the sector.

> Leadership of the **Sabotage Section**, the main one of the ten composing our staff, falls to an active officer, extraordinarily dynamic, extremely intelligent and clever, having a creative imagination, adaptability, and a real vocation for his assignment. He must conduct his missions so that all types of sabotage are exploited to the fullest; if possible hitting new objectives daily.

> **The Recruiting Section** obtains personnel to fill out our ranks or replace our losses. It will list names of young volunteers separating them into three groups. In the first group will be those who are to replace our casualties; in the second, those who can serve as machete men or demolition agents; and the third group, used only for the construction of fortifications and other such tasks.

> The officer in charge of training will supervise the training in handling firearms and close order drill, literacy courses for peasants, and all educational and cultural programs of the guerrilla unit.

The **Armament Section** is concerned with the maintenance of the unit's weapons; with the shotguns of the shotgunners serving with our forces as well as with our hand guns. It will keep lists of instructors and armorers and their assistants, providing for the acquisition of replacement parts needed to maintain our arms in good repair.

The **Munitions Section** is in charge of everything pertaining to the guerrilla unit's ammunition. It trains civilians who are to pass cartridges on to the guerrillas, and furthermore maintains small caches of cartridges and spare parts so that in no encounter will the guerrilla be without munitions.

The **Quartermaster Section**, because of its vital importance, will be the province of one of the most responsible men in the unit. This section sees to it that food is never lacking for the troop, rationing intelligently whatever it has, and assuring by its negotiations, orders, and purchases the feeding of the unit.

The **Sanitation chief** doesn't have to be a doctor or nurse, although it would be helpful if he were. This section has the responsibility for keeping a complete stock of medicine, and whatever else is needed to bring our comrades back to health. This includes the addresses of doctors and nurses in our sector who will either voluntarily treat our men or who will be forced to do so when called upon.

The man in charge of **Propaganda** will make known all our successful exploits in newspapers and magazines throughout the country; and if that is not possible, then by means of letters, mimeographed bulletins, etc. This publicizing of our military accomplishments will raise the morale of our people and wear down the morale of our enemies.

In Combat

15. What physical training should a guerrilla hare before going on missions?

He will engage in even longer marches until reaching a total of fifteen hours duration with only a short rest of ten minutes every four hours; besides, he will practice night marches of seven hours, at least.

16. How should one move about in the field at night?

One should walk as though riding a bicycle, lifting the feet high each step in order not to trip over stones, tree trunks, or other objects in your way. Use your compass at least every hour to check your directions. If you have no compass you can orient yourself by the polar star whose location you will learn in our manuals. On starless nights you can get your bearings from the trees. In our countries, the north side of live trees has either no bark or the bark is thin and worn.

17. How should guerrillas treat one another?

Everyone should be friendly or at least cooperative. Practical jokes and tricks are considered bad taste. They cause enmity among the men, weaken the unit's strength, and therefore are forbidden in our organization.

18. How can one orient himself during the day?

By means of the sun. Stand pointing your right arm and side toward the place the sun has risen. This arm points toward the east; the opposite side is the west; in front of you, north; and at your back, south.

19. When in the field we come upon a house or peasant's hut, how should we proceed before entering for the first time?

Only two of our number will go in; the others will let the occupants of the building know they are surrounded in case they are enemies or intend to betray us. When a careful search of the house has been made and the possibilities of betrayal or the hiding of enemies in the house have been ruled out, the rest of the guerrilla unit can enter after lookouts have been placed on the hills overlooking the road along which enemies might come. While we are inside, we will not let anyone leave, for he might warn an enemy. The recruiting officer will be in charge of interrogating the owner and discerning his true feeling toward us. Afterwards he will be asked to help as an informal agent, or as a farm guerrilla. If he refuses, showing open sympathy for the enemy cause, he will be made to leave the area; for it is impossible in an area where the guerrilla unit is operating to allow freedom of movement to individuals who might be working against US. Once he has been told to leave his house or farm we will attach all of his property without any compensation. All his belongings will become the property of the armed forces of popular liberation.

20. What shall we do with the young men who wish to join our unit?

The recruiting section will process them one by one, investigating their merits and deciding whether w c can accept them as fellow-soldiers in our revolutionary struggle. In case we can, they are trained to be farm guerrillas; if we have the weapons and the need for more people, they can be taken in as regular guerrillas after receiving the proper training. I personally trained Calixto Sánchez's guerrilla leaders who later landed in Oriente, and in Cabonico, and whose initial operation was a complete success. Not a cartridge was lost; only one boat, which got stuck on the beach. Many times in my classes I emphasized that those who did not voluntarily offer to join up might be accepted one at a time, searched, and given a rigorous interrogation by the recruiting officer to decide who should be assigned to our elite, to the regulars (the less inspired) or to the third section--the unreliables. But we never accept people merely because they claim to be on our side. Calixto Sánchez's leaders did not follow this warning, one which I learned well in a hundred encounters with the enemy. When a group of soldiers dressed as peasants came up shouting, "Viva Fidel Castro!", our people received them with open arms. The soldiers then drew their pistols from hiding and arrested our men saying they were surrounded by many others in the mountains. Our guerrillas, new at the tricks of war, were stricken by fear, the disease that all unseasoned troops are subject to. The rest is well known. They were taken prisoners and that butcher Colonel Cowley assassinated all that were with Calixto Sánchez. Cowley, in turn, was later brought down by a heroic shotgunner of the 26th of July Movement. Only the seven men in Calixto Sánchez' advanced guard, commanded by Héctor Cornillot, survived this encounter and later most of them joined the Sierra Maestra units.

21. What should the guerrilla unit do after an amphibious landing?

Once on the beach, we march toward the highest ridge offering concealment. Of course this is after hiding in the most appropriate places all the heavy materiel we have unloaded. If we succeed in moving inland in secrecy, we carry along our materiel to hide in even safer places in the highlands.

22. Can you tell me some of the assignments in which volunteers of both sexes can assist?

Here are some of the missions they can undertake:

1. To form a small platoon of attendants for each guerrilla unit.

2. To provide pairs of people to serve as scouts in front and on the flanks.

3. To provide liaison pairs to give proper personnel status reports to the command post.

4. To act as runners to maintain contact with the flanks.

5. To provide large platoons to comb (clean) the enemies from our zone of control. This job must be done frequently.

6. Other platoons can ask the loan of hammers, nails, saws, picks, shovels, hoes, barbed wire, food, canteens, empty bottles and tin cans, and typewriters that the commander requires.

7. Others may compile a list of volunteers, both men and women, of the proper age to give service.

8. To form political groups to inquire of the political leanings of the people in our zone.

9. To select individuals who are ready and able to make our status reports, plans, selected scale maps, detailed operational information, to keep guerrilla service records, speeches to the people, etc.

10. Printers, typists, mimeographists, and others may work in the propaganda section.

11. To form brigades of propagandists of our revolutionary ideas to carry out meetings in plazas and other places.

12. To form police platoons, in which women should participate, to impose order and to prevent robbery, pillage, violations, and abuses.

13. To provide and guard storage for our material.

14. Women will also be used to bring complete information from cities not yet dominated by us. By sending many of them to the same place without their knowing that they have the same mission, more complete and cross-checked information will be gained.

15. To form water carriers and quartermaster personnel and distributers of provisions from the women.

16. Women can be used to form a corps of nurses and helpers.

17. To form the sections of carrier pigeons.

18. To establish a report-carrying section using trained dogs.

19. Cooks.

20. Cook's helpers.

21. Wood carriers for the kitchen.

22. Kitchen dishwashers.

23. Water carriers for the kitchen.

24. Seamstresses.

25. Clothes ironers.

26. Laundresses.

27. Registerers of home residents (preferably women).

28. Bathkeepers.

29. Typists, for consignment to sections that ask for them.

30. Separation, storage, and control of captured enemy clothing.

31. Hospital personnel.

32. To form units of saboteurs of trains, highways, bridges, wire communications, etc.

33. To make groups of slingers and throwers of incendiary bombs.

34. Teams of sling instructors.

35. To provide picked groups designated to prepare incendiary bottles, filling them with gasoline and capping them, so that they will be ready at the proper time.

36. From the most intelligent and brave women, to form sowers of fear.

37. Statisticians.

38. To form a group of carpenters to make sawhorses, barbs, fence stakes, trench floors when the ground is wet, grenade boxes, frames to mount rails in trenches, etc.

39. As groups to collect rails for fortification works.

40. To carry the rails to the place of their use.

41. To form recruiting parties to bring people from villages not yet controlled.

42. To form espionage and counterespionage sections.

43. To make up flag and signal communication sections.

44. For fortification works, using whatever workshops that are available.

45. To provide day and night relief teams.

46. To form cavalry with whatever animals are available among the people.

47. Enemy aircraft spotters.

48. Basket carriers to carry dirt from the trenches.

49. Arms cleaners.

50. Cold steel weapons (cutting weapons) storage.

51. Providers of horse rations.

52. Investigation of traitors.

53. Food storage.

54. Throwers of incendiaries against vehicles on the roads.

55. Personnel to set up and improve airfields.

56. Tree cutters.

57. Keeper of the "operations diary."

58. Correspondents.

59. Letter carriers.

60. Tool keepers.

23. What is the first offensive action that a recently formed guerrilla unit should take?

Our first action, as soon as we reach our sector, is to cut in as many places we can, all roads and railroads so that our enemies will only be able to travel on foot. We must force them into infantry roles. Because of their inferior training, lack of morale, because they are armed forces at the service of the oligarchic enemies of the people, and because of their lack of fighting spirit they should be very inferior to our forces who, with greater nobility and efficiency of personnel, are in better condition than the enemy. We should not become panic-stricken finder any circumstances, even though the enemy might throw thousands of men at us. We will have a better chance to inflict casualties on him. It would be more dangerous to our guerrilla team of fifteen men if they assigned twenty-five soldiers to hunt us down. This is worse than having a thousand after us. Always remember that Sandino fought against the Americans for seven years without once being cornered in spite of his pursuers' many thousands of perfectly trained men with motorized units and dozens of radios beaming concentric rings around the Sierra de Segovia where our hero was fighting. After seven years of fruitless pursuit they had to grant him a truce on his own terms. Augusto César Sandino, the Nicaraguan patriot, was assassinated a short time after leaving the Segovia highlands.

24. What should we do with the peasants who wish to join us?

The recruiting officer will organize them into two different divisions. Into the first one will be put fighting men whom we trust completely, and into the second will go those who can be utilized on secondary combat tasks such as water carriers, wood cutters for the mess units, and porters for long marches. To those individuals who display an avid desire for combat and have unquestionable backgrounds will be issued machetes and incendiary bombs. They will march along with our unit as members of machete and bomb squads.

25. When should we do battle with the enemy?

This is the prime question for a guerrilla unit. The answer should be learned by heart and always put into practice. The perfect guerrilla, that is the one who best serves the daily interests of the peoples' revolutionary cause, is one who never invites the enemy to do battle. Nor does he accept challenge to fight the enemy who hopes to meet us where he would hold the advantage. Every good guerrilla

8

should attack by surprise, in skirmishes and ambushes, and when the enemy least suspects any action. When the soldiers load and prepare to repel our attack, we should all fade out of sight and redeploy in safer places. Obviously, in all actions we try to inflict the heaviest possible casualties. We will never lose visual contact with our enemies; that is, we will accompany them from afar keeping within field glass range so that we are constantly aware of their position. If we do not fire into their quarters every night we are not performing our duty as guerrillas. A good guerrilla is one who looks after his men not exposing them to enemy fire; he makes sure they cannot see his troops with camouflage and skillful tactics. He hounds the enemy day and night, carrying on "minuet" tactics. That is, he advances when the enemy falls back; retreating to our right when the enemy plans to encircle us on that flank. We always keep the same distance from the enemy forces: some 800 to a thousand yards by day, sending two or three of our sharpshooters up as close as possible during the night to pester them, and thus bringing about the highest number of casualties.

26. How should a police headquarters be attacked?

If the headquarters is built in the center of a lot one hundred yards wide by fifty in length, there will be fifty yards between the building and the fence surrounding it. First, we have to take the adjacent buildings with our fire force the garrison to take cover, waiting for reinforcements and outside help. Once in possession of a neighboring building and setting our riflemen around the headquarters so that no one can escape, we will begin our plan of attack as follows:

In the building we have taken, we will dig a tunnel toward the center of the headquarters. Once we have the first shaft and the tunnel begun, we put two men with pick axes shoulder to shoulder digging a six-foot- high tunnel. Each one digs out a cubic yard of earth. They then withdraw while the dirt is quickly removed by others with shovels and baskets. When one side of the tunnel is clear of loose dirt, the shovel and basket men withdraw and the pick men begin again. All the workers thus have a break and can perform their tasks with greater efficiency. The tunnel bores away underground, just wide enough to allow two to work without interference. All work as fast as possible; the supervisor relieves the men when they seem to be slowing down.

It is next to impossible for reinforcements to reach the garrison by day so it will probably surrender. If it does not do so soon, it should be blown up--first with the object of taking it over; secondly, as a lesson for other police headquarters to surrender quickly. To hasten the job, not only one tunnel will be dug, but many leading under the headquarters. We do not know what kind of earth we will encounter in any one tunnel, nor whether the first mining attempt will be successful. A second or third bombing may be needed.

If on igniting the charge we discover that the blast is not underneath the building, our soldiers, ready and waiting, should be sent into the tunnel to reach the garrison from the crater or at least to occupy the crater. It has to be somewhere near the building and as such serve as a good place to attack the building from.

For these operations we need the following teams: strong men for the picks, shovelmen, and basketmen; those to handle the lanterns and other tunnel lights; those who will shore up the tunnel after it is dug; and finally those who will set the charge, as well as soldiers to race down the tunnel after the explosion.

Before setting off one tunnel explosion under the headquarters, all other tunnel activities from other buildings must be halted so as to safeguard our comrades.

We have to be prepared at all times for a counterattack from the garrison itself as well as by the army, keeping a 24-hour guard posted. We also will make beforehand the necessary preparations for accommodating the wounded, prisoners, and the dead resulting from the attack. One man will be assigned to take care of all equipment we might capture. All enemy survivors will be given a thorough interrogation to learn what should be done with them.

If, after the first explosion the garrison still does not surrender, we keep up work intensively on the other tunnels as well as in the first one. After an unsuccessful first attempt we should be able to correct the angle for the next try. Up to the time of the second bomb, the first crater can be used to pin down the occupants of the building from close by.

After the headquarters has been taken, the teams we have utilized in the tunnel operations will be sent on to other targets to do similar work.

When all garrisons in our zone have fallen, these specialists will be given jobs in our corps of engineers. The leaders of tunneling operations will at all times inform the general staff of their progress.

As a closing note to this section keep in mind that from all of the world's famous prisons, men have escaped by digging their way under walls and past sentinels.

27. What should be done before attacking from a tunnel?

If it is not possible to achieve a surprise attack, an intense psychological campaign should be carried out making use of emissaries, wives of the besieged, local bigwigs, and enemy prisoners taken in previous attacks.

28. How is a guerrilla column on the march made up?

The guerrillas cover their flanks (right and left sides), an advance party (those preceding the main body) and rear guard (protecting from behind) utilizing peasants who volunteer (as they all should) to help us, as well as troops from the guerrilla unit itself.

29. What should appear on service records?

The dates and places where each guerrilla has fought in addition to his rating as a soldier in each action, and whether he received any distinction for his performance. It is important to be precise in keeping service records so that promotions can be given to the most valuable men.

30. How can you make a hand grenade?

Take an empty condensed milk can, dry it thoroughly inside; put in a dynamite cap, nails or small pieces of iron; press smoothly so no sparks are produced; be careful not to jar or hit it. Continue inserting other dynamite caps and more shrapnel, tamping gently each time until the can is full. A wooden or metal cover is then placed over the can after the contents have been compressed as much as possible. The cover should be pierced to allow a fuse with a percussion cap at its end to make contact with the dynamite. On lighting the fuse, the percussion cap is exploded, which in turn ignites the dynamite in the grenade.

31. How can you make a land mine?

Take a length of pipe, seal it at one end by welding or screwing on a pipe cap. Fill it with dynamite and cover the open end, leaving a small hole in the cap for the fuse. Insert a tube about 1/8 of an inch thick with a percussion cap in the end contacting the dynamite. In the other end of the fuse, place a wad of cotton impregnated with potassium chlorate and sugar. Another wad of cotton, and next to this a little glass chamber containing sulfuric acid are next inserted into the fuse tube, making certain the glass receptacle is well-sealed to keep the acid inside. Next, stick in a length of metal or wood for a plunger that can slide easily down the tube to break the glass. This releases the acid which forms a chemical reaction with the sugar and potassium chlorate, producing a flame to ignite the fuse and percussion cap. Set the mine in a road with a board attached to the plunger. The first vehicle or pedestrian to pass over ignites the charge.

32. How do you make a time bomb?

Use the same system as for the land mine, adapting to the fuse a connection to drive the plunger with an alarm clock, whose alarm bell of course will be removed.

33. How do you make a delayed action fire bomb?

Take a small bottle and fill it with sulfuric acid; then cap it with a wad of cloth or newspaper (one page). The paper is attached to the bottle with a rubber band. Cut off the end of the cover that sticks out. This is done so that the acid is not wasted in this material. Then take another small bottle with a slightly larger mouth so that the top of the first bottle can fit into it. Into the second bottle put twelve tablespoons of potassium chlorate and four of ordinary sugar. Mix up the sugar and potassium chlorate. Now set the first bottle upside down in the mouth of the second. The acid eats through the paper or cloth and on reaching the potassium chlorate and sugar, produces a large multicolored and long-lasting flame. If we have taken the precaution to set the bottles next to inflammable materials, we are assured of a good blaze.

34. What happens if acid and glycerine are used instead of acid alone?

The action of the bomb can be delayed up to five or six days depending upon the amount of glycerine to acid. Experiments with different mixtures should be made to establish formulae for various time durations.

35. How can you obtain the maximum delay?

By putting the sulfuric acid into a covered bottle with a siphon in the top which reaches well below the level of the acid inside. The acid slowly evaporates on contact with the air and consequently fills the siphon with vapor. Later, the vapor condenses and drips onto the chlorate in an adjoining bottle, producing the combustion. A bomb like this can be set to explode weeks or months later.

36. What is the principle of the military time fuse?

Military time fuses which can produce results days, weeks, or even months later are used by all modern armies. The triggering device consists of a plunger on a compressed spring. Acid eats at the wire compressing the spring. When the wire is cut through, the released spring drives the plunger into a priming tube containing the combustible acid, touching off the bomb. This was the type bomb used by the anti-Nazis against Hitler in 1943. The attempt to blow up the German dictator's plane failed due to the discovery of the bomb by vigilant crew members.

37. How do you make an incendiary bomb?

Incendiary bombs should be used by the masses to insure our victory. Every man, woman, and child should know how to handle them. By converting everyone into a combatant and hurling thousands of incendiary bombs against the defenders of tyranny, no enemy can stand before us and victory will certainly be ours. An incendiary bomb can be made with any kind of a bottle, a rag fuse, and gasoline. These can easily be found in any town. Fill the bottle with gasoline and put in a piece of cloth; any size will do so long as it reaches the bottom and has a bit sticking out to light. The bottle is closed with a cork stopper, paper, cloth, or can even be left open. Light the fuse and throw the bottle at the target. When it breaks open on striking the hard surface, the gas is spilled out and ignites. There is first a huge flame and small explosion which cannot hurt the thrower, even, though he is close try. The flame lasts a few minutes depending upon the amount of gasoline in the bottle. The bottle with its lighted fuse, whether uncovered or not, never explodes. It makes no difference whether the bottle is open or covered, the gas fills a third, a half, or two-thirds of the container, or whether the bottle is carried around for hours. It will not explode in your hands. We emphasize this so that the future bomb throwers will know that only those on the receiving end can be injured by this bomb. It is advisable to cover the bottle, however, so that upon being thrown no gasoline spills out on the ground, but all of it hits the objective. Suggested training exercises include using a bottleful of water to begin with, though actually lighting the fuse each time. Using a thick glass bottle, like the Coca-Cola ones, practice throwing as far as possible over soft earth. Plowed earth makes a good "range" to practice on. Thus you can use the same bottle many times over, practicing daily for accuracy and distance. Later, practice with different-size bottles to achieve versatility. In actual combat conditions thin-walled bottles are best as they require less energy to smash on reaching the target. Incendiary bombs can be used to good effect at night since their flames illuminate the enemy objective and thus help make the bomb thrower's position less visible. When attacking the military garrison in a town, the revolutionaries should proceed as follows:

Everyone at a predetermined time will appear on all the surrounding flat-roofed buildings. Five minutes later everyone lets fly with a rain of incendiary bombs against all the walls of the building, trying to hit doors and windows. Revolutionaries in the streets also hurl the bombs they can against the walls at the same time, trying for the same prime targets. Especially those in the streets will throw rocks and shoot at doors, balconies, and windows.

If the police or soldiers come out they will he riddled with bullets, rocks, and bombs by the whole populaces, and especially by those on rooftops. Outnumbered like this, not one garrison can hold out.

If the garrison is constructed of wood, fire bombs can be used to good effect no matter where they hit; but even in this case doors and windows should be the prime targets. Even uncapped bottles can be used without having to light the fuses first, if a good blaze is already going. Gasoline can even be thrown in cans and earthen pots. It is well to have our revolutionaries also practice with slings, as do shepherds and country folk, so they can hurl gasoline bombs with them. You make a sling with a piece of rope two yards long, into the middle of which you attach a can, piece of heavy cloth, or leather pouch (from a handbag, etc.) where the missile is placed. We then tie one end of the rope to the right wrist. Put the gasoline bottle into the pouch (or even into a partially unwoven and widened section in the middle of the rope). Grabbing both ends of the rope, swing the sling with the bottle around your head (like hill people do all over when they want to hit a hog, bull, or horse) until you build up speed. When ready, take aim and release the free end of the rope, sending your projectile smashing into its target. That is, if you have been practicing!

Time spent in practicing bomb-throwing with a sling is really worthwhile. You can become invaluable to the revolutionary cause as a precision bomb marksman being much more valuable than the hand thrower.

Another way of throwing fire bombs is with large launchers similar to the slingshots used by children in hunting birds. The elastic bands of course must be heavier and more powerful.

Patriots in towns should become skilled in throwing fire by hand, sling and slingshots; then engage in contests with one another. It goes without saying that all this is to be carried out under the utmost secrecy and with the least possible noise so as not to arouse the suspicions of the police.

These marksmen will be in the front line when the revolution comes.

When the day of the revolution comes, these units should attack the town garrison, the houses, and other places the enemy is holding out. If all the enemy strongholds in your town are immediately smashed, the whole bomb squad should report together immediately afterwards at other localities where their services are needed. If all the objectives (town garrisons, barracks, forts, etc.) are successfully taken, the Revolutionary Command will assign them to the highways to attack, from a distance, all vehicles moving through the area. These operations are best carried out in daylight and from ambush. Other bomb throwers should be ready to defend the first attackers should enemy parties pursue them from the highway.

Ideally, every revolutionary, man woman, or child (over twelve), should know how to wield incendiary bombs. To achieve this goal, not one day should go by without our practicing with water-filled bottles.

In order to prepare for the eventual battle of liberation from the forces of oppression, exploitation, and the bourgeois dictatorship, all revolutionaries should continue collecting all the empty bottles they can (even buying them), as weld as storing gasoline, old rags, and matches so that when the crucial day arrives nothing will be wanting. Empty cans and cardboard boxes, well lined with paper so that no liquid comes through, should be kept. Wooden boxes can be made if bottles and pots are unavailable.

Teamwork makes for efficient fire bomb attacks. Comrades should aid the bomb thrower; some filling the containers with gasoline; others sticking in fuses; others closing the bottles with corks, paper, or rags; and still others lighting the fuses.

In a fire bomb attack, our people should be well hidden so that the police or soldiers, when driven from their refuges by the heat and flames, can be fired upon with rifles, pistols, and rocks and given the warm reception they deserve.

If, after the bombardment, there still remain inside the garrison enemies who have not surrendered, then platoons of volunteer machete men should rush in, being careful to divide up the rooms to be attacked. Some will only go down the main corridors, others into the rooms on the left, others into the ones on the right. As soon as they have eliminated the enemy occupants of the rooms, they should cut a hole no more than a yard from the floor to let their comrades in the next room know they are in command there and not to use fire bombs against them. When one side of a garrison is in our power, the revolutionaries should come out into the street to help the others attacking the other sides. This draws enemy fire and attention from our forces inside and thus shares the burden of the siege.

38. How can communication be organized between various guerrilla sectors?

Portable walkie talkie radios are used by wireless experts these days for communication among groups in the field. It is understandable that for guerrillas who have to scale high mountains and engage in long marches heavy communication equipment is out! We cannot count on vehicles to carry the equipment, nor even on hand generators which are also heavy. For our operations we are limited to only the lightest of apparatus, working off dry cells. Even these need to be replaced. The 114 mc. (two-meter) band is the best. On the air, keep your messages clear and to the point to guarantee speed and security in communication. Groups in the field should communicate with each other directly and privately. Each group should carry a small transmitter-receiver and maintain contact on a previously determined wave length adjusted on their sets by means of a crystal oscillator. Other groups intercommunicate with one another in the same way. If various groups gather in one place they can contact a shelter or supply depot over the same system to acquire more and better equipment and aid. When making preliminary incursions in unfriendly territory it is not advisable to complicate this basic system of communications. Radio sets can be acquired or even built, and tested before their being put to use. Sets measuring $2 \frac{1}{2}$ x $3 \frac{1}{2}$ x 10 inches powered by a 3-volt A battery and a 90-volt B battery have a 15-mile range and, in favorable conditions, twice as far. The sets are delicate, precision-made instruments, and should be handled with care.

39. How should guerrillas report current developments to their superiors?

Each guerrilla leader should report such happenings on three different sheets. One of them furnishes valuable personnel information; another lists the materiel on hand at the moment of its signature; and the third concerns political military information from the sector. This last report might include the latest rumors, enemy troop movements, new men who have joined us, data on informers and spies, etc. These three parts are sent to the chiefs of the personnel section, the materiel and armament section, and the intelligence section, respectively.

40. How should guerrillas in neighboring sectors communicate with one another?

They should report their strength and the state of their supplies. These reports should be delivered verbally and in person by liaison officers of the utmost confidence. The officers should also have the authority from their superiors to set the day and hour for combined operations, including their own and another or possibly two other units.

41. Should reports be made in code?

It is advisable to code messages which might be captured by the enemy. Usually duplicate messages are sent, cast in special language. Two men, or better yet, two boys start out at different times with the same message. These runners should be natives of the region, clever fellows and fleet of foot.

42. What is the complement of a guerrilla company?

The tactical unit designated as the company contains one hundred men including the commander, a captain. A company has four lieutenants, each commanding a section. Including their lieutenants in command, the first three sections each contain twenty-five men, except for the last. The captain is the twenty-fifth member of the fourth company. Each section has two sergeants who in turn command a platoon apiece of eleven men. Each platoon has two corporals who command squads of five men each. In the squads a second corporal assists the corporals.

43. *What is the complement of a battalion?*

A battalion has five companies. In the fifth company are the cooks, helpers, mechanics, barbers, tailors, cobblers, office personnel, and all those who because of the nature of their work are relieved of instruction and daily activities. Of course even this company reports for duty when the guerrilla war has attained the magnitude approaching a civil war. In other respects, the fifth company is like any other.

44. *Is it necessary for all guerrilla companies to keep this same complement?*

In order to have complete and precise control over all units it is indispensable. If all the units are the same size you can at all times know your total strength. The quartermaster, for example, must know that three companies contain exactly 300 men, etc., without having to make any calculations. All units can then contribute equally in whatever they are called on to perform. An undermanned company could not be expected to obtain the same results as one fully staffed. Also important: no guerrilla wants to be held back in his career for having been associated with an ineffectual outfit.

45. *When your complement is full and you still have extra men, what do you do with them?*

Report the fact at once to your immediate superior so that he can order the men sent to other units as yet undermanned. If, after all units have been brought up to full strength you still have extras, then new units can be made up with the additional men.

46. *If we have said previously that the ideal guerrilla unit in the interest of nobility is composed of fifteen men, why are we now talking of companies of one hundred?*

Because this organization has nothing to do with combat operational necessities. A captain can command a hundred men, hut does not have to use all of them together. On certain occasions, for example, in the siege of an army or police garrison defended by a small detachment, it is a good idea to use the whole guerrilla company for the assault. The captain who operates in certain sectors assigned to him by the Guerrilla Staff has his platoons of twelve men trebled to be perfect guerrillas; he will sometimes utilize groups of twenty-five men commanded by lieutenants.

47. *What is the best procedure for replacing battle casualties?*

The captain should have in some strategic site, out of the enemy's range, if possible, a training base where new guerrillas spend all their time undergoing intensive training, including the memorization of this manual and other necessary information. After having tested these trainees, a ranking will be made according to each man's knowledge, aptitudes, and intelligence section report. As necessary, to fill vacancies, the new men are then sent to active units. After reporting to the captain in charge they are given their permanent assignments.

48. What are close order and extended order drill?

Close order drill is a type of exercise designed to instill habits of discipline in the troops. The guerrilla must surrender his own will completely to the one in command, no matter who it may be. While close order drill is part of the training of armies all over the world, it is no longer employed in combat. It is merely a preliminary form of exercise and does produce good results. Extended order drill is used in the field to deploy troops in the various positions of combat formations.

49. If while on the march, in camp, or at any other time you are fired upon by the enemy, what is your first move?

The first thing to do is to hit the ground and as best you can lie facing the direction the shots are coming from. Then space yourself as far as possible from your comrades who will be doing the same. Thus if the enemy fire misses the one aimed at, there is no possibility of a lucky hit on another man. After this choose the best protection within reach and take cover. If you are a captain or in command of a smaller unit, order your men to take cover as well. Do not counterattack, but try to find some way out of the ambush as quickly as possible. If the fire is too heavy and the enemy is not cutting down our men, because of lack of morale, or in fear of our return fire (which will probably be the case), you might sit tight and wait for nightfall. A daylight retreat would probably cost you too many casualties. After dark, slip out of the trap.

50. What shall we do with our dead and wounded in the field?

If we have time, we will bury our dead, first seeing to it that our wounded are removed from the scene of combat; and when possible, taken to where our comrades can administer medical treatment. If there is no time nor possibility for burial of the dead, we must face the necessity of leaving them. When absolutely imperative, we leave a dead companion; but never one who is wounded.

51. What should we do so as not to lose visual contact with the enemy?

When you withdraw, leave one or two men (better one than two) to keep an eye on the enemy. These observers should never open fire on the enemy, but rather do nothing to let him know he is being watched. When the enemy makes camp for the night one of the observers should report the enemy position so that some of our men can be sent to harass them during the night.

52. If the enemy continues marching during the night, what should we do?

In that case we will follow him, keeping him in sight. The party we send out to follow him should stick as close as possible to him, maintaining harassing tactics as he marches. If the enemy later makes camp or stops to rest or eat, we continue annoying him.

53. How many men should the harassing party contain?

Very few--perhaps two or three. The rest of our men should get their sleep. Our snipers, taking care not to be surrounded, will spend the night firing into the enemy. We will cover both of our flanks while they are resting, so that the snipers can do their job without unexpected risks. This harassment should be carried out every night without fail. You would not he doing your duty if you overlook it.

54. What is the difference between a spy and a counterspy?

Espionage and counterespionage are arts which all guerrillas should become proficient in, since wars are not won only by using one's head, but also by using one's foot in tripping up the enemy as often as possible. A spy is a peasant working for us who accompanies the enemy troop pretending to be their friends and selling them anything they need. The type of article sold or his profits or losses are of no consequence. The important thing is that he become friendly with as many of the enemy, of all ranks as possible. He should never ask them for any information whatsoever, but rather report everything, every movement, he sees; shout the equipment the enemy has; information on their delays, etc. Women are invaluable in this role. That is after they have had the proper training. Their reports should be brought in by intermediaries, and in code. If the information is of extreme urgency, by oral message. A counterspy is one who works with the enemy forces, or is a volunteer in the ranks of the oppressors. Once in their confidence, he goes to work for us, keeping us up to date with firsthand intelligence information. In wartime, counterespionage is of greater service than simple espionage.

55. How is a secret society formed?

A secret society is always formed with a maximum of three members. A fourth member is never admitted, but one can operate with two members. Experience has shown that anything can be done with three agents; any more get in each other's way. Besides if we have the misfortune (and it is to a certain extent inevitable) to have one of our cells infiltrated by a spy, the most that are lost to us are two agents. This does not represent too great a risk nor expense. We must abolish those cells containing eight to ten where each member is in turn the leader of another cell with ten or twelve members, and so on.

56. How does the sabotage section operate?

A secret society will never be given more than one mission. Giving the cells many of them has always produced poor results. Each society should choose a special name for identification purposes, such as José Antonio Galán, Antonio Nariño, or names of other martyrs to our cause. The sabotage section will assign but one mission to each such cell. This way they will have ample opportunity to do a good job.

57. Does only the sabotage section have secret societies?

No. The Intelligence Section can and should have their information gathering sub-organizations, but these never engage in sabotage.

58. How many types of guerrillas are there?

Two types: Field troops and farm troops.

59. What are farm troops?

Farm troops are those who work as farm hands, apparently neutrals politically, who operate periodically, perhaps two or three times a month. They get their arms from the cache, carry out a night mission, then return to the farm and go to work the next day as though nothing had happened. If questioned, they know nothing of the operation, but all say they have seen a few armed men at a distance whom they thought to be guerrillas.

60. How can you blow up sizable buildings, barracks, etc?

The easiest, surest, and least dangerous way to blow up big barracks or buildings like the Presidential Palace is by digging a tunnel ending just below the center of the building.

61. How do you dig the tunnel?

First one must select a house in the neighborhood. It doesn't matter if the house is not too close to the objective. It might be more dangerous if the house is not close since the larger the distance to the objective the bigger the risk, but distance might help in order to ensure the operation without arousing suspicions. Once the house is obtained the tunnel can be started from it, but before anything else is done canned food should be acquired and kept in the house. Food should be enough for the four or five men who are to dig the tunnel, however these men should not give the impression of being the tenants of the house. On the first day a shaft has to be made in one of the rooms of the house reaching farther down if the building to blow is very big, and less if the building is not as heavy. Introduce in the shaft a log shaped like an E without the middle line, the one in between, the log looks then like a C with the top and the bottom straightened. The top arm of the log must be oriented toward the objective and consequently the parallel bottom arm will equally point toward the objective. The tunnel must be started in this direction and only one man will work in the shaft since it has to be narrow in order to avoid earth slides. When this man has dug out enough earth, a second man will remove it with a shovel and a third man will take it out of the tunnel with a basket. This operation will go on until the tunnel has become long enough.

62. What do you do with the earth removed?

When the blasting takes place within a city it is hard to take the earth out of the house without being noticed since in these cases you have to handle a great deal of earth. The best way to handle it is by simulating in the house a business that requires loading and unloading operations. This way sandbags can be taken to an unnoticeable place or preferably cast into the river, the sea, etc.

63. How long does it take to dig a tunnel?

When the earth is of average hardness a man can remove a cubic meter of earth per hour. It is easy to determine how long it will take to cover the distance between the house and the objective.

64. How do you estimate the distance to the objective?

An exact calculation requires a comrade with some knowledge of trigonometry and of how to resolve triangles. Otherwise you will have to use your eyes and discuss repeated measurements with other comrades until the estimate is as accurate as desired.

65. How much dynamite has to be placed below the building to blow it out?

It depends on how heavy the building is but it is better not to under estimate the amount. Let's say that it is safe to use 500 to 1000 kilos of dynamite.

66. How do you go about blasting?

A technician should be in charge of the operation, but everybody should know that the dynamite will only blow by means of a fulminant detonator inserted into the load and in contact with a fuse that will carry the fire from afar. To ensure the blasting it is better to use two different detonators and two fuses, and if one fails use the other.

67. How do you place the detonator in the dynamite?

Pick up a sharp stick and make a hole in the dynamite, then place the detonator in the hole. Don't ever use metalic tools to open the hole unless you want to go to heaven instead of fighting in the guerrilla.

68. How do you attach the fuse to the detonator?

The fuse is introduced in the open side of the detonator and is fixed with special pliers (crimpers) pressing evenly around the open side of the detonator, which prevents loosening of the fuse and failure of the blasting. If pliers are not available at blasting time bite the detonator, it is not dangerous, it is the most common method among guerrilla men.

69. What would happen if dynamite burns or is exposed to fire?

It doesn't blast, it is just consumed as a melting sugar lump.

70. How do you light the fuse?

With a cigarette, and if there are two fuses both must be burned simultaneously.

71. How can you achieve a sympathetic blasting?

The formula for sympathetic blasting is $S=0.9 \times K$ (Kilos). The number of kilos of the load multiplied by 0.9 will give us the distance in meters (ms) from where to blast the other bomb. If the bomb weighs 23 kilos, multiplying 23 by 0.9 the result will be 20.7. Any bomb exploding within this distance will make the other blow up, but if eve increase the distance no matter how well prepared the bomb is it will not explode.

72. What precautions must the chief of the force have in mind before the blasting is ordered?

He must send an officer to every tunnel to make sure that nobody is still there, he will also make sure that each man knows what to do the minute he hears the blast. He will make a speech to encourage speed in the assault and will indicate that shameful acts during the attack will be severely punished.

73. What else should be kept in mind for after the explosion?

Before lighting the fuse the chief will announce to the troop that the blasting time has arrived and, immediately after the explosion, all our fighters will approach the building to be taken from all sides, taking advantage of the confusion that will necessarily follow the explosion. This attack must be carried out fast for better results.

74. What is to be done with used cartridges?

We better keep them, we can always find an officer or a sergeant among the enemy who will exchange them for new ones to make friends with us. He can very well say that then were used by his own troops in order to turn them in and get a resupply; besides, we must not keep the enemy informed about the state of our supply by letting them know how many shots were made.

75. If our fighters could take advantage of a plain to build an airfield, how would they go about it?

The terrain must be cleared of stones, holes straightened, and hills made even. The field selected must be 1000 meters long and some 400 meters wide. Close obstacles like trees, telegraph poles, etc., must be removed.

76. How can the field be made available for the use of our planes?

First it will be convenient to send our side information about the existence of the field, and a chart of it indicating its exact dimensions and location in a chart at a scale of 1/10,000; if possible send also a photograph. When we get news of the day and hour in which our planes will land on the field, right on that day logs and branches of trees will be placed around the perimeter of the field. As the plane appears on the horizon at the fixed hour, signals will be made with a whistle or a flag and the logs set on fire so that the plane may find the field, determine its limits as pointed out by the fires and find out the direction of the wind, since landing must be made always against the wind. As soon as the plane has landed all the fires will be put out, things transported by the plane unloaded and the plane itself pushed by hand to the extreme end of the field where it will be again facing the wind. Only then, if the pilot requests it, which he shouldn't, a single fire will be set to indicate the direction of the wind in case the pilot cannot determine it himself by using a handkerchief. The pilot will see to it that the plane does not remain longer than necessary in order to prevent identification. If there were any mountains around the field we will place a machine gun on the top to harass enemy airplanes that might appear on the horizon.

77. What is to be done if the plane must land during the night for security reasons?

At the day and the hour which the plane will be directly over the field we will light the fires and keep somebody minding the fires so that they are burning constantly to let the pilot know where to land. Night landing is usually very dangerous for the pilot, since even with a good compass precise positioning over the field is always hard to achieve due to the winds which might deviate the plane without allowing the pilot to find the field. To prevent this from happening, landing may be fixed at an hour that will allow the pilot some visibility. Landings will be accordingly fixed for one hour before dawn, unless repeated utilization of the field by the same pilot makes disorientation improbable, in which case landing may be fixed for an earlier time. After a night landing, whistles or a shot will indicate that it is time to put the fires out. If it is still dark after unloading and the plane must leave, fires will be started again all along the runway for good orientation and put out when the plane has been for fifteen minutes in the air

78. How does a plane take off and land?

Always facing the wind.

79. How will our men be busy when there is no immediate task?

They will relax during the day, wash their feet daily and take care of their toenails since feet and legs are the engines of the guerrilla. They will study the maps of the region, memorizing the names of all nearby villages, and their population and some of the names of the people, they will identify on a blank chart all rivers, tributary rivers, springs, reservoirs, and wells. They will learn the distances between different points within that sector and the location of bridges and sewers that might be used for train sabotage. In other words they must learn by heart whatever piece of information might be helpful to carry on the war or to facilitate the tasks of other sections of the militia.

80. How are they given such training?

They are first enlisted as bomb and machete men and will go with us on the marches. Beginning as scouts and carriers of water and ammunition for the guerrilla, then they will take over the watching as sentinels while the fighters rest and will be given rifles for the moment in the capacity of fighters for the first time. Then they will be employed in assaults on the police headquarters or refuges of counterrevolutionary forces, etc. Finally when new rifles captured from the enemy are available, they will be given the rifles and promoted to guerrilla fighters.

81. What is the standard procedure to administer capital punishment to traitors?

They must be given an opportunity to defend themselves, and as in the army, the regular procedures of a court martial will be followed.

82. What are we supposed to do with sick comrades?

When a comrade is sick we will leave him with a family that can be trusted if they make themselves responsible for his cure and protection. They will be better off hiding in some place other than peasant huts even though attended by the peasants.

83. What is understood by the term resupply storage?

Weapon and ammunition officers will keep their supplies hidden in secret places or buried close to peasants' huts. Since it is better not to keep all the eggs in the same basket lest they be broken, resupply storages will be dispersed in strategic sectors so that we may have recourse to the supplies regardless of our position at any moment.

84. What is the attitude of the fighters with regard to peasants?

All food taken from them must be paid for at a good price, thanks must be repeatedly expressed and peasants made aware that they are helping their own cause. Our men will try to repair things in the house such as beds, closets, tables that might be ruined. They will help the peasant in fencing his lot or in sowing or clearing the fields, and in so doing they will clearly show our sympathy and attract the peasants to our cause so that we may eventually request their help any time.

85. How is the defense of a town taken from the enemy organized?

In order to organize this defense, the town must be rearranged to take the configuration of a complex of fortifications by opening connecting passages between adjacent houses. These passages must be small, letting only one man crouching go through, so that if it is an enemy he can be easily disposed of and if it is a friend he may go through with only the relative discomfort of bending his knees. Once all the houses are connected, those facing the street where the enemy will attack first will have in the front several holes like small vents from which to shoot. These openings will be made at a level higher than the regular stature of a man so that even bullets that occasionally go through them will not hit our men. Of course, in order to shoot from these openings one must be standing on a chair.

86. What will be our attitude toward the population of the town?

We will try to convince them gently that they must evacuate their houses, that it is an imperative of the war to fortify them. If this can't be achieved peacefully then they will be evacuated by force as an imperative of war.

87. What will be done with the furniture?

All the furniture, good or bad, will be used to connect houses of separate blocks. Blocks must be connected by barricades made of furniture, stones, bricks, etc.

88. What about military defensive organization?

The chief of highest seniority or rank will appoint his deputies for different sectors of the captured town that is to be defended. Every chief responsible for a sector will see to it that houses and blocks are prepared to conform to the specified defense configuration.

89. What will be the role of the groups operating in the vicinity of the town during the course of the enemy attack?

They will be in constant activity, striking at the rear guard of the besiegers and most of all their supply sources.

90. How can we slow down the capture of entire blocks by the enemy?

We will have parapets in the corners of all the roofs and firing from them will deny the enemy access to the houses. We will also have in the houses dry husk and rags impregnated with cheap oil. If a house is taken, the husk set on fire will have the effect of a smoke grenade stopping enemy advance.

91. How long can we keep defending a town in this manner?

It may last for years. This was the type of defense put into practice during the defense of the University district of Madrid; Franco's troops never went through it.

92. What if the enemy completely cuts the water supply to the town?

It was presupposed that the activity of the outside guerrillas would make it frightful and unacceptable for the enemy to maintain a protracted assault; however, if after all there is no other alternative the best way to escape is to break through the enemy lines in the middle of the dark and flee to the hills.

93. Which must be the main concern of the fighter while in the hills?

His main concern must be the care of his gun, since the weapon is his friend and protector, his means of survival. The rifle must be kept clean and oiled, especially when you are out in the country, marching by dusty paths where guns easily get dirty.

94. Who can be properly called a hill fighter?

He who is in open and declared rebellion against the oligarchy, against bourgeois dictatorship, against the people's enemies; in other words, all regular soldiers in the guerrilla who wage war against oppression and exploitation.

95. Which is the maximum time for a guerrilla to remain in the same place?

Three days is the longest they can stay in one particular place. On the third day they must start toward a position far away from their previous one.

96. What qualifications make a perfect guerrilla fighter?

To correctly handle a gun, a rifle, a machine gun, and a revolver. To be able to fight with a knife and fence with a stick. To be able to throw a knife well and hit a distant target. Horseback riding, bicycle driving, automobile driving. Making and using bombs. Know how to take and develop pictures. Know how to use the phone. Typing. Chart designing. An elementary knowledge of topography. Know how to read a chart and interpret contour data. Know how to whistle loudly. Practice in climbing ramparts and walls using ropes or human towers. Practice in twelve-hour marches through rugged hills with slight descents. Swimming, rowing, motor boat driving. Practice in climbing trees and telegraph poles rapidly. Familiarization with piston-engine parts. Know how to start a car with a crank, how to reach the fuel tank, how to fill the tires of a car or a bicycle, how to change the tires fast. Know the Morse Code. Know how to start the propeller of a light plane. Extreme tolerance to all religions. Finally to be courageous, daring, cunning, to anticipate needs and dangers, to avoid ties with things or persons; to love danger.

23

97. Are all those conditions indispensable to become a guerrilla fighter?

Those are qualities of the perfect fighter only and are only achieved at the peak of the fighter's performance. Take Pancho Villa for instance. He was an outstanding fighter and nevertheless he was an illiterate. However, all those qualifications must be required as an ideal, as they are required by military academies to graduate officers who can defend the fatherland in case of aggression.

98. What items should be on hand for the guerrilla?

The perfect guerrilla must have: Combat boots for the men. Thick socks. Pants reinforced with inside and back patches. Thick and resistant belts that eventually can be joined together as the links of a chain and be used in crossing rivers, climbing walls and obstacles . . . They are called "tails" by the fighters. Coats according to the weather (jackets). Compasses. Good watches. Knives and folding knives. Scissors to cut hair. Scissors to cut nails (especially toenails). Soap for clothes washing. Guns, submachine guns. Grenades. Combat binoculars. Medicaments proper for the guerrilla in the first aid kit. Pliers with oilskin handle (you can also use a thin pipe to cover the handle). Hatchets to cut wood. Razors and blades. Flashlights. Forehead lights (of the kind that can be attached to the head, as the miners do). Batteries for all these lights. Three-corner files. Saws. Threads and fishhooks for fishing. Lighters. Hammocks. Wirecutters.

99. Isn't that too much to be carried by the fighters?

For sure, but it can be taken by the irregulars who always accompany the guerrilla. The list is just a catalog of things that we should have on hand at one time or another and that eventually will all be needed, but that doesn't necessarily mean that all of them should be taken in every raid.

100. What precautions must we have before attacking a village?

In order to attack a village we must know first all the details about it. Some of the most important details will be: Whether or not it has telegraph or telephone communications. Whether or not there are troops guarding communication centers. If there are no troops to guard them, where (how far) is the closest communication center. How many civilians have rifles. Whether there is in the village an amateur radio transmitter. Names of traitors and executioners, domiciles of the best known oppressors of patriotic and revolutionary agents. Location of the railway or road bridges closest to the village and size of the guard. Distance to the closest airfield. Timetable of trains passing through the village and of trucks or buses of lines regularly serving the town. Analysis of the topography of the local area and all other useful data that could possibly be collected. Once the information has been gathered, the data should go into the Staff Section (Operations), which will prepare plans for the assault based on the information received. Assaults are possible without all these requirements, hut this is the technical approach that will give us the highest probability of success.

101. Once plans have been prepared how does the operation develop?

The precise time must be fixed. Each special task is assigned to a special team. These teams must operate fast and with decision, without being concerned with the development of the operations of the other teams, or with the failure of these operations. A team will cut telegraph and telephone communications at the entrance of the village, another team will cut them at the exit. Since individuals possessing weapons are known, a team guided by friendly villagers will break into the houses of these individuals and take their weapons along. The addresses of these individuals and the order in which these searches will take place will be specified in the lists that the chief of the team will be given. Other teams will pick up squealers, spies, or traitors. All these things must be done "electrically," that is, in the least possible time, and the faster we do it the greater the success will be, even in terms of convincing the enemy about our great discipline and morals. This way hopeful revolutionaries will see with their own eyes that our organization can do the job. After the operation is finished we will leave the village by car, meeting the vehicles in predetermined places where they will be waiting with their engines running.

102. What will be the mission of a guerrilla chief in an area under control?

He will organize, under the guidance of the recruiting officer, several groups with the following purposes:

- A. One unit will "comb" the area, that is, it will inspect all the houses and sectors where there are enemies of the people's cause. This same unit will carry out the requisition of all the combat elements that we may need by searching wherever they may be.

- B. Another unit will intensify propaganda for our cause in that area. Both units will be integrated by honest personnel who are notoriously incapable of stealing or of abusing those whom they dislike or their personal enemies.

103. What will be the punishment for those who commit abuses?

Those who would dare to steal in these circumstances, or perpetrate abuses or infractions, must immediately undergo a drum-head court martial and after conviction sent to the firing squad without wasting any time.

104. How will the execution take place?

It will take place at an hour that will permit attendance of a large audience. It will be publicly announced and dramatically set. An officer will address the crowd explaining that the man to be shot is guilty of rape, murder, theft, or of any other shameful and anti-revolutionary action that he may have accomplished. He will use this occasion to emphasize the honesty of the People's Army and praise it in the most laudatory terms insisting upon the fact that shameful acts against the dignity of the people will never be left without their rigorous and well deserved punishment.

105. What is the most important advice that should never be overlooked in the marches?

Combat marches will mainly be accomplished during the night, especially when our purpose is to be seen again far away from our previous position and let the enemy think that there are two different units in operation. During the day we will be sleeping, studying, or occupied in activities proper of the guerrilla, such as care of the weapons, distribution of ammunition, care of the feet, study of the map of the region, attention to the business of the sections of the guerrilla, memorizing names of nearby villages or individuals living within the sector, etc. Don't forget the names of the ranches that we visit, etc. But in marching during the night it is a must that we walk in the most absolute silence and without smoking; otherwise, the entire unit might be destroyed.

106. How must we proceed in a surprise enemy attack after taking shelter?

First we will try not to answer enemy fire, and then, even if they seem to be fewer than we are, wait until the day is over to retreat. If we were superior in number, we engage them for a short time and cause them some casualties. If the situation is not clear it is better to disappear because it might be a decoy or a stratagem to surround us with superior forces. The course of action must be resolved by the chief of the guerrilla who knows by heart that our tactic is not to engage the enemy but to hit and run.

107. Is the purpose of such skirmishes to cause casualties or to cause psychological effects?

Our aim is to destroy enemy morale, keeping their mercenaries from relaxing. If a troop does not sleep during the night they are worthless during the day and slow in the marches. Therefore, the enemy will not be left in peace for a single night.

108. Shall we take turns in this mission?

Of course. This is a mission that should be shared by all members of the guerrilla for several reasons: They must all share the honor of harassing the enemy, our fighters must acquire more experience in this type of actions to improve their morale, and finally it is known that when a soldier does not shoot he gets more and more paralyzed, rusty, useless.

109. Which is the most vulnerable part of a camp?

Kitchens, stables, dispensary, etc. These are points that cants be defended and where the combat morale would be lower.

110. How will we keep weapons in a peasant hut?

It would be a big mistake to keep it in boxes in the hut itself. They must be buried in boxes with an inner cover of zinc, in other words the box will be patched up inside with straightened oil or gasoline cans that will be nailed to the box. The weapons should be wrapped in rags if time permits. Then the box must be closed tight and well hidden in the hole. There you have your hideaway.

111. How deep should the boxes be buried?

Always rather deep to prevent any soldiers from digging around the field close to the house and finding the boxes there (although even this is improbable).

112. How far from the house should the boxes be buried?

Rather far, between 30 and 60 meters from the house, and the place will be known only by the man who buried them and two other fighters; one of whom must always be from the weapons section and the other will, in each case, be from a different one.

113. What should be done from time to time in order to prevent rifles from getting rusty when in use?

They should be examined by the weapons expert that always goes along with the guerrilla and in all cases the fighter must take care of his own weapon with love and dedication, for it is an insurance policy for his life and for those who are in his company.

114. How many times a week do the chiefs of the sections report to the commander of the guerrilla?

Twice a week, during the stops in the marches, the commander will call his chiefs of sections away from the rest of the comrades in a spot called the "office," close to a rock or a tree; the commander will talk with each one of his chiefs separately, starting with the Information chief. The commander will ask as many questions as he thinks fit. One after the other all the chiefs will be examined by the commander about the status of every branch in his section and about the efficiency of their activities.

115. What basic knowledge should the guerrilla fighter possess?

They all should have an idea of plotting, plot reading, contour interpretation and be able to reproduce at a different scale a map of installations or facilities such as schools, court buildings, police stations, barracks, etc.

116. If we have on hand a map of Colombia, for instance, and we want to change it from a scale of 1:300,000 to a scale of 1:5000, what is the best way to do it?

Since the quotient of 300,000 divided by 5,000 is 60 it would be very hard and bothersome to enlarge the map on a paper 60 times larger than the original, and besides, many parts of the map would not be of interest for guerrilla operations. Therefore, it is better to design first a map four times larger covering only the part in which we are interested. The new scale would be 1:75,000 (300,000 divided by 4). After this we place within a square the zone of operations and enlarge it to 1:15,000; finally by a similar operation we enlarge only a concrete part of the zone of operations making it three times larger or to a scale 1:5,000. Instead of a direct 60 times enlargement, we enlarged a part of the original 4 times, a part of that 5 times and then a portion of the latter 3 times. The scale is: $4 \times 5 \times 3 = 60$ times larger.

117. What shall we do with the maps 1:75,000 and 1:15,000 that were made and will not be used?

Give them to the Operations Section which can certainly use them.

118. What does the fraction 1:100,000 mean in a map scale?

It means that every meter on the map will represent 100 kilometers in reality, that is, 100,000 meters on the ground.

119. What is the best scale for maps used in guerrilla operations?

The best scale is 1:10,000 or 1:5,000.

120. What acts of sabotage can be accomplished by isolated patriots?

Those who don't feel that they have the courage to get together and form secret societies and those who don't trust anybody around but still would like to cooperate by means of individual action may carry out the following tasks:

a. If they are working at a post office they could slow down service or send official communiques to the wrong place or in the wrong direction, always avoiding the possibility of being suspected.

b. If they work on a phone board they can boycott the service and slow it down.

c. Mailmen may pick out letters addressed to important personalities in the regime and open them by steam, learning about their contents. If they contain intelligence data they will pass these data to the Intelligence Service.

d. Phone operators will try not to miss a word of interesting conversations and will communicate all useful information to our movement. The operator should do this by telephone without disclosing her name.

e. Those who work in garages will put emery powder in the oil of automobiles used by mercenaries or by officials who are against the people. If emery were not available they may use sand or pulverized rocks, etc.

f. If they work in garages belonging to the armed forces or in official maintenance depots, they will ruin the supplies, hide the tools, misuse gas either in engine tests or by washing their hands often, always trying to throw away some.

g. If they are government chauffeurs they will try to ruin the tires with nails if they can do it in the garage or by driving close to the sidewalk to scratch their sides or by driving over rocks.

h. Schoolteachers will talk to their pupils about the greatness of progress, of beautiful ideals, about love among human beings and solidarity among nations, looking after each other even within the moral slavery in which they find themselves.

i. Everybody will pass on gossip about the exploitation that the people suffer, the increasing prices of essential goods, and complain about the miserable life they are leading.

j. Workers will ask for leave affecting sickness, and request increases in salaries or try to manufacture defective articles especially if the factories are managed by a few enemies of the working class.

k. Wherever there are no water or light meters people will leave faucets open and the lights on.

l. Government employees will not brief or correct their subordinates, instead they will criticize all orders from above and emphasize the defects of their superiors. They will use their time as much as possible in telephone conversations, coffee breaks, reading newspapers, will change the sense of documents, cause disorder, break the furniture, break machines, etc.

m. When opportune, they will change personnel, reprimand those who are friendly with the regime, and at the same time they will appear to be the most fanatic supporters of the government and of the people's enemies. They will ruin the urinaries, bathrooms, water, light, and gas installations, not only in the public offices but in cafes, casinos, theaters, etc. The best way to destroy a urinary is by throwing cotton packages and newspapers mixed with nails and wire into it.

n. In larger offices they will let loose rats and feed them with cheese until they adapt to the place and can operate by themselves. They will also try to blow the light bulbs in the offices and try to cause a short-circuit.

o. While traveling on the train or other public means of transportation they will cut the seats with razors or scissors, etc. In the stadiums or other games they will protest and disturb the peace by yelling against the authorities, the police, etc.

p. In the streets they will try to stop the traffic by going against the traffic regulations.

q. On the anniversaries of traditional commemorations that are not celebrated by the bourgeois government they will be in the streets marching past military, government, and police offices, in a silent protest against the arbitrary government of the oligarchy. They must also go to the plazas where there are statues of freedom heroes and circle around until their presence is noticed, and attract other demonstrators provoking police intervention. Then they will all start booing the police, manifesting indignation. They will convene crowds large enough to break police ranks and to expand and shrink like an accordion, rushing like gigantic waves toward the enemy only to disperse in the collision and to reorganize, forming other waves to clash with police trucks or armored cars or army tanks or "steel helmets." If there is opportunity and impunity they must boo the most prominent figures of the bourgeois landowners and the dictatorship, yelling "down with them," and encouraging revolt so as to form a massive clamor, howling and wild. The idea is to cause methodically the greatest disorder possible. If political debates with mercenaries take place try to keep the opponent surrounded by comrades, especially if he is a police official, and try to out-yell him and out-act him.

121. What is to be done if the police or the troops open fire on the people?

If in a street fight either the police, the armed forces, or the "steel helmets" open fire against the crowd, the next day all our friends and comrades in the work must be induced not to go to work so that a protest may be transformed into a revolutionary general strike. If this end is achieved all efforts will be directed toward generalization of the strike so that business will stop and nobody will dare to work in the factories. To this end we must recruit the help of all our friends and use coercion and energetic measures upon shy and cowardly people.

122. How do we use rumors?

We echo all sorts of rumors and fibs to discredit the most prominent figures of the oligarchy, including presidents elected in a referendum, and we "improve" these rumors. They may also discredit chiefs of police, army, or secret police.

123. How should we react in the event of a vehicle collision?

When we are present at the scene of a collision and one of the drivers is a government driver we must direct the indignation of the people against him.

124. What shall we do if a fire starts?

If a fire starts we will attempt to interfere with the work of the firemen. We will make a call from a distant place from which escape is easy and give the firemen a wrong address. (This refers to fires due to sabotage of government facilities or of offices of prominent figures in the regime.)

125. How can we use vacant apartments?

If we can get vacant apartments for rent belonging to persons in favor of the regime we will throw gasoline, or any other inflammable on hand, into them and set them on fire, escaping only after the fire starts.

126. How do you spoil gasoline?

To sabotage gasoline it is sufficient to put some water or sugar in it.

127. How do we sabotage a machine or a car?

To sabotage a car it is enough to take a small part essential to make it run; it is better to pick out parts that cannot be easily found in the store and must be ordered. Summing up, all efforts must be directed to paralyze regular work, whether in government offices or in private factories, especially wherever it may affect influential figures. We must never give a peaceful moment to the representatives of the criminal bourgeois dictatorship. We will never stop until we see the ultra-reactionary dictatorship collapse violently and lose the power that it held for so long at the expense of the people, while the peasant and working majorities and the middle classes suffered misery and hunger and strains and worries.

128. How do we distribute the troops in order to defend a village?

The village itself must be divided into four zones, each one under a responsible chief who will operate independently from the other but keeping them posted as to the steps that he takes so that they can all depend on each other although they are all subordinate to the commander of the village.

129. How will the troop itself be divided by its commander?

The troop will be classified in three categories: roof shooters, balcony shooters, and window shooters.

130. How will the sentries on the roofs react in case of an air attack?

They will get out of reach of the planes' machine guns, but if a plane is flying low they will open fire against it trying to be always under the cover of walls or old ramparts and aiming at the bushing of the propeller.

131. How do we keep the doors of the houses?

The front doors will be all locked and if possible blocked so that the only way to get in the house is by destroying them.

132. Should we remove all doors within the house?

All the doors within the house must be removed or pulled away except those of the rooms where food and ammunition are kept.

133. How do we arrange the houses that make one block?

They must all be connected by passages made in the separation walls; these passages should not be higher than one meter or wider than sixty centimeters so that people can go through one by one only and stooping. This prevents the enemy from entering a house ready to attack the defenders.

134. What shall we do about women and children living in these houses?

Women, children, and elderly people will be evacuated. Some women, useful old men, and children over sixteen will be allowed to stay if they want to fight for the revolutionary cause. These women and old men will be used in the many jobs and arrangements that defense requires, such as preparation of the blocks, recruiting, encouraging those who don't dare to fight, and especially distributing the ammunition, because at fighting time all men should be shooting and it should be these women who take care of providing men with ammunition.

135. What will people evacuated from the houses be allowed to take with them?

All their private things, except weapons, ammunition, even if it is only shot cartridges, knives, hatchets, picks, bottles, gasoline, alcohol, or anything that might be helpful in the battle.

136. What shall we do with requisitioned food and ammunition.

They will be kept in a room of the house especially adapted for this purpose, the food in one room and the ammunition in another together with everything useful in battle. Those who guard the food will be made aware that they will be responsible for every crust of bread, and will not take anything for themselves unless they want to be accused of disobedience, irresponsibility, cheating their comrades, and of faults against the ethics of revolutionary war.

137. Who will guard the rooms where food and ammunition are kept?

Both rooms will preferably be guarded by women who can be trusted with this task, since men will be dedicated to missions requiring more strength or to missions of more responsibility and risk.

138. What kinds of communications will be maintained?

Communications from house to house and between the zones and the headquarters of the defense. Communications may be verbal, but it is better if it is done by writing. Other communications will be conducted by means of flags or other signals previously agreed upon, such as cloth hanging from the balconies, etc. We will also have to manage to establish communications with the guerrillas in the hills.

139. What kind of discipline will be maintained in the confusion originated by our occupation of private quarters?

We will be more severe with our own men than with the population. We will shoot right away those who trespass or steal for their own pocket, and severely punish those who beat, insult, or humiliate civilians who refuse to give up their houses or goods or who don't understand our explanations for breaking into their houses. Our troops will take whatever is necessary without cruelty or insults and will evacuate people from their homes only as a necessary imposition due to the war.

140. How do we attend to the care of the wounded?

The wounded from all houses will be gathered in a house well adapted to their care that will be as far as possible from enemy fire. Since all the houses will be connected, as we said before, it will be possible to transfer the wounded from house to house and from block to block to the point where they will be attended.

141. What shall we do if the enemy takes one house in the village?

We will defend the next house room by room.

142. What if they take several blocks?

We will defend the village block by block until there is none left. It should be clear that this is only done in the phase of open war with the enemy and not in the phase of guerrilla warfare in which this type of combat is never admissible.

143. What is our answer to those who argue against this phase of combat saying that we are destroying the fatherland?

We will contend that the best way to destroy the country is by allowing the enemies of the people from all the parties of the oligarchy to eat it up and give it up to Yankee imperialism. We will tell them that the shame of living under oppression, under the dictatorship of the bourgeoisie, is worse than to fight for the fatherland and for true freedom, even if we have to start reconstruction from scratch. Finally we will tell them that we prefer to build the new walls at the expense of the blood of our brothers rather than leave the old walls to serve as prisons for the eternal seclusion of the workers, the peasants, the students, the employees, etc.

144. Is it convenient to have in our cause people working in counterespionage?

Undoubtedly. Persons in the villages who do this job will render a better service than anybody who would give us fifty machine guns.

145. Should the counterspy take part in the battles against our own men?

A counterspy should take part in such battles, but his role should be to show off as much as possible without causing great damage to the guerrilla, without really hurting anybody.

146. What services can be rendered by a counterspy who serves as an officer in the enemy ranks?

He may give us details about the strength of each one of the units that follow us, names of their officers, material at their disposal, maps of the places where they are assigned, information about the morale of these troops, ammunition supply, and movements planned, etc. etc. One of the best services that he may render is to assign pickets to engage a guerrilla in places previously agreed upon among us or to leave a small garrison in a particular place, impeding their defense by leaving them short of ammunition or by leaving in command a cowardly sergeant or corporal who would waste the ammunition, or by moving their soldiers some place so that we may attack them in this way at a determined place and time, etc. An officer of the enemy forces that collaborates with us is more useful than ten of our own officers fighting the enemy. For this reason those who work in counterespionage must always volunteer to participate in actions against us or in outfits set for repression of guerrillas, such as the so-called "peace guerrillas" organized by the dictators Laureano Gómez and Rojas Pinilla, among others, in Colombia, etc.

In the Occupied Zones

147. What precautionary steps must we take after occupation of enemy territory?

Small units will be formed with men that don't move as fast as the other fighters because of injuries, wounds, physical defects, or exhaustion. These units will "comb" the area. In these circumstances all the sections will be able to work efficiently, without rush or fears. The Information Section will gather information as necessary, Operations will interrogate the peasantry about bridges and sewers and will mark them in their maps, Sabotage will increase its manpower with as many men as they please, instructing and training the men to form new secret societies. Recruitment will take care of the necessary propaganda to add new men to the guerrilla, checking with Information before making the selection. Training will carry out its mission by establishing camps, selecting and stimulating the instructors who will produce good hill fighters and refined technicians in the specialty of demolition, etc. Armament will make an inventory of the material at the disposal of the different sections, requesting and accepting from the General Staff orders to the effect that all units report to our rearguard facilities for armament inspection and repairs. Ammunition will be able to select good places for their secret storage and make a status report on their supply and will also from time to time dig out their stocks and expose them to the sun. Supply can take care of the purchases of food and will catalogue requisitioned items in their storage. All of which will be done more accurately now that the pressure of the fight is overcome. Finally, Health and Propaganda will carry out their own missions.

In Victory

148. What will be done by the chief when he sees that victory is coming?

He will carefully attempt to separate those who will volunteer to fight at the last moment from those who are truly our own men. He will attempt to keep a good record of his own men and of those who in the last minute jumped onto the bandwagon of revolution. These new volunteers will be registered in a card with complete information and two pictures, and will be requested to sign their service record which will be passed to purging committees for verification.

149. What will be the attitude of the chief toward the indignation of the masses and their intentions of revenge against the agents and spies of the people's enemies and the mercenaries of the dictatorship?

He will strongly and effectively oppose those attitudes, because it is prescribed that all persons suspected of being war criminals will have the right of self defense, and especially because there were many among the enemy who were secretly doing counterespionage, and risked their lives for the victory of our cause.

150. What is the greatest danger that eve face after victory over the bourgeois dictatorship and over the oppression and exploitation of the various oligarchical regimes?

The greatest danger is dissipation of our victory. The forces of evil and oppression, the historical legions of the reactionary classes never give up. They are like snakes that always fight back even after we step on their poisonous throats; they crawl and crouch, only to get ready to jump over the people. They never give up, they always resist, they are always trying to stifle us. Some of these snakes are the politicians in the clergy who hardly deserve the name of Christian or Catholic. They are the ones who want to do in Latin America as they did in Spain where they achieved complete domination of the Spanish people after a horrible mass murder. And thus they preach in the Dominican Republic the slogan of God and Trujillo and have the shamelessness of repeating in our countries, exhausted by exploitation, misery, oppression, injustice, and by bourgeois reactionary dictatorships that call themselves democratic, that heaven is for the poor in spirit. This clergy, by nature reactionary and always meddling in politics, should not call itself Christian or Catholic since their only ambition is to use religion as a cover to justify the oppression of the minorities and to deceive the peasants and workers by forming the so-called "Christian" and "Christian Democratic" or "Social Christian" parties, or others with equal pretensions, and to them they preach that there should neither be hatred nor aversions, that God will judge humanity, that the conquerors should be lenient with the defeated. It must be noted well that these individuals, who don't believe in God or in anything of the kind, mean "the God Capitalism," "the God Exploitation" when they mention God. They never preach these things when it is the enemy who have their feet on our neck, but no sooner is the war over when they will tell you this and more. They will go around shrieking to stop our fight against reaction if it ever makes a show; and so they will carry out their "missions" for the benefit of the enslavers. But you, revolutionary son of the people, beware of "the incense of the sacristy." Watch over and mind your own victory, never let clerical winds impress or hypnotize you, beware of those who in all nations dominated by capitalism or imperialism adulate and support exploiters and oppressors: beware lest those hypocrites of the "kyrieleison" undermine your heroic and well-deserved victory..

EPILOGUE

TIP Most people click the numbers or operators using the mouse, but if you want to use the numeric keypad for entering numbers and operators, press the Num Lock key on the keyboard. Remember to turn Num Lock off again when you are done.

The Edit and View menus contain simple options. Use Edit to copy and paste, and use View to change from the Standard calculator to the Scientific calculator and back again. As usual, Help is also available.

Performing Standard Arithmetic Calculations

To add, subtract, multiply, divide, and perform other standard arithmetic operations, follow these steps:

1. Enter the first number to be used in the calculation.

2. Click the operator: + to add, – to subtract, * to multiply, / to divide, sqrt to take a square root, % for a percentage of another number, or 1/x to calculate the reciprocal.

3. Enter the next number.

4. Continue to click operators and numbers in the order you want them entered.

5. Click = to get the final answer.

If you want to store a number while you are doing another calculation, use the calculator's memory function. After entering or calculating the first number you want to store, click MS. Do whatever intermediate calculations you want, then click M+ to add result of your calculations to the contents of the calculator's memory. When you are ready to look at the result in the calculator's memory, click MR. You can then clear the calculator's memory by clicking MC.

TIP To subtract a number from the current contents of the calculator's memory, enter the number, click the +/– button to make the number negative, click M+, and then click MR.

Performing Scientific Calculations

To perform scientific calculations such as logarithms, follow these steps:

1. Choose View ≻ Scientific. The Calculator window (shown next) expands to include additional buttons.

CALCULATOR

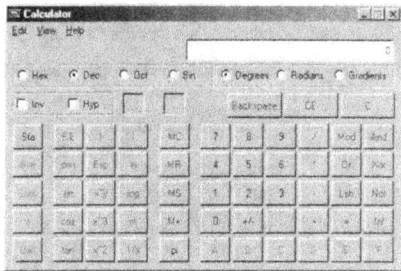

2. Choose a number system in the upper left: Hex (hexadecimal), Dec (decimal), Oct (octal), or Bin (binary).

3. Enter the first number and then click an operator.

4. Continue to enter numbers and operators.

5. Click = for the final result.

TIP To get help on a calculator key, right-click it and then click What's This.

Performing Statistical Calculations

You can apply statistical functions—such as Average, Sum, and Standard Deviation(s)—to a set of numbers entered in Scientific view with the Sta and Dat buttons. Clicking the Sta button opens the Statistics Box, which contains the set of numbers entered with Dat—you must click Sta before you can use the other statistical functions.

Follow these steps to enter a set of numbers that you can then use with the statistical functions:

1. Choose View ➢ Scientific and then enter your first number in the display area.

2. Click Sta to open the Statistics Box and then click Dat to enter the data.

3. Enter the rest of the numbers by typing them over the first number, clicking Dat after each entry.

4. Click Sta to display the statistical data being entered and then click the RET button to return to the calculator.

CAPTURING A SCREEN IMAGE

5. When you finish entering your number set, click the statistics functions you want. The results are shown in the display area.

The buttons in the Statistics Box perform the following functions:

RET Exits the Statistics Box.

LOAD Displays the selected number from the Statistics Box in the Calculator display area.

CD Clears the selected number.

CAD Clears all numbers in the Statistics Box.

Capturing a Screen Image

You can capture screen images—either the entire screen or only the active window—by copying the image to the Clipboard and then pasting it into a document you are creating. Use the following techniques:

- To capture the image of the active window and place it on the Clipboard, press Alt+Print Screen.
- To capture the image of the whole screen, press Print Screen.
- To paste the captured image, open the appropriate document, move the insertion point to the desired location, and choose Edit ➢ Paste or press Ctrl+V.

CD Player

The CD Player from previous versions of Windows has been replaced by the Windows Media Player, an all-purpose player that plays not only CDs but also sound files and video clips.

See Also Windows Media Player, Sounds and Multimedia

CLIPBOARD

Character Map

Displays the set of characters available with a given font. Using the Character Map feature, you can copy special characters not available on a regular keyboard and then paste them into a document. Follow these steps:

1. Choose Start ➢ Programs ➢ Accessories ➢ System Tools ➢ Character Map to open the Character Map dialog box.

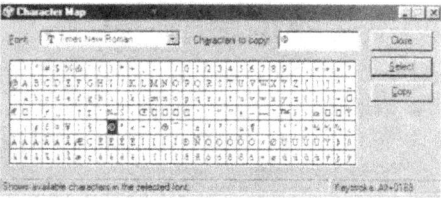

2. Click the Font text box to display a list of the currently available fonts and then select the font you want. Its characters are displayed in the main window.

3. Select the individual characters to copy by clicking a character and then clicking Select (or double-clicking the character). The Characters to Copy box displays the selected characters. Repeat steps 2 and 3 to select more characters.

4. Click Copy.

5. Open the document into which the characters are to be inserted, move the insertion point to the correct location, and choose Edit ➢ Paste, or press Ctrl+V.

TIP To ensure that a character is really the one you want, you can enlarge it by clicking it and holding down the left mouse button.

Clipboard

A temporary storage place for data. You can use the Cut and Copy commands as well as the Windows screen capture commands to place data on the Clipboard. The Paste command then copies the data from the Clipboard to a receiving document,

CLIPBOARD VIEWER

perhaps in another application. You cannot edit the Clipboard contents. However, you can view and save the information stored in the Clipboard by using the Clipboard Viewer, or you can paste the contents of the Clipboard into Notepad.

WARNING The Clipboard holds only one piece of information at a time, so cutting or copying onto the Clipboard overwrites any existing contents. In Microsoft Office 2000 programs, however, there is an enhanced Clipboard that enables you to cut or copy multiple selections.

See also Clipboard Viewer, Notepad

Clipboard Viewer

Provides a way to view and save the contents of the Clipboard. The Clipboard Viewer is not included with the normal Windows installation, but you can add it using the Add/Remove Programs applet.

Choose Start ➢ Programs ➢ Accessories ➢ System Tools ➢ Clipboard Viewer to open the Clipboard Viewer window. It displays the contents of the Clipboard.

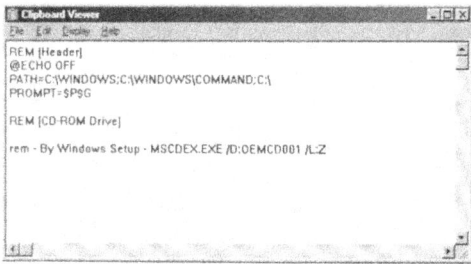

The Clipboard Viewer contains the following menus:

File Allows you to save and open a Clipboard file. Files are saved as .CLP files, which other applications, such as Notepad, cannot easily read.

Edit Allows you to clear the contents of the Clipboard with the Delete command.

COMMUNICATIONS

Display Lets you define the current Clipboard contents as text, picture, graphics, or other objects.

See also Add/Remove Programs, Clipboard

Closing Windows

Closing an application program window terminates the operations of that program. You can close windows in a number of ways:

- Click the Close button in the upper-right corner of the program title bar.
- Click the Control Menu icon (the icon to the left of the program name in the title bar) and then choose Close, or simply double-click the Control Menu icon.

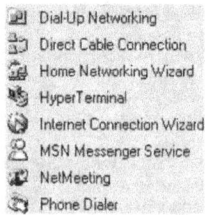

- Choose File ➢ Close or File ➢ Exit within the application.

If the application is minimized on the Taskbar, right-click the application's icon and choose Close, or press Alt+F4.

See also Shut Down

Communications

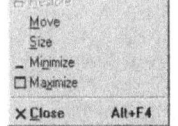 Choose Start ➢ Programs ➢ Accessories ➢ Communications to access all the Windows Me communications tools, including Dial-Up Networking, Direct Cable Connection, HyperTerminal, Home Networking Wizard, and the MSN Messenger Service.

- Dial-Up Networking
- Direct Cable Connection
- Home Networking Wizard
- HyperTerminal
- Internet Connection Wizard
- MSN Messenger Service
- NetMeeting
- Phone Dialer

See also Internet Connection Wizard, Dial-Up Networking, Direct Cable Connection, HyperTerminal, Home Networking Wizard, Modem, Phone Dialer, NetMeeting, MSN Messenger Service

COMPRESSED FOLDERS

Compressed Folders

A new feature in Windows Millennium edition that enables you to view the contents of compressed archives, such as Zip files, as if they were regular folders. This eliminates the need to use a separate unzipping utility program. After you open a compressed archive as a folder, you can simply drag the files out of the folder to extract copies of them.

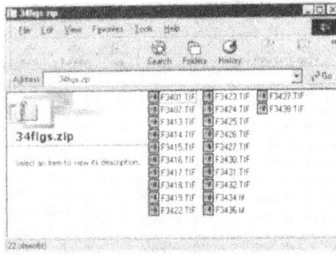

If you double-click a compressed archive file and its content does not appear in a folder, as above, the Compressed Folders feature may need to be enabled in Windows. To do so, use Add/Remove Programs in the Control Panel to add that Windows component.

See Also Add/Remove Programs

Connecting to the Internet

See   Internet Connection Wizard

Control Panel

Provides a way to establish settings and defaults for all sorts of important Windows features. To access the Control Panel, choose Start ≻ Settings ≻ Control Panel.

COPYING FILES AND FOLDERS

To open an applet, double-click it, or click once on its icon to select it and then choose File ≻ Open.

WARNING Other programs sometimes put icons for their settings in the Control Panel, so the icons you see on your screen may not be exactly the same set as shown here.

The Control Panel window is similar to all folder windows, and you can modify it through the View menu. For detailed information about each Control Panel applet, consult its entry in this book.

See also Folder Options, Printers, Start, Taskbar

Copying Files and Folders

When you copy a file or a folder, you duplicate it in another location and leave the original in place. In Windows, you can copy files and folders in three ways.

Using Drag-and-Drop

To use the drag-and-drop method, you must have both the source and the destination folders open on the Desktop. Press and hold the Ctrl key while holding down the left mouse button and drag the file or folder from one location to another. When the file or folder is in the correct place, release the mouse button and then release the Ctrl key.

D

COPYING FILES AND FOLDERS

WARNING Be sure to hold down Ctrl. If you do not, the file or folder will be moved rather than copied.

Using the Edit Menu

The Edit menu in My Computer, Explorer, or any folder window provides a Copy and Paste feature. Follow these steps to use it:

1. Select the file or folder you want to copy.
2. Choose Edit ≻ Copy.
3. Find the destination file or folder and open it.
4. Choose Edit ≻ Paste.

You will see the name of the file in the destination folder.

TIP You can select multiple files or folders to be copied by holding down Ctrl and clicking them one after the other. If the files to be copied are contiguous, you can also use Shift to select them.

Using the Right Mouse Button

Right-clicking a file or a folder opens a pop-up menu that you can use to perform a number of functions, including copying. To copy using the right mouse button, follow these steps:

1. Locate the file or folder you want to copy, and right-click to open the pop-up menu. Select Copy.
2. Open the destination folder, click the right mouse button, and select Paste.

You will see the name of the file in the destination folder.

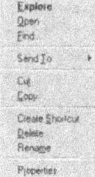

See also Explorer

COPYING FLOPPY DISKS

Copying Floppy Disks

To copy a floppy disk, use the Copy Disk command. The disks used must be of the same type—for example, a 3 1/2" high-density disk must be copied to another 3 1/2" high-density disk. Any information on the receiving disk is replaced with the data from the source disk. Follow these steps:

1. Open My Computer or Explorer and find the floppy disk drives you want to copy to and from.

TIP You can copy to and from the same drive. You will be prompted when to change from the source to the destination disk.

2. Right-click the drive to open the pop-up menu.
3. Choose Copy Disk to open the Copy Disk dialog box:

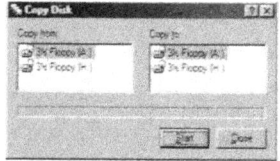

4. On the left, select the disk to copy from. On the right, select the disk to copy to.
5. Click the Start button to begin the copying process. The progress bar at the bottom of the dialog box indicates how much information remains to be copied.
6. If you are working with a single drive, you will be prompted to swap source and target disks as the copy proceeds.

You will see the message "Copy Completed Successfully" when the program has completed copying. Get ready to copy another disk, or click the Close button to close the Copy Disk dialog box.

See also My Computer, Explorer

 CREATING NEW FOLDERS

Creating New Folders

Sooner or later, you will want to add a new folder to a disk or to another folder, and you can do so in Explorer. Follow these steps:

1. In Explorer or My Computer, select the disk or folder in which you want to place a new folder.
2. Choose File ➢ New ➢ Folder. A new folder is added to the disk or the folder you indicated with the name "New Folder" highlighted (shown next).

3. Type a new folder name, something that will act as a reminder as to the files it contains, and press Enter.

You can also right-click the blank part of the Windows Explorer file pane to open a pop-up menu from which you can choose New ➢ Folder.

 TIP If you would rather bypass Explorer altogether, you can create a new folder on the Desktop by double-clicking My Documents and then choosing File ➢ New ➢ Folder. Give the folder a new name and then drag it to the Desktop.

See also Explorer

Windows ® *Me for Me!*

DATE/TIME

Date/Time

Date/Time

The clock that appears in the right corner of the Taskbar displays the system clock, which not only tells you the time, but also indicates the time and date associated with any files you create or modify.

To set the clock, follow these steps:

1. Double-click the time in the Taskbar, or choose Start ➢ Settings ➢ Control Panel and then double-click Date/Time to open the Date/Time Properties dialog box (shown next).

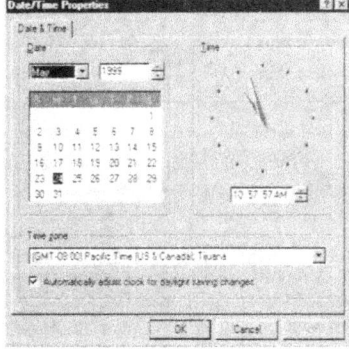

2. To change the time, either drag across the numbers you want to change beneath the clock and type the new time, or highlight the numbers and click the up and down arrows to increase or decrease the values.

3. To change the date, click the drop-down arrow to select the month, use the up and down arrows to change the year, and click the appropriate day of the month.

TIP At any time, you can place the mouse pointer on the time in the Taskbar to display the complete date.

DIV

DELETING FILES **AND FOLDERS**

You can also change the time zone, and automatically adjust the clock for daylight savings time, in the Date/Time Properties dialog box.

> **NOTE** To vary the format of the date and time displayed in the Taskbar, select the Regional Settings applet in the Control Panel.

See also Regional Settings, Taskbar

Deleting Files and Folders

You can delete a file or a folder in several ways. First, select the file or folder you want in My Computer or Windows Explorer, and then do one of the following:

- Choose File ≻ Delete. After you confirm that you want to delete the file or folder, Windows sends it to the Recycle Bin.

- Press the Delete key on the keyboard and verify that you want to delete the selected file or folder. Windows then sends it to the Recycle Bin.

- Right-click the file or folder to open the pop-up menu. Select Delete and then verify that you want to delete the selected file or folder. Off it goes to the Recycle Bin.

- Position the My Computer or Explorer window so that you can also see the Recycle Bin on the Desktop; then simply drag the selected file or folder to the Recycle Bin.

> **NOTE** If you accidentally delete a file or folder, you can choose Edit ≻ Undo Delete, or retrieve the file or folder manually from the Recycle Bin. You cannot retrieve a deleted file or folder if the Recycle Bin has been emptied since your last deletion.

> **TIP** To delete a file without placing it in the Recycle Bin, select the file and then press Shift+Delete. You cannot recover the file if you do this. You will be asked to confirm the deletion.

See also Explorer, Recycle Bin

DESKTOP THEMES

Desktop

What you see on the screen when you first open Windows. Initially, it contains a set of icons arranged on the left, plus the Taskbar with the Start button across the bottom. As you work with Windows and load application programs, other objects such as dialog boxes and messages boxes are placed on the Desktop.

You can also change the appearance of the Desktop by right-clicking it and selecting Properties. This allows you to change display properties for the Desktop background and screen savers. You can also change the monitor type, font types and sizes, and colors for objects on the screen.

See also
Active Desktop, Control Panel, Display, Folder Options

Desktop Themes

 Lets you select a Desktop theme. A theme provides a unified look for your computer's Desktop, icons, font styles, wallpaper, and sounds.

Desktop Themes Choose Start ➢ Settings ➢ Control Panel and then double-click Desktop Themes to open the Desktop Themes dialog box. To select a new theme, follow these steps:

1. In the Desktop Themes dialog box, choose from the list of themes.

2. Confirm that you want to use all the options listed in the Settings area of the dialog box, or clear the check mark from any options you don't want to use.

3. Click OK to apply the selected portions of the new theme.

4. You can choose Save As to store your newly created theme under a new name so that you can select it again in the future (shown on next page).

Click Screen Saver for a preview of the screen saver used in your chosen theme, and click Pointers, Sounds, Etc. for a preview of the icons, mouse pointer, and sounds used in the theme.

See also Display, Screen Saver

DEVICE MANAGER

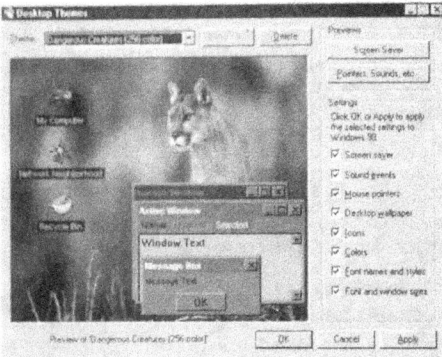

Device Manager

Provides a list of the hardware installed on the PC, along with the status of each device, and offers easy access to each device's properties.

To display the Device Manager, choose Start ➤ Settings ➤ Control Panel and then double-click System, or right-click the My Computer icon on the desktop and choose Properties. Then click the Device Manager tab.

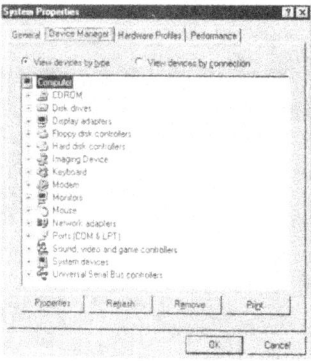

DEVICE MANAGER

Devices are listed by category. Click a plus sign next to a category to see a list of installed devices in it, or double-click the category itself.

Devices that are functioning normally appear on the list with a small icon and their name. If a device has a problem, a yellow circle with an exclamation point appears next to its icon. If a device has been disabled, a red X appears through its icon.

Viewing a Device's Properties

If you are not sure whether a device is functioning, or if you need to know what resources or settings it is using, view its properties. To view a device's properties from Device Manager:

1. Click the device to select it. (You may need to open a category to display the individual device names.)

2. Click Properties. A Properties box for that device opens.

3. Use the controls in the Properties box to specify how the device operates.

Each device has different properties, but almost all have at least a General, Driver, and Resources tab, described in the following sections.

General Tab

The General tab reports the device status. It also contains the following checkboxes:

Disable in This Hardware Profile Disables the device, freeing up any system resources that were allocated to it. Leave this unmarked in most cases.

Exists in All Hardware Profiles Makes the installed device available in all hardware profiles, if multiple profiles have been set up. Leave this marked in most cases.

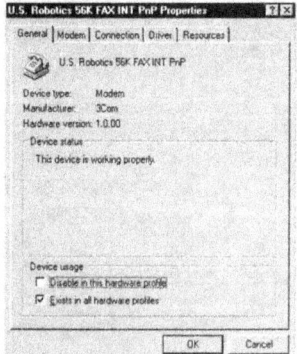

**VICE MANAGER**

See Also Profiles

Driver Tab

The Driver tab reports information about the driver files in use for the device. It also contains two buttons:

Driver File Details   Provides detailed information about the driver files for the device.

Update Driver   Runs the Update Device Driver Wizard to locate and install a better driver for the device, if possible. There is usually no need to do this unless you are having a problem with the device.

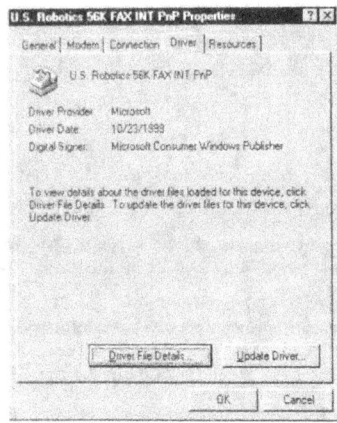

Resources Tab

The Resources tab provides information about the system resources the device is using, such as IRQ (Interrupt Request), I/O Address, and DMA channels. Windows automatically assigns these resources to most devices, so there is no need to change them.

DEVICE MANAGER

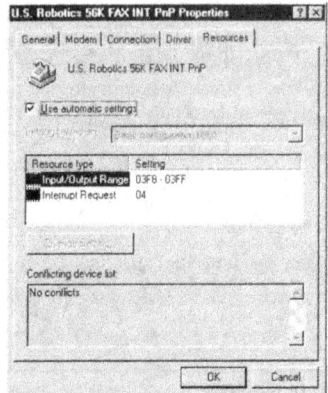

If you do need to change the resource assignments for a device, do the following:

1. Deselect the Use Automatic Settings checkbox.

2 Choose a different configuration from the Settings Based On drop-down list. Find one that reports No Conflicts in the Conflicting Device List section of the dialog box.

3 If none of the settings in step 2 reports No Conflicts, it may be possible to change an individual setting. Click the line in the Resource Type column for the conflicting device, then click Change Setting.

4 Choose a different setting for that resource, and then click OK.

5 Click OK to close the Properties box for the device.

NOTE A message may appear in step 3 that the resource cannot be modified. In that case, check the Conflicting Device List to see what device is at the other end of the conflict, and try to change its setting instead.

DX

DIAL-UP NETWORKING

Removing and Redetecting a Device

Sometimes a problem with a device can be solved by removing the device from Device Manager and letting Windows redetect it. When it is redetected, Windows assigns available resources to it, often eliminating a former resource conflict.

To remove and redetect a device in Device Manager, follow these steps:

1. From Device Manager, select the device to remove.
2. Click Delete.
3. If a confirmation box appears, click Yes.
4. Click Refresh. The device should be redetected.

If the device is not redetected, restart the PC; it might be redetected upon startup.

Dial-Up Networking

 Connects your computer to remote computers, called *servers*, using your modem. This is the most common kind of connection for using the Internet. You use dial-up networking to connect your PC to your Internet Service Provider (ISP).

In addition, if you travel as a part of your job, you can connect your laptop computer to your corporate network right from your motel room. If you telecommute, you can connect to headquarters from your home office. Once you make the connection, you can share information with other computer users back at the office and, at least temporarily, become part of the network to which the remote server is connected.

You can automate the dialing process if you dial one number often, and you can also save telephone numbers and other dialing specifications in a special Dial-Up folder you can use again and again.

NOTE Dial-Up Networking supports conventional modems, ISDN (Integrated Systems Digital Network) connections, and a null-modem connection between serial ports.

www.ingramcontent.com/pod-product-compliance
Lightning Source LLC
Chambersburg PA
CBHW050044230526
45470CB00004B/1406